FINANCIAL MATHEMATICS

SECOND EDITION

REN-RAW CHEN

GLOBAL SOCIAL SCIENCES INSTITUTE

Published by Global Social Science Institute

Library of Congress Control Number 2009935642 Chen, Ren-Raw Financial Mathematics / Ren-Raw Chen ISBN 9780982546017

Contents

Preface

Recently, there are a growing number of text books and MS curricula on mathematical finance. These books and curricula focus on how to apply mathematics in finance. These texts and curricula are designed to help mathematicians, physicists, and engineers to understand finance. As a result, the languages and methodologies used in these books and curricula are not suitable for the finance people to learn necessary mathematics. This book is doing precisely that. This book is written to teach finance students the necessary mathematics to do financial modeling.

My undergraduate major was International Business. I have always loved math. But I have learned no math in my college years. When I took math classes in Illinois, I found it to be absolutely fascinating. Nevertheless, it took me almost two years to learn the necessary math to do what I do today. The reason for such slow progress was that the math classes were not designed for finance students. I took at the math department with math students. I had great fun with my math classmates but I also wasted much valuable time learning the math that was completely unrelated to finance.

When I started teaching Financial Modeling II (which is equivalent to continuous time finance) at Rutgers in 2001, I needed to explain math to my students whose undergraduate majors were economics and finance. Then I realized how important it was to teach the opposite of mathematical finance – financial mathematics. Hence, for 8 years, I have been teaching financial mathematics – the necessary mathematics for a finance major to do modeling.

To do financial modeling, we need certain mathematical rigor, but not at the level a solid mathematician would desire. For example, we write as . It is not correct, but it works. Many of my fine colleagues at the math department at Rutgers would voucher for me on such a sloppy but working mathematical maneuvering. I teach change of measure with a dart-shooting example. I do forward and backward equations as an extension of the binomial model. It is sloppy but it works. I found that students could very quickly pick up necessary knowledge to build financial models. The chapters (lessons) in the book are taken from my past teaching notes of the Quantitative Finance (Financial Modeling), MBA (Analysis of Fixed Income

Securities), and Ph.D. (Continuous Time Finance) courses at Rutgers University, National Taiwan University, University of Pittsburgh, and Fordham University. I must thank my former students from these classes who constantly challenged me and helped improve the contents of the book.

Quantitative finance has come a long way. Today, quantitative finance is so broad that no single book can contain all materials and no single author can be the expert of all the materials. This book seeks the balance between math (e.g. Shreves book) and finance (Huang and Litzenbergers book), breadth (Hulls book) and depth (Rebertos term structure book).

The coverage of the book is to consider the widest exposure to various modeling techniques possible. I spend a lot of effort on term structure theories because this is an area where many quantitative finance tools lie. I should note that many product details are missing from this book because, after all, it is a methodology book. For that information, readers are referred to other books.

Ren-Raw Chen

New York, New York

Acknowledgements

I first need to thank my wife, Hsing-Yao, for her support for all these years. Without her taking care of all the work at home, I would not have achieved not only this book, but any of my research. Her unconditional love and support is the main reason I could accomplish anything at all. I am also so obliged to my former students and co-authors Dr. Shih-Kuo (Seiko) Yeh (Chung-Hsing University), Dr. Xiaolin Cheng (Morgan Stanley), Dr. Bo Liu (Citi Bank), and Dr. Hsuan-Chu Lin (Cheng-Kung University) and my colleague Dr. Xu Bai (Morgan Stanley) for very helpful comments and suggestions. In this second edition, I am particularly grateful for my Fordham MSQF students who pointed out a number of typographical mistakes. Also their questions toward the contents of the earlier edition have changed the how the materials are presented in this edition. Last but not least, I must thank Dr. Phelim Boyle and Dr. Louis Scott for their caring and support for all these years. Without their patience, encouragement, and kind guidance, this book would never be possible.

Words for the Second Edition

In this second edition, I corrected mistakes in the first edition (Errata sent out to the previous edition buyers). I also added some new materials, including (i) a new risk management chapter (Chpater 15) where I discussed market risk, credit risk, and CVA; (ii) a new liquidity quantification chapter (Chapter 16) that is what Basel III demands; and (iii) enhancements of various chapters (Chapter 4 adds more discussions on Brownian motion; Chapter 12 adds forward rate version of the short rate models).

Notation

Throughout this book, I try to keep notation as consistent as possible so that when readers jump in the middle of the book it is still readable. Inevitably, notation becomes quite complicated in some places. Yet the basic principle applies. For example, σ should always represent volatility. It could carry subscripts often times to help represent volatilities of different but parallel assets. When necessary, other symbols will be introduced for volatility and those should be temporary.

The following is a set of notation that is used consistently throughout the paper:

$\Phi(t, T, s)$ = futures price of a zero coupon bond $P(t, T)$
$\Psi(t, T, s)$ = forward price of a zero coupon bond $P(t, T)$
$P(t, T)$ = discount bond price at time t of \$ 1 at time T
$Q(t, T)$ = survival probability between t and time T
$r(t)$ = instantaneous risk-free rate
$f(t, T)$ = continuous time forward rate at time t (over T and $T + dT$)
$f(t, T, s)$ = discrete forward rate

Note that the discrete forward price is model-free as $\Psi(t, T, s) = \frac{P(t,s)}{P(t,T)}$ and the discrete forward rate is $f(t, T, s) = \frac{1}{\Psi(t,T,s)} - 1$.

The continuous time forward rate is also model-free is $f(t, T) = -\frac{\partial \ln P(t,T)}{\partial T}$.

Although not as consistently appearing as the above notation, the following is also frequently used:

$B(t, \underline{T}; c) = c\Sigma_{j=1}^{n} P(t, T_j) + P(t, T_n)$ = risk-free coupon bond price

where $\underline{T} = < T_1, \cdots, T_n >$. In the Vasicek and Cox-Ingersoll-Ross models in particular:

α = reversion speed
μ = reversion level
σ = volatility
λ = market price of risk

Conditional expectations (conditional on t) under various measures are described as:

$\mathbb{E}_t^{\mathbb{P}}[\cdot] = \mathbb{E}_t[\cdot] =$ physical expectation
$\mathbb{E}_t^{\mathbb{Q}}[\cdot] = \hat{\mathbb{E}}_t[\cdot] =$ risk neutral expectation
$\mathbb{E}_t^{\mathbb{F}(T)}[\cdot] = \tilde{\mathbb{E}}_t^{(T)}[\cdot] =$ forward expectation

It is noted that $\mathbb{E}[\cdot]$ is an unconditional expectation. When $t = 0$, then the expectation is by definition unconditional. Variance and covariance are symbolized as $\mathbb{V}$ and $\mathbb{K}$. An indicator function is often used and is $\mathbb{I}$.

Matrices are represented by bold capital letters such as $\mathbf{X}$ and vectors are represented by underlined small-case letters such as $\underline{x}$. In most of the situations vectors are vertial. Any variable with a $\sim$ on the top represents a random variable, such as $\tilde{X}$.

Finally, we use a few acronyms frequently throughout the book:

- RND $\rightarrow$ Radon-Nykodym derivative
- PDE $\rightarrow$ partial differential equation
- KBE $\rightarrow$ Kolmogorov back equation

Chapter 1

Introduction

1.1 Introduction

Financial mathematics is a broad area and many authors have written it in different ways. This book takes an approach that mimics my learning experience. I graduated with an undergraduate International Business degree that had very little math at the time. Throughout my graduate study I struggled my way through mathematics so that I can do research in derivatives. I have taken courses that would have qualified me a master degree in mathematics. However, at the end, I found that all I need is a very limited set of mathematics. While I had enjoyed my time spent in the math department of the University of Illinois, I felt that I could have learned the needed math more efficiently. This is the motivation of this book.

This book is written in a way that any finance student can start learning needed mathematics to start his/her research in derivatives asset pricing. This book is organized in such a way that only the absolutely necessary derivations are provided. In spirit, this book mimics the style of Hull's book.

1.2 Contents

To equip with necessary mathematics, one must start with stochastic calculus that begins with Browanian motion. Then it comes naturally Ito's lemma. Change of measure is a necessary tool so we need to touch upon measure theory. Since option pricing relies on continuous rebalancing, we must learn stochastic processes and martingales. The fundamental economic theory that supports this mathematics is Merton's (and Cox, Ingersoll, and Ross) Intertemporal Capital Asset Pricing Model (ICAPM) and Arrow-Debrus complete market argument via state pricing theory.

It is also necessary to study a number of derivative pricing models. Usually we start with the Black-Scholes model. Then it expands to exotic option pricing models and term structure theories.

Many option pricing models do not have closed-form solutions so we must study numerical methods such as Monte Carlo simulation and finite different methods.

1.3 Coverage

This book aims at how to prepare a financial engineering student to deal with Wall Street interview questions. Hence, the coverage of this book contains only necessary materials in quantitative financial modeling. This is not to say that the coverage of this book is exhaustive. For example, this book does not cover hard core econometrics and advanced statistics that have become increasingly important for hedge funds. This book covers mainly derivatives materials.

This book starts with State Price Theory. This is the first option pricing theory (contingent claim pricing theory) in the literature. In late 50's and early 60's, Gerard Debreu and Kenneth Arrow already recognized that many securities have contingent payoffs and they derived a theory for pricing such securities. The well known Arrow-Debreu security had therefore been born. The concept of Arrow-Debreu securities is not only useful in theory, it is widely used in building lattice models. Lattice models need forward induction in calibration and backward induction in pricing. In complex situations like term structure modeling, the use of Arrow-Debreu securities is very efficient. Readers can find excellent discussions in a series of works by John Hull and Alan White.

Then I talk about the Intertemporal Asset Pricing Model (or ICAPM). There are two most relevant papers in ICAPM, one by Robert Merton in 1973 in Econometrica and the other Cox, Ingersoll, and Ross in 1985 in Econometrica. CIR's work was actually completed in 1977. Merton has a long series of background works on intertemporal portfolio selection that are all relevant. The ICAPM, particularly the CIR model, lays the foundation for contingent claim pricing when the underlying asset is not publicly traded. This chapter is particularly useful for those who transfer to finance from science or engineering areas. Many such people can do the math well but make fatal mistakes in building financial models. I feel responsible to point out some important issues in microeconomics. This chapter should be read carefully.

Then I introduce the basic stochastic calculus in Chapter 4. This chapter contains the absolutely necessary math materials for doing contingent claim pricing. All the definitions and theorems are used throughout the book. As I mentioned in the Preface, the math materials introduced in the book is by no means rigorous, and yet they are correct.

Chapter 5 is the famous risk neutral pricing. As many of you already know, the most amazing result of the Black-Scholes model is not its closed-form formula, but its independence of risk preference. Prior to Black and Scholes, many microeconomists had already derived explicit option formulas (e.g. Case Sprinkle and Paul Sameulson). Unfortunately they all failed to recognize no-arbitrage hedging. While Fisher Black and Myron Scholes themselves did not recognize the theory at the time (as their method to arrive at the solution is to solve a partial differential equation), many insightful scholars (Sharpe (1976), Cox-Ross-Rubinstein (1979), Cox-Ross (1976), Harrison-Pliska (1979), Harrison-Kreps (1978), Duffie-Huang (1985), among numerous others) gradually over time filled in the details and completed the theory.

Risk neutral pricing theory (or alternatively called equivalent martingale measure theory) states that there is this unique probability measure under which all assets earn the risk-free return. As a result, if the market is complete (i.e., all underlying assets are publicly traded, vaguely speaking), we can price contingent claims by no-arbitrage. Hence, no-arbitrage is another term for the existence of the equivalent martingale measure.

From Chapter 6 till the end of the book is pretty much all about various models. By no means these are all the models the readers should study. I chose these models by the following principle:

- It must be widely used by the industry, or
- It must represent a useful pricing technique

For example, Fourier transform is used widely in the financial industry. So I included all important models that use such technique. Monte Carlo is a must tool. So I introduced how Monte Carlo should be used in simulating the CIR model. Multi variate distributions are important. So I introduce the Geske model.

Finally, I hope you enjoy the book and find its coverage efficient and useful. Of course, any comments are extremely welcomed.

Chapter 2

State Price Theory

This chapter leverages heavily upon Ingersoll's Theory of Financial Decision Making. State price theory is the most fundamental theory in asset pricing. The first people who discovered state securities were Gerard Debreu and Kenneth Arrow. That is why they are also known as Arrow-Debru securities.

2.1 Introduction

The State Price Theory is the earliest pricing theory of derivatives. We see many basic principles of derivatives pricing in the State Price Theory. The Theory provides an evaluation of a "State Contingent Claim" where \$1 is paid if a particular economic state is realized. As a result, it is apparent that the theory can only apply to discrete states (even though there is a continuous time analogy).

2.2 Notation

We start the state price theory by defining a state payoff. Assuming in a one-period economy where there are N securities and J states. Define a state payoff as the dollar cash flow received from a security (i) at a state (j). As a result, the entire possibilities of state payoff can be summarized in the following matrix:

$$\mathbf{S} = \begin{bmatrix} S_{11} & S_{12} & \cdots & S_{1J} \\ S_{12} & S_{22} & \cdots & S_{2J} \\ \vdots & \vdots & \ddots & \vdots \\ S_{N1} & S_{N2} & \cdots & S_{NJ} \end{bmatrix} \tag{2.1}$$

where S_{ij} represents the payoff at state j for security i. In other words, each row represents random payoffs of a given security (say i) and each column represents the payoffs of different securities if a given state (say j) is realized.

Define similarly the return matrix as:

$$\mathbf{R} = \begin{bmatrix} R_{11} & R_{12} & \cdots & R_{1J} \\ R_{12} & R_{22} & \cdots & R_{2J} \\ \vdots & \vdots & \ddots & \vdots \\ R_{N1} & R_{N2} & \cdots & R_{NJ} \end{bmatrix} \tag{2.2}$$

where $R_{ij} = S_{ij}/v_i$ and v_i is the current value of security i. We write the value matrix as follows:

$$\underline{v} = \begin{bmatrix} v_1 \\ v_2 \\ \vdots \\ v_N \end{bmatrix} \tag{2.3}$$

The initial wealth, W_0, is the amount of money spent in N securities:

$$W_0 = \sum_{i=1}^{N} n_i v_i \tag{2.4}$$

where n_i is the number of shares bought for the i-th security. Hence, the portfolio weight of the i-th security can be written as:

$$w_i = \frac{n_i v_i}{W_0} \tag{2.5}$$

and

$$\sum_{i=1}^{N} w_i = 1 \tag{2.6}$$

The portfolio returns in various states are:

$$\begin{bmatrix} w_1 & w_2 & \cdots & w_N \end{bmatrix} \times \begin{bmatrix} R_{11} & \cdots & R_{1J} \\ \vdots & \ddots & \vdots \\ R_{N1} & \cdots & R_{NJ} \end{bmatrix} \tag{2.7}$$
$$= \begin{bmatrix} \sum_{i=1}^{N} w_i R_{i1} & \sum_{i=1}^{N} w_i R_{i2} & \cdots & \sum_{i=1}^{N} w_i R_{iJ} \end{bmatrix}$$

Now we are ready to introduce two important definitions. These definitions seem trivial in finance. But we need these formal definitions in order to derive models with proper rigor.

Definition – Stochastic Dominance. Portfolio A $\begin{bmatrix} w_{a1} & w_{a2} & \cdots & w_{aN} \end{bmatrix}$ is said to dominate another portfolio B $\begin{bmatrix} w_{b1} & w_{b2} & \cdots & w_{bN} \end{bmatrix}$ if:

$$\begin{aligned} &\begin{bmatrix} w_{a1} & w_{a2} & \cdots & w_{aN} \end{bmatrix} \times \begin{bmatrix} R_{11} & \cdots & R_{1J} \\ \vdots & \ddots & \vdots \\ R_{N1} & \cdots & R_{NJ} \end{bmatrix} \\ &\geqslant \begin{bmatrix} w_{b1} & w_{b2} & \cdots & w_{bN} \end{bmatrix} \times \begin{bmatrix} R_{11} & \cdots & R_{1J} \\ \vdots & \ddots & \vdots \\ R_{N1} & \cdots & R_{NJ} \end{bmatrix} \end{aligned} \tag{2.8}$$

□

Note that the resulting matrix is a 1 by J vector. "$\geqslant$" means all components in the LHS are larger than those in the RHS.

Definition – Arbitrage. An arbitrage exists if

$$\sum_{i=1}^{N} w_i \leqslant 0 \tag{2.9}$$

and

$$\begin{bmatrix} w_1 & w_2 & \cdots & w_N \end{bmatrix} \times \begin{bmatrix} R_{11} & \cdots & R_{1J} \\ \vdots & \ddots & \vdots \\ R_{N1} & \cdots & R_{NJ} \end{bmatrix} \geqslant 0 \tag{2.10}$$

□

Ever since the Black-Scholes discovery, "no arbitrage" has served the basis for modern financial models. The above definition can be interpreted easily as "no free lunch" – no out-of-pocket investment should yield no return (i.e. not all states have positive returns). An alternative definition for no free lunch is that any risk-free investment should yield the risk-free return. In the rest of this book, the two definitions are used inter-changably.

2.3 State Security

Define a state security (a.k.a. Arrow-Debreu security, state contingent claim, or pure security) as a security that pays \$1 only at one state (e.g. j-th state) and nothing else. Then the payoff matrix for J state securities is:

$$\mathbf{I} = \begin{bmatrix} 1 & 0 & \cdots & 0 \\ 0 & 1 & \cdots & 0 \\ \vdots & \vdots & \ddots & \vdots \\ 0 & 0 & \cdots & 1 \end{bmatrix}_{J \times J} \tag{2.11}$$

It is an identity $J \times J$ matrix. And we write the current price vector as:

$$\underline{\pi} = \begin{bmatrix} \pi_1 \\ \pi_2 \\ \vdots \\ \pi_J \end{bmatrix} \tag{2.12}$$

where π_j represents the price of \$1 paid if state j is realized. In other words, each element in (2.12) is the price of a j-state contingent claim. Finally, we introduce an important lemma about state security.

Lemma There exists a non-negative state price vector if and only if there exist no arbitrage opportunities.

□

2.4 Relationship Between S and I Under Complete Markets

Under complete markets, that is $J = N$, there is a relationship between regular securities prices and state securities prices. In other words, we can replicate state securities using regular securities. As an example, we suppose we want to replicate the first state security using regular securities. In order to do so, we form a portfolio of regular securities with the following shares:

$$\begin{bmatrix} n_{11} & n_{12} & \cdots & n_{1J} \end{bmatrix} \tag{2.13}$$

each of which, say i where $i \in [1, \cdots, J]$, pays S_{ij} at state j. The payoff matrix of these regular securities hence is:

$$\begin{aligned} &\begin{bmatrix} n_{11} & n_{12} & \cdots & n_{1J} \end{bmatrix} \times \begin{bmatrix} S_{11} & S_{12} & \cdots & S_{1J} \\ S_{12} & S_{22} & \cdots & S_{2J} \\ \vdots & \vdots & \ddots & \vdots \\ S_{J1} & S_{J2} & \cdots & S_{JJ} \end{bmatrix} \\ &= \begin{bmatrix} \sum_{i=1}^{J} n_i S_{i1} & \sum_{i=1}^{J} n_i S_{i2} & \cdots & \sum_{i=1}^{J} n_i S_{iJ} \end{bmatrix} \end{aligned} \tag{2.14}$$

For this portfolio of regular securities to produce the payoff of a state security, the shares defined in (2.13) must be chosen in such a way that the state payoffs equal those of the first state security:

$$\begin{bmatrix} 1 & 0 & \cdots & 0 \end{bmatrix} \tag{2.15}$$

In other words:

$$\begin{bmatrix} n_{11} & n_{12} & \cdots & n_{1J} \end{bmatrix} \times \begin{bmatrix} S_{11} & S_{12} & \cdots & S_{1J} \\ S_{12} & S_{22} & \cdots & S_{2J} \\ \vdots & \vdots & \ddots & \vdots \\ S_{J1} & S_{J2} & \cdots & S_{JJ} \end{bmatrix} = \begin{bmatrix} 1 & 0 & \cdots & 0 \end{bmatrix} \tag{2.16}$$

Hence, to solve for the vector of shares, we simply inverse the matrix S:

$$\begin{bmatrix} n_{11} & n_{12} & \cdots & n_{1J} \end{bmatrix} = \begin{bmatrix} 1 & 0 & \cdots & 0 \end{bmatrix} \times \begin{bmatrix} S_{11} & S_{12} & \cdots & S_{1J} \\ S_{12} & S_{22} & \cdots & S_{2J} \\ \vdots & \vdots & \ddots & \vdots \\ S_{J1} & S_{J2} & \cdots & S_{JJ} \end{bmatrix}^{-1} \tag{2.17}$$

which is the first row of $\mathbf{S}^{-1}$. The cost of the portfolio is:

$$\sum_{i=1}^{J} n_i v_i = \pi_1 \tag{2.18}$$

Similarly, to reproduce the j-th state security, the numbers of shares in regular securities are equal to the j-th row of $\mathbf{S}^{-1}$. For convenience, to express the general case where all state securities are included we use the following matrix notation:

$$\mathbf{N} \times \underline{v} = \mathbf{S}^{-1} \times \underline{v} = \underline{\pi} \tag{2.19}$$

It is straightforward to go from state securities to regular securities. Let the number of shares of state securities be:

$$\begin{bmatrix} h_{i1} & h_{i2} & \cdots & h_{iJ} \end{bmatrix} \tag{2.20}$$

To mimic the payoffs of the i-th regular security, we simple hold shares of the state securities as follows:

$$\begin{bmatrix} h_{i1} & h_{i2} & \cdots & h_{iJ} \end{bmatrix} = \begin{bmatrix} S_{i1} & S_{i2} & \cdots & S_{iJ} \end{bmatrix} \tag{2.21}$$

Hence,

$$\begin{bmatrix} S_{11} & S_{12} & \cdots & S_{1J} \\ S_{12} & S_{22} & \cdots & S_{2J} \\ \vdots & \vdots & \ddots & \vdots \\ S_{J1} & S_{J2} & \cdots & S_{JJ} \end{bmatrix} \times \begin{bmatrix} 1 & 0 & \cdots & 0 \\ 0 & 1 & \cdots & 0 \\ \vdots & \vdots & \ddots & \vdots \\ 0 & 0 & \cdots & 1 \end{bmatrix} = \begin{bmatrix} S_{11} & S_{12} & \cdots & S_{1J} \\ S_{12} & S_{22} & \cdots & S_{2J} \\ \vdots & \vdots & \ddots & \vdots \\ S_{J1} & S_{J2} & \cdots & S_{JJ} \end{bmatrix} \tag{2.22}$$

The price of this portfolio is:

$$\sum\nolimits_{j=1}^{J} \pi_j h_{ij} = \sum\nolimits_{j=1}^{J} \pi_j S_{ij} = v_i \tag{2.23}$$

for the i-th security. In matrix notation for all securities, we have

$$\underline{v} = \mathbf{H} \times \underline{\pi} = \mathbf{S} \times \underline{\pi} \tag{2.24}$$

Note that this relationship yields a very convenient and useful result:

$$\sum\nolimits_{j=1}^{J} \frac{\pi_j S_{ij}}{v_i} = 1 \tag{2.25}$$

for all i. Recall that $R_{ij} = S_{ij}/v_i$, and hence the above result implies:

$$\sum\nolimits_{j=1}^{J} \pi_j R_{ij} = 1 \tag{2.26}$$

for all i. In a matrix form:

$$\begin{bmatrix} R_{11} & R_{12} & \cdots & R_{1J} \\ R_{12} & R_{22} & \cdots & R_{2J} \\ \vdots & \vdots & \ddots & \vdots \\ R_{J1} & R_{J2} & \cdots & R_{JJ} \end{bmatrix} \times \begin{bmatrix} \pi_1 & \pi_2 & \cdots & \pi_J \end{bmatrix} = \begin{bmatrix} 1 \\ 1 \\ \vdots \\ 1 \end{bmatrix} \tag{2.27}$$

2.5 Insurable State

Definition A state is said to be insurable if there exists a portfolio (of regular securities) that can replicate the payoff of the state security.

□

Hence, we can see from above that the states that are insurable must have a supporting **S** matrix that is full rank (i.e. invertible).

2.6 Risk-free Rate

If we buy a share of every state security, then we are guaranteed to receive \$1, regardless which state is realized; hence the portfolio is riskless and should yield the risk-free return.

$$\sum_{j=1}^{J} \pi_j = \frac{1}{1+r_f} \tag{2.28}$$

where r_f is the risk-free rate.

2.7 Price By Arbitrage And Risk Neutral Pricing

The basic economic law says that:

$$v_i = \frac{\mathbb{E}[\tilde{S}_i]}{k_i} \tag{2.29}$$

where k_i is the discount rate for the i-th security. Define:

$$\hat{p}_j = \pi_j(1+r_f) \\ = \frac{\pi_j}{\sum_{j=1}^{J}\pi_j} \tag{2.30}$$

Hence it is true that

$$\sum_{j=1}^{J}\hat{p}_j = 1 \tag{2.31}$$

and $\hat{p}_j$'s can be regarded as probabilities. Note that:

$$\hat{\mathbb{E}}[\tilde{S}_i] = \sum_{j=1}^{J}\hat{p}_j S_{ij} = \sum_{j=1}^{J}\frac{p_j S_{ij}}{\sum_{j=1}^{J}p_j} = v_i(1+r_f) \tag{2.32}$$

Or we can write it as:

$$v_i = \frac{\hat{\mathbb{E}}[\tilde{S}_i]}{1+r_f} \tag{2.33}$$

Hence, $\hat{p}_j$'s are risk-neutral probabilities as the expected value using $\hat{p}$ yields the risk-free rate. Also note that, interestingly,

$$\hat{\mathbb{E}}[\tilde{R}_i] = \sum_{j=1}^{J}\hat{p}_j R_{ij} = \sum_{j=1}^{J}\frac{\pi_j R_{ij}}{\sum_{j=1}^{J}\pi_j} = 1+r_f \tag{2.34}$$

This holds regardless of i! Next, we define the state price per unit as:

$$q_j = \frac{\pi_j}{p_j} \tag{2.35}$$

where p_j is the actual probability for the j-th state. Then,

$$\mathbb{E}[\tilde{q}\tilde{S}_i] = \sum_{j=1}^{J}p_j q_j S_{ij} = \sum_{j=1}^{J}\pi_j S_{ij} = v_i \tag{2.36}$$

So we know that q is like a discount factor. We can also show that

$$\mathbb{E}[\tilde{q}\tilde{R}_i] = \sum_{j=1}^{J}p_j q_j R_{ij} = \sum_{j=1}^{J}\pi_j R_{ij} = 1 \tag{2.37}$$

Rewriting the above,

$$\begin{aligned} 1 &= \text{cov}[\tilde{q}, \tilde{R}_i] + \mathbb{E}[\tilde{q}]\mathbb{E}[\tilde{R}_i] \\ \mathbb{E}[\tilde{R}_i] &= \frac{1 - \text{cov}[\tilde{q}, \tilde{R}_i]}{\mathbb{E}[\tilde{q}]} \\ &= (1 + r_f) - (1 + r_f)\,\text{cov}[\tilde{q}, \tilde{R}_i] \end{aligned} \tag{2.38}$$

The last line is obtained because:

$$\mathbb{E}[\tilde{q}] = \sum_{j=1}^{J} p_j q_j = \sum_{j=1}^{J} \pi_j = \frac{1}{1 + r_f} \tag{2.39}$$

Equation (2.39) is quite significant in asset pricing. $\tilde{q}$ is the stochastic discount factor which can be interpreted as the Marginal Rate of Substitution (MRS) in microeconomic theory. Equation (2.39) states that the expected value of the MRS yields the risk-free discount. The theory on MRS is not limited to one period setting as the state price theory introduced here. The theory on MRS is beyond the scope of this book. Interested readers can find such materials in standard microeconomic text books.

2.8 Portfolio Problem

Any portfolio problem starts with a utility maximization objective (using state prices π_j and number of shares h_j) as follows:

$$\max \mathbb{E}[\tilde{U}] = \sum_{j=1}^{J} p_j U(h_j) \tag{2.40}$$

subject to $\sum_{j=1}^{J} h_j \pi_j = W_0$.

Note that the wealth at time-1 is $W_1 = h_j$ where state j is realized. Form Lagrangian as follows:

$$\mathcal{L} = \sum_{j=1}^{J} \pi_j U(h_j) + \lambda \left[W_0 - \sum_{j=1}^{J} h_j \pi_j \right] \tag{2.41}$$

The first order conditions:

$$\begin{aligned} 0 &= \frac{\partial L}{\partial \lambda} = W_0 - \sum_{j=1}^{J} h_j \pi_j \\ 0 &= \frac{\partial L}{\partial h_j} = \pi_j \frac{\partial U(h_j)}{\partial h_j} - \lambda \pi_j \end{aligned} \tag{2.42}$$

The first condition yields the budget constraint, and the second gives:

$$\frac{p_j}{\pi_j}U'(h_j) = \lambda \tag{2.43}$$

Given this result, we obtain, for the return on any arbitrary regular security:

$$\begin{aligned}\mathbb{E}[U'\tilde{R}_i] &= \sum\nolimits_{j=1}^{J} p_j U'(h_j) R_{ij} \\ &= \lambda \sum\nolimits_{j=1}^{J} \frac{\pi_j S_{ij}}{v_i} \\ &= \lambda\end{aligned} \tag{2.44}$$

and also, for the return on the state security:

$$\begin{aligned}\mathbb{E}\left[\frac{U'(h_j)}{\tilde{\pi}}\right] &= \sum\nolimits_{j=1}^{J} p_j U'(h_j)\frac{1}{\pi_j} \\ &= \lambda\end{aligned} \tag{2.45}$$

They both are equal to the shadow price.

Chapter 3

Intertemporal CAPM

Intertemporal Capital Asset Pricing Models (ICAPM) tie closely to the state price theory introduced in the previous chapter. They serve the foundation of derivative pricing theories. Without proper understanding of ICAPM, financial modellers might develop models that violate equilibrium. While there are a number of ICAPMs in the economic literature, we only demonstrate the two most relevant ones - Merton and Cox-Ingersoll-Ross.

3.1 Intertemporal Portfolio Selection

3.1.1 Definition

Define $P_i(t)$ as price for stock i at time t and $\tilde{z}_i(t) = \frac{\tilde{P}_i(t+1)}{P_i(t)}$ as the (random) return of stock i. The symbol "tilde" on top of a variable represents a random variable. There are $n+1$ assets where $i = 0$ represents the risk-free asset and $i = 1, \cdots, n$ are risky assets. We assume, for simplicity, that the risk-free return is constant over time – $z_0(t) = z_0$ for all t. As a utility maximizer, we maximize the following objective function:

$$\max_{C,B}\{\mathbb{E}[U(C(0), C(1), \cdots, C(T-1), B(T))]\} \tag{3.1}$$

where $C(t)$ is consumption and B is bequest. Under the von Neumann-Morgenstern utility assumption, the above utility maximization problem becomes (additive):

$$\max_{C,B}\left\{\mathbb{E}\left[\sum_{t=1}^{T-1} U(C(t), t) + B(W(T), T)\right]\right\} \tag{3.2}$$

where the bequest on the terminal wealth is the last utility. The budget constraint is given as:

$$\left.\begin{aligned} I(t) &= W(t) + Y(t) - C(t) \\ &= \sum\nolimits_{i=0}^{n} P_i(t)N_i(t) \end{aligned}\right) \tag{3.3}$$

where W represents wealth, Y represents wage income, and C as defined earlier represents consumption. Equation (3.3) states that the amount at the current time t that can be used for future investment (i.e. I) is equal to the current wealth and the wage income made at current time minus the money spent in consumption. The second line of equation (3.3) is to use the the total investment income in risky stocks where P represents the price and N represents shares. As indicated earlier, $i = 0$ represents the risk-free asset and there are n risky stocks (i.e. $i = 1, \cdots, n$).

Define the portfolio weight as:

$$x_i(t) = \frac{P_i(t)N_i(t)}{I(t)} \tag{3.4}$$

Then, the wealth in the next period is:

$$\begin{aligned} \tilde{W}(t+1) &= \sum\nolimits_{i=0}^{n} N_i(t)\tilde{P}_i(t+1) \\ &= I(t)\sum\nolimits_{i=0}^{n} \frac{N_i(t)P_i(t)}{I(t)}\frac{\tilde{P}_i(t+1)}{P_i(t)} \\ &= I(t)\sum\nolimits_{i=0}^{n} x_i(t)\tilde{z}_i(t) \\ &= I(t)\tilde{z}^*(t) \end{aligned} \tag{3.5}$$

where

$$\begin{aligned} \tilde{z}^*(t) &= \sum\nolimits_{i=0}^{n} x_i(t)\tilde{z}_i(t) \\ &= x_0(t)z_0 + \sum\nolimits_{i=1}^{n} x_i(t)\tilde{z}_i(t) \\ &= \left[1 - \sum\nolimits_{i=1}^{n} x_i(t)\right] z_0 + \sum\nolimits_{i=1}^{n} x_i(t)\tilde{z}_i(t) \\ &= z_0 + \sum\nolimits_{i=1}^{n} x_i(t)(\tilde{z}_i(t) - z_0) \end{aligned} \tag{3.6}$$

Intuitively, equation (3.6) indicates that the next period wealth (which is random) is the current investment times the next period portfolio return (which is random). The portfolio return is further expressed as the risk-free rate plus the portfolio risk premium.

At $T-1$, we maximize (note that the choice variable changes from B to x since portfolio decision determines the terminal wealth, and hence bequest).

$$\max_{C,x}\{U(C(T-1),T-1)+\mathbb{E}_{T-1}[B(W(T),T)]\} \tag{3.7}$$

subject to:

$$\begin{aligned}\tilde{W}(T) &= I(T-1)\tilde{z}^*(T-1)\\ &= \{W(T-1)+Y(T-1)-C(T-1)\}\\ &\quad\left(z_0+\sum\nolimits_{i=1}^{n} x_i(T-1)(\tilde{z}_i(T-1)-z_0)\right)\end{aligned} \tag{3.8}$$

The first order conditions are (with respect to $C(T-1)$)

$$U_{C(T-1)}+\mathbb{E}_{T-1}[\tilde{B}_{W(T)}\tilde{W}_{C(T-1)}]=U_{C(T-1)}-\mathbb{E}_{T-1}[\tilde{B}_{W(T)}\tilde{z}^*]=0 \tag{3.9}$$

which results in (marginal utility of consumption equals to marginal utility of bequest):

$$U_{C(T-1)}=\mathbb{E}_{T-1}[\tilde{B}_{W(T)}\tilde{z}^*] \tag{3.10}$$

where

$$U_{C(T-1)}=\frac{\partial U}{\partial C(T-1)}\tilde{B}_{W(T)}=\frac{\partial \tilde{B}}{\partial W(T)} \tag{3.11}$$

and (with respect to x_i):

$$\mathbb{E}_{T-1}[\tilde{B}_{W(T)}\tilde{W}_{x_i}]=\mathbb{E}_{T-1}[\tilde{B}_{W(T)}I(T-1)(\tilde{z}_i(T-1)-z_0)]=0 \tag{3.12}$$

which can be simplified as:

$$\mathbb{E}_{T-1}[\tilde{B}_{W(T)}(\tilde{z}_i(T-1)-z_0)]=0 \tag{3.13}$$

Multiplying (3.13) by x_i and summing across i, we obtain:

$$
\begin{aligned}
\sum_{i=1}^{n} \mathbb{E}_{T-1}[\tilde{B}_{W(T)} x_i(T-1)(\tilde{z}_i(T-1) - z_0)] \\
&= \mathbb{E}_{T-1}\left[\tilde{B}_{W(T)} \sum_{i=1}^{n} x_i(T-1)(\tilde{z}_i(T-1) - z_0)\right] \\
&= 0
\end{aligned}
\tag{3.14}
$$

which is:

$$
\begin{aligned}
\mathbb{E}_{T-1}[\tilde{B}_{W(T)}(\tilde{z}^* - z_0)] = 0 \\
\mathbb{E}_{T-1}[\tilde{B}_{W(T)}\tilde{z}^*] = z_0 E_{T-1}[\tilde{B}_{W(T)}]
\end{aligned}
\tag{3.15}
$$

This equation relates to risk-neutral pricing. It states that the expected compounded bequest is identical to the expected bequest compounded at the risk-free rate. From (3.10), it follows that

$$
U_{C(T-1)} = z_0 \mathbb{E}_{T-1}[\tilde{B}_{W(T)}] \tag{3.16}
$$

Now, define $J(W(T-1), T-1)$ as the indirect utility function such that

$$
\begin{aligned}
J(W(T-1), T-1) &= \max_{C,x}\{U(C(T-1), T-1) + \mathbb{E}_{T-1}[B(W(T), T)]\} \\
&= U(C^*(T-1), T-1) + \mathbb{E}_{T-1}[B(W^*(T), T)]
\end{aligned}
\tag{3.17}
$$

The indirect utility function is an already optimized utility function and is a function of wealth as opposed to the original utility function that is a function of consumption. Taking a partial derivative with respect to wealth, we obtain

$$
\begin{aligned}
J_{W(T-1)} &= \frac{\partial J(W(T-1), T-1)}{\partial W(T-1)} \\
&= \frac{\partial U(C^*(T-1), T-1)}{\partial C^*(T-1)} \frac{\partial C^*(T-1)}{\partial W(T-1)} + \mathbb{E}_{T-1}\left[\frac{\partial B(W(T), T)}{\partial W(T)} \frac{\partial W(T)}{\partial W(T-1)}\right] \\
&= 0 \cdot \frac{\partial C^*(T-1)}{\partial W(T-1)} + \mathbb{E}_{T-1}\left[B_{W(T)} z^*\right] \\
&= \mathbb{E}_{T-1}\left[B_{W(T)} z^*\right] \\
&= U_{C(T-1)}
\end{aligned}
\tag{3.18}
$$

This is a particularly interesting result that states that the marginal utility of consumption is the same as the marginal utility of wealth. Now we will perform backward induction and perform Bellman optimization.

Moving backwards, at $T-2$, we get

$$\begin{aligned} J(W(T-2),T-2) &= \max_{C,x}\{U(C(T-2),T-2) + \mathbb{E}_{T-2}[U(C(T-1),T-1) \\ &+ \mathbb{E}_{T-1}[B(W(T),T)]]\} \\ &= \max_{C,x}\{U(C(T-2),T-2) + \mathbb{E}_{T-2}[U(C(T-1),T-1) \\ &+ B(W(T),T)]\} \\ &= \max_{C,x}\{U(C(T-2),T-2) + \mathbb{E}_{T-2}[J(W(T-1),T-1)]\} \end{aligned} \tag{3.19}$$

subject to the budget constraint (3.8) which is:

$$\begin{aligned} \tilde{W}(T-1) &= (W(T-2) + Y(T-2) - C(T-2)) \\ &\left(z_0 + \sum\nolimits_{i=1}^{n} x_i(T-2)[\tilde{z}_i(T-2) - z_0]\right) \\ &= (W(T-2) + Y(T-2) - C(T-2))z^* \end{aligned} \tag{3.20}$$

The last line of (3.19) is a result of Bellman Dynamic Optimization Principal. The first order conditions are (with respect to $C(T-2)$):

$$\begin{aligned} U_{C(T-2)} + \mathbb{E}_{T-2}[\tilde{J}_{W(T-1)}\tilde{W}_{C(T-2)}] = U_{C(T-2)} - \mathbb{E}_{T-2}[\tilde{J}_{W(T-1)}\tilde{z}^*] = 0 \\ U_{C(T-2)} = \mathbb{E}_{T-2}[\tilde{J}_{W(T-1)}\tilde{z}^*] \end{aligned} \tag{3.21}$$

Then, similar to (3.16), we reach the following result

$$\mathbb{E}_{T-2}[\tilde{J}_{W(T-1)}(\tilde{z}_i - z_0)] = 0 \tag{3.22}$$

Repeated substitutions imply that the above result applies to any arbitrary t.

3.2 CAPM in Continuous Time

Define the following process for returns:

$$\zeta_i(t) = z_i(t) - 1 = \frac{P_i(t+h) - P_i(t)}{P_i(t)} \tag{3.23}$$

with

$$\begin{aligned} \mathbb{E}[\zeta_i] &= \alpha_i h \\ \mathbb{E}[\zeta_i \zeta_j] &= \sigma_{ij} h \end{aligned} \tag{3.24}$$

Note that the budget constraint is:

$$\begin{aligned} W(t+h) &= I(t) \sum\nolimits_{i=1}^{n} x_i(t) \tilde{z}_i(t) \\ V(t) &= \frac{W(t+h)}{I(t)} - 1 = \sum\nolimits_{i=1}^{n} x_i(t) \tilde{\zeta}_i(t) \end{aligned} \tag{3.25}$$

such that:

$$\begin{aligned} \mathbb{E}[V(t)] &= \sum\nolimits_{i=1}^{n} x_i(t) \alpha_i(t) h \\ \mathbb{E}[V(t)^2] &= \sum\nolimits_{i=1}^{n} \sum\nolimits_{j=1}^{n} x_i(t) x_j(t) \sigma_{ij} h \end{aligned} \tag{3.26}$$

Rewrite the budget constraint as:

$$\begin{aligned} \tilde{W}(t+h) &= (W(t) - C(t)h) \sum\nolimits_{i=0}^{n} x_i(t) \tilde{z}_i(t) \\ &= (W(t) - C(t)h) \sum\nolimits_{i=0}^{n} x_i(t)(1 + \tilde{\zeta}_i(t)) \\ &= (W(t) - C(t)h) \sum\nolimits_{i=0}^{n} x_i(t) \tilde{\zeta}_i(t) + W(t) - C(t)h \end{aligned} \tag{3.27}$$

Hence,

$$\tilde{W}(t+h) - W(t) = (W(t) - C(t)h) \sum\nolimits_{i=0}^{n} x_i(t) \tilde{\zeta}_i(t) - C(t)h \tag{3.28}$$

Take the expected value of the first and second moments and get:

$$\begin{aligned}\mathbb{E}[W(t+h)-W(t)] &= (W(t)-C(t)h)\sum\nolimits_{i=0}^{n} x_i(t)\alpha_i(t)h - C(t)h\\ &= W(t)\sum\nolimits_{i=0}^{n} x_i(t)\alpha_i(t)h - C(t)h\\ &= W(t)\left(x_0 rh + \sum\nolimits_{i=1}^{n} x_i(t)\alpha_i(t)h\right) - C(t)h\\ &= W(t)\left(rh + \sum\nolimits_{i=1}^{n} x_i(t)(\alpha_i(t)-r)h\right) - C(t)h\end{aligned} \tag{3.29}$$

and

$$\mathbb{E}[(W(t+h)-W(t))^2] = \sum\nolimits_{i=1}^{n}\sum\nolimits_{j=1}^{n} x_i(t)x_j(t)\sigma_{ij}W(t)^2 h \tag{3.30}$$

Define (known as the indirect utility function):

$$J(W(t),t) = \max \mathbb{E}_t\left[\int_t^T U(C(s),s)ds + B(W(T),T)\right] \tag{3.31}$$

By the result of portfolio selection above, the problem can be written as:

$$J(W(t),t) = \max \mathbb{E}_t\left[\int_t^{t+h} U(C(s),s)ds + J(W(t+h),t+h)\right] \tag{3.32}$$

Expand $J(W(t+h),t+h)$ around $W(t)$ and t to get:

$$\begin{aligned}&J(W(t+h),t+h)\\ &= J(W(t),t) + J_W[W(t+h)-W(t)] + \frac{1}{2}J_{WW}[W(t+h)-W(t)]^2 + J_t h\end{aligned} \tag{3.33}$$

Substitute this result back to (3.32) to get,

$$\begin{aligned}0 &= \max \mathbb{E}_t\left[\int_t^{t+h} U(C(s),s)ds + J_W[W(t+h)-W(t)] + \frac{1}{2}J_{WW}[W(t+h)-W(t)]^2 + J_t h\right]\\ &= \int_t^{t+h} U(C(s),s)ds + J_W\left[W(t)\left(r+\sum\nolimits_{i=1}^{n} x_i(t)(\alpha_i(t)-r)\right) - C(t)\right]h\\ &+ \frac{1}{2}J_{WW}\sum\sum x_i(t)x_j(t)\sigma_{ij}W(t)^2 h + J_t h\end{aligned} \tag{3.34}$$

Apply the mean value theorem on the utility function,

$$\int_t^{t+h} U(C(s),s)ds = U(C(t^*),t^*)h \tag{3.35}$$

where $t < t^* < t+h$. Substitute the result back into (3.32) to obtain:

$$\begin{aligned} 0 = U(C(t^*),t^*)h + J_W \left[W(t)\left(r + \sum_{i=1}^{n} x_i(t)(\alpha_i(t) - r)\right) - C(t)\right] h \\ + \frac{1}{2} J_{WW} \sum\sum x_i(t)x_j(t)\sigma_{ij} W(t)^2 h + J_t h \end{aligned} \tag{3.36}$$

Canceling h on both sides and as $h \to 0, U(C(t^*),t^*) \to U(C(t),t)$, we have,

$$\begin{aligned} 0 = U(C(t),t) + J_W \left[W(t)\left(r + \sum_{i=1}^{n} x_i(t)(\alpha_i(t) - r)\right) - C(t)\right] \\ + \frac{1}{2} J_{WW} \sum_{i=1}^{n} \sum_{j=1}^{n} x_i(t)x_j(t)\sigma_{ij} W(t)^2 + J_t \end{aligned} \tag{3.37}$$

The first order conditions (with respect to C) are:

$$\begin{aligned} 0 &= U_C - J_W \\ U_C &= J_W \end{aligned} \tag{3.38}$$

and (with respective to x_i),

$$\begin{aligned} 0 &= J_W W(\alpha_i - r) + J_{WW} \sum_{i=1}^{n} x_i \sigma_{ij} W^2 \\ &= J_W(\alpha_i - r) + J_{WW} \sum_{j=1}^{n} x_j \sigma_{ij} W \end{aligned} \tag{3.39}$$

In matrix notation (underscore represents vector),

$$0 = J_W \begin{bmatrix} \alpha_1 - r \\ \alpha_2 - r \\ \vdots \\ \alpha_n - r \end{bmatrix} + J_{WW} W \begin{bmatrix} \sigma_{11} & \sigma_{12} & \cdots & \sigma_{1n} \\ \sigma_{21} & & & \\ \vdots & & & \\ \sigma_{n1} & & & \sigma_{nn} \end{bmatrix} \begin{bmatrix} x_1 \\ x_2 \\ \vdots \\ x_n \end{bmatrix} \tag{3.40}$$

$$= J_W(\underline{\alpha} - r\underline{1}) + J_{WW} W \Omega \underline{x}$$

Hence, to solve for the optimal weights, we have:

$$\begin{aligned}
-\frac{J_W}{J_{WW}W}(\underline{\alpha} - r\underline{1}) &= \Omega\underline{x} \\
K(\underline{\alpha} - r\underline{1}) &= \Omega\underline{x} \\
K\Omega^{-1}(\underline{\alpha} - r\underline{1}) &= \underline{x}^* \\
x_i^* &= K\sum\nolimits_{j=1}^{n} v_{ij}(\alpha_j - r)
\end{aligned} \tag{3.41}$$

where $K = -\frac{J_W}{J_{WW}W}$ is the relative risk aversion. Note that K disappears in x_i^*/x_j^* and hence the ratio of two portfolio weights is independent of risk preference. This is the famous separation theorem.

Given a particular K, the optimal portfolio is $x_i^* = K\sum_{j=1}^{n} v_{ij}(\alpha_j - r)$. The market has a certain endowment (supposed to be the sum of all individual wealth):

$$\underline{x}^M = K^M\Omega^{-1}(\underline{\alpha} - r\underline{1}) \tag{3.42}$$

Then

$$\begin{aligned}
\Omega\underline{x}^M &= K^M(\underline{\alpha} - r\underline{1}) \\
\sigma_{iM} &= K^M(\alpha_i - r) \\
\sigma_M^2 &= K^M(\alpha_M - r) \\
\beta_i &= \frac{\alpha_i - r}{\alpha_M - r}
\end{aligned} \tag{3.43}$$

We verified the famous CAPM in the continuous time framework.

3.3 Intertemporal CAPM – Merton Model

In a multi-period framework, $z_i(t)$ is a stochastic process. In other words, the efficient frontier changes over time. In a simplest setting, we assume that the change of the frontier is driven by a single factor. In other words, all assets are intertemporally perfectly correlated and yet correlated freely within a period. To see that, let the optimal utility function, $J(W(t), \pi(t), t)$, be a function of two state variables, wealth and some state variable that drives intertemporal movements of the risky returns.

Define the state variable behavior (dynamics) as follows:

$$\begin{aligned}\mathbb{E}[\pi(t+h)-\pi(t)] &= \mu h \\ \mathbb{E}[(\pi(t+h)-\pi(t))^2] &= \sigma_\pi^2 h\end{aligned} \tag{3.44}$$

Taylors series expansion on the indirect utility function,

$$\begin{aligned}J(W(t+h),\pi(t+h),t+h) &= J(W(t),\pi(t),t)+J_t h \\ +J_W[W(t+h)-W(t)] &+ \frac{1}{2}J_{WW}[W(t+h)-W(t)]^2 \\ +J_\pi[\pi(t+h)-\pi(t)] &+ \frac{1}{2}J_{\pi\pi}[\pi(t+h)-\pi(t)]^2 \\ &+J_{\pi W}[W(t+h)-W(t)][\pi(t+h)-\pi(t)]\end{aligned} \tag{3.45}$$

The last two terms are new, due to the intertemporal risk. The last term can be expanded as follows:

$$\begin{aligned}&[W(t+h)-W(t)][\pi(t+h)-\pi(t)] \\ &= \left[(W(t)-C(t)h)\sum_{i=0}^n x_i(t)\tilde{\zeta}_i(t)-C(t)h\right][\pi(t+h)-\pi(t)] \\ &= (W(t)-C(t)h)\sum_{i=0}^n x_i(t)\tilde{\zeta}_i(t)[\pi(t+h)-\pi(t)]-C(t)h[\pi(t+h)-\pi(t)]\end{aligned} \tag{3.46}$$

which, after taking expectation, leads to:

$$\mathbb{E}\{[W(t+h)-W(t)][\pi(t+h)-\pi(t)]\} = W(t)\sum_{i=0}^n x_i(t)\sigma_{i\pi}h \tag{3.47}$$

Then, (3.34) becomes:

$$\begin{aligned}0 = \max\mathbb{E}_t\Big\{ &\int_t^{t+h} U(C(s),s)ds + J(W(t),\pi(t),t)+J_t h \\ +J_W[W(t+h)-W(t)] &+ \frac{1}{2}J_{WW}[W(t+h)-W(t)]^2 \\ +J_\pi[\pi(t+h)-\pi(t)] &+ \frac{1}{2}J_{\pi\pi}[\pi(t+h)-\pi(t)]^2 \\ &+J_{\pi W}[W(t+h)-W(t)][\pi(t+h)-\pi(t)] \Big\}\end{aligned} \tag{3.48}$$

By a similar arrangement that leads to (3.37), we achieve the following:

$$
\begin{aligned}
0 = U(C(t),t) + J_t + J_\pi \mu + \frac{1}{2} J_{\pi\pi}\sigma_\pi^2 + J_{\pi W} W(t) \sum_{i=0}^{n} x_i(t)\sigma_{i\pi} \\
+ J_W \left[W(t)\left(r + \sum_{i=1}^{n} x_i(t)(\alpha_i(t) - r) \right) - C(t) \right] \\
+ \frac{1}{2} J_{WW} \sum_{i=1}^{n} \sum_{j=1}^{n} x_i(t) x_j(t) \sigma_{ij} W(t)^2
\end{aligned}
\tag{3.49}
$$

The first order conditions (similar to (3.38)) are:

$$
U_C = J_W \tag{3.50}
$$

and

$$
0 = J_{\pi W}\sigma_{i\pi} + J_W(\alpha_i(t) - r) + J_{WW} W(t) \sum_{j=1}^{n} x_j(t)\sigma_{ij} \tag{3.51}
$$

which can be further shown as:

$$
-\frac{J_{\pi W}}{J_{WW} W}\sigma_{i\pi} - \frac{J_W}{J_{WW} W}(\alpha_i(t) - r) = \sum_{j=1}^{n} x_j(t)\sigma_{ij} \tag{3.52}
$$

In a matrix form, we can write the above equation as:

$$
\begin{aligned}
-\frac{J_{\pi W}}{J_{WW} W}
\begin{bmatrix} \sigma_{1\pi} \\ \sigma_{2\pi} \\ \vdots \\ \sigma_{n\pi} \end{bmatrix}
- \frac{J_W}{J_{WW} W}
\begin{bmatrix} \alpha_1(t) - r \\ \alpha_2(t) - r \\ \vdots \\ \alpha_n(t) - r \end{bmatrix}
=
\begin{bmatrix} \sigma_{12} & \sigma_{13} & \cdots & \sigma_{1n} \\ \sigma_{21} & \sigma_{22} & \cdots & \sigma_{2n} \\ \vdots & & \ddots & \vdots \\ \sigma_{n1} & \sigma_{n2} & \cdots & \sigma_{nn} \end{bmatrix}
\begin{bmatrix} x_1(t) \\ x_2(t) \\ \vdots \\ x_n(t) \end{bmatrix} \\
-\frac{J_{\pi W}}{J_{WW} W}\underline{\sigma_{i\pi}} - \frac{J_W}{J_{WW} W}(\underline{\alpha(t)} - r\underline{1}) = \Omega \underline{x(t)}
\end{aligned}
\tag{3.53}
$$

Solving the equation for the vector of portfolio weights, $\underline{x(t)}$, we get:

$$
-\frac{J_{\pi W}}{J_{WW} W}\Omega^{-1}\underline{\sigma_{i\pi}} - \frac{J_W}{J_{WW} W}\Omega^{-1}(\underline{\alpha(t)} - r\underline{1}) = \underline{x^*(t)} \tag{3.54}
$$

or equivalently, each individual component is:

$$x_i^* = -\frac{J_{\pi W}}{J_{WW}W}\sum_{j=1}^{n} v_{ij}\sigma_{j\pi} - \frac{J_W}{J_{WW}W}\sum_{j=1}^{n} v_{ij}(\alpha_j - r) \tag{3.55}$$

Note now that x_i^*/x_j^* is no longer independent of preference. This is an important result!

Define:

$$\begin{aligned} D &= -\frac{J_W}{J_{WW}W}\underline{1}'\Omega^{-1}(\underline{\alpha} - r\underline{1}) \\ H &= -\frac{J_{\pi W}}{J_{WW}W}\underline{1}'\Omega^{-1}\underline{\sigma}_{i\pi} \end{aligned} \tag{3.56}$$

Then:

$$\begin{aligned} \underline{1}'\underline{x}^* &= -\frac{J_{\pi W}}{J_{WW}W}\underline{1}'\Omega^{-1}\underline{\sigma}_{i\pi} - \frac{J_W}{J_{WW}W}\underline{1}'\Omega^{-1}(\underline{\alpha(t)} - r\underline{1}) \\ 1 - x_0 &= H + D \end{aligned} \tag{3.57}$$

Substitute D and H back to (3.55) to get:

$$\underline{x}^* = H\underbrace{\frac{\Omega^{-1}\underline{\sigma}_{i\pi}}{\underline{1}'\Omega^{-1}\underline{\sigma}_{i\pi}}}_{\underline{h}} + D\underbrace{\frac{\Omega^{-1}(\underline{\alpha} - r\underline{1})}{\underline{1}'\Omega^{-1}(\underline{\alpha} - r\underline{1})}}_{\underline{d}} \tag{3.58}$$

Note that $\underline{1}'\underline{h} = \underline{1}'\underline{d} = 1$ so both $\underline{h}$ and $\underline{d}$ are two "portfolios" of risky stocks, and H and D are "weights" in the two portfolios. Note that $1 = x_0 + H + D$ is how investment is allocated among three portfolios – risk-free asset, hedging portfolio and market (tangent) portfolio respectively. It can be proved that h has the maximum absolute correlation with the state variable. This is left as an exercise for the reader.

Note that (3.53) is for a particular investor, say k:

$$-\frac{J_{\pi W}}{J_{WW}}\underline{\sigma}_{i\pi} - \frac{J_W}{J_{WW}}(\underline{\alpha} - r\underline{1}) = b^{(k)}\underline{\sigma}_{i\pi} + a^{(k)}(\underline{\alpha} - r\underline{1}) = W^{(k)}\Omega\underline{x}^{(k)} \tag{3.59}$$

where

$$a^{(k)} = -\frac{J_W}{J_{WW}}$$
$$b^{(k)} = -\frac{J_{\pi W}}{J_{WW}}$$

For any given security i, summing over all individuals (k), we obtain the entire market as follows:

$$\begin{aligned}
\sigma_{i\pi}\sum_{k=1}^{K} b^{(k)} + (\alpha_i - r)\sum_{k=1}^{K} a^{(k)} &= \sum_{k=1}^{K}\sum_{j=1}^{n} W^{(k)}\sigma_{ij}x_j^{(k)} \\
&= \sum_{j=1}^{n}\sigma_{ij}\sum_{k=1}^{K} W^{(k)}x_j^{(k)} \\
&= \sum_{j=1}^{n}\sigma_{ij}\frac{\sum_{k=1}^{K} W^{(k)}x_j^{(k)}}{\sum_{k=1}^{K} W^{(k)}}\sum_{k=1}^{K} W^{(k)} \\
&= \sum_{j=1}^{n}\sigma_{ij}\left(x_j^M\right)\sum_{k=1}^{K} W^{(k)} \\
&= \sigma_{iM}\sum_{k=1}^{K} W^{(k)}
\end{aligned} \tag{3.60}$$

Re-arrange terms yields:

$$(\alpha_i - r) = \sigma_{iM}\frac{\sum_{k=1}^{K} W^{(k)}}{\sum_{k=1}^{K} a^{(k)}} - \sigma_{i\pi}\frac{\sum_{k=1}^{K} b^{(k)}}{\sum_{k=1}^{K} a^{(k)}} = \sigma_{iM}A - \sigma_{i\pi}B \tag{3.61}$$

This equation is important as it already demonstrates how expected excess returns (or risk premium: $\alpha_i - r$) can be decomposed into the covariance with the market and the covariance with the hedging portfolio. Aggregate (3.61) over market weights to arrive at:

$$\begin{aligned}
\sum_{i=1}^{n} x_i^M(\alpha_i - r) = \sum_{i=1}^{n} x_i^M\sigma_{iM}A - \sum_{i=1}^{n} x_i^M\sigma_{i\pi}B \\
\alpha_M - r = A\sigma_M^2 - B\sigma_{\pi M}
\end{aligned} \tag{3.62}$$

Similarly, aggregate (3.61) over hedging portfolio weights to arrive at:

$$\begin{aligned}
\sum_{i=1}^{n} h_i(\alpha_i - r) = \sum_{i=1}^{n} h_i\sigma_{iM}A - \sum_{i=1}^{n} h_i\sigma_{i\pi}B \\
\alpha_h - r = A\sigma_{\pi M} - B\sigma_\pi^2
\end{aligned} \tag{3.63}$$

With these two equations, we can solve for the two unknowns, A and B, as follows:

$$\begin{bmatrix} A \\ B \end{bmatrix} = \begin{bmatrix} \sigma_M^2 & -\sigma_{\pi M} \\ \sigma_{\pi M} & -\sigma_\pi^2 \end{bmatrix}^{-1} \begin{bmatrix} \alpha_M - r \\ \alpha_h - r \end{bmatrix}$$
$$= \frac{1}{-\sigma_\pi^2\sigma_M^2 + \sigma_{\pi M}^2} \begin{bmatrix} -\sigma_\pi^2 & \sigma_{\pi M} \\ -\sigma_{\pi M} & \sigma_M^2 \end{bmatrix} \begin{bmatrix} \alpha_M - r \\ \alpha_h - r \end{bmatrix} \tag{3.64}$$
$$= \frac{1}{-\sigma_\pi^2\sigma_M^2 + \sigma_{\pi M}^2} \begin{bmatrix} -\sigma_\pi^2(\alpha_M - r) + \sigma_{\pi M}(\alpha_h - r) \\ -\sigma_{\pi M}(\alpha_M - r) + \sigma_M^2(\alpha_h - r) \end{bmatrix}$$

Substituting this back to (3.61), we get,

$$(\alpha_i - r) = \sigma_{iM}\frac{-\sigma_\pi^2(\alpha_M - r) + \sigma_{\pi M}(\alpha_h - r)}{-\sigma_\pi^2\sigma_M^2 + \sigma_{\pi M}^2} - \sigma_{i\pi}\frac{-\sigma_{\pi M}(\alpha_M - r) + \sigma_M^2(\alpha_h - r)}{-\sigma_\pi^2\sigma_M^2 + \sigma_{\pi M}^2}$$
$$= (\alpha_M - r)\frac{-\sigma_\pi^2\sigma_{iM} + \sigma_{i\pi}\sigma_{\pi M}}{-\sigma_\pi^2\sigma_M^2 + \sigma_{\pi M}^2} + (\alpha_h - r)\frac{\sigma_{\pi M}\sigma_{iM} - \sigma_{i\pi}\sigma_M^2}{-\sigma_\pi^2\sigma_M^2 + \sigma_{\pi M}^2} \tag{3.65}$$

If the hedging portfolio and the market are orthogonal, that is $\sigma_{\pi M} = 0$, then we achieve a "two-factor CAPM" as follows:

$$(\alpha_i - r) = (\alpha_M - r)\frac{\sigma_{iM}}{\sigma_M^2} + (\alpha_h - r)\frac{\sigma_{i\pi}}{\sigma_\pi^2} \tag{3.66}$$

where the first factor can be interpreted as the market factor as in the standard "one-factor" CAPM and the second factor is the hedging factor. Many finance text books try to relate this result to Ross' APT (Arbitrary Pricing Theory) where Ross proposes a linear, multi-factor, return generating function. Indeed, Ross' linear return generating function is like the "market model" in the standard single factor CAPM and Merton's ICAPM is like its equilibrium result. However, note that the two models are not directly mathematically connected.

3.4 Intertemporal CAPM – CIR Model

The CIR model is essentially the same as the Merton model but it is written more in the language of derivative pricing. CIR start with a joint process for the production technologies η and the state variables Y:

$$\begin{bmatrix} d\eta_1 \\ \vdots \\ d\eta_n \\ dY_1 \\ \vdots \\ dY_k \end{bmatrix} = \begin{bmatrix} \alpha_1(Y,t) \\ \vdots \\ \alpha_n(Y,t) \\ \mu_1(Y,t) \\ \vdots \\ \mu_k(Y,t) \end{bmatrix} dt + \begin{bmatrix} g_{1,1} & \cdots & g_{1,n} & g_{1,n+1} & \cdots & g_{1,n+k} \\ \vdots & \ddots & \vdots & \vdots & \ddots & \vdots \\ g_{1,n} & \cdots & g_{n,n} & g_{n,n+1} & \cdots & g_{n,n+k} \\ s_{1,1} & \cdots & s_{1,k} & s_{1,k+1} & \cdots & s_{1,n+k} \\ \vdots & \ddots & \vdots & \vdots & \ddots & \vdots \\ s_{k,1} & \cdots & s_{k,k} & s_{k,k+1} & \cdots & s_{k,n+k} \end{bmatrix} \begin{bmatrix} dW_1 \\ \vdots \\ dW_n \\ dW_{n+1} \\ \vdots \\ dW_{n+k} \end{bmatrix} \tag{3.67}$$

We label the upper-left corner of the variance-covariance matrix, i.e. g_{ij} where $i = 1, \cdots, n$ and $j = 1, \cdots, n$, **G** and the lower-right corner, i.e. s_{ij} where $i = 1, \cdots, k$ and $j = k+1, \cdots, k+n$, **S**. Then **GG'** is positive definite and **SS'** is non-negative definite. The upper-right and lower-left of the variance-covariance matrix describe the dependency between the state variables and the production returns. If those are 0, then the two are independent.

Then, CIR hypothesize the securities process (state contingent claims):

$$\begin{bmatrix} dF_1 \\ dF_2 \\ \vdots \\ dF_k \end{bmatrix} = \begin{bmatrix} F_1\beta_1 - \delta_1 \\ F_2\beta_2 - \delta_2 \\ \vdots \\ F_k\beta_k - \delta_k \end{bmatrix} dt + \begin{bmatrix} F_1h_{1,1} & F_1h_{1,2} & \cdots & F_1h_{1,n+k} \\ F_2h_{2,1} & F_2h_{2,2} & \cdots & F_2h_{2,n+k} \\ \vdots & \vdots & \ddots & \vdots \\ F_kh_{k,1} & F_kh_{k,2} & \cdots & F_kh_{k,n+k} \end{bmatrix} \begin{bmatrix} dW_{n+1} \\ dW_{n+2} \\ \vdots \\ dW_{n+k} \end{bmatrix} \tag{3.68}$$

and we symbolize the matrix **H**$= [h_{ij}]$. Define the utility function as in Merton:

$$\mathbb{E}_t \left[\int_t^T U(C(s), Y(s), s) ds \right] \tag{3.69}$$

subject to the wealth constraint:

$$\begin{aligned} dW = & \left[\sum\nolimits_{i=1}^{n} a_i W(\alpha_i - r) + \sum\nolimits_{i=1}^{n} b_i W(\beta_i - r) + rW - C \right] dt \\ & + \sum\nolimits_{i=1}^{n} a_i W \left(\sum\nolimits_{j=1}^{n+k} g_{ij} dW_j \right) + \sum\nolimits_{i=1}^{k} b_i W \left(\sum\nolimits_{j=1}^{n+k} h_{ij} dW_j \right) \\ \equiv & \, W\mu(W)dt + W \sum\nolimits_{j=1}^{n+k} q_j dW_j \end{aligned} \tag{3.70}$$

where a_i for $i = 1, \cdots, n$ is the portfolio weight in the production processes and b_i for $i = 1, \cdots, k$ is the portfolio weight in the state contingent claim.

This constraint is identical to Merton:

$$J(W, Y, t) = \max \mathbb{E}_t \left[\int_t^T U(C(s), Y(s), s) ds \right] \tag{3.71}$$

and (equations (11a) - (11e) in the CIR paper):

$$\psi_C = U_C - J_W = 0 \tag{3.72}$$

$$C\psi_C = 0 \tag{3.73}$$

$$\psi_a = (\alpha - r1)WJ_W + (\mathbf{GG}'a + \mathbf{GH}'b)W^2 J_{WW} + \mathbf{GS}'WJ_{WY} \leqslant 0 \tag{3.74}$$

$$a'\psi_a = 0 \tag{3.75}$$

$$\psi_b = (\beta - r1)WJ_W + (\mathbf{HG}'a + \mathbf{HH}'b)W^2 J_{WW} + \mathbf{HS}'WJ_{WY} = 0 \tag{3.76}$$

Equilibrium requires that:

$$\sum a_i = 1 \tag{3.77}$$

and

$$b_i = 0 \tag{3.78}$$

Like Merton, the goal is to obtain a_i, β_i, and r. With $b_i = 0$, (3.74) and (3.75) determine a_i given r. With a_i and r determined, (3.76) is a linear system in β_i. This is the Merton result. To solve for the equilibrium interest rate, we study the following problem:

$$\max a'\gamma + a'Da \tag{3.79}$$

subject to:

$$a'\underline{1} = 1 \tag{3.80}$$

and $a \geqslant 0$ where (* represents optimal)

$$\begin{aligned} \gamma &= \alpha W J_W^* + \mathbf{GS}' W J_{WY}^* \\ D &= {}^1\!/_2 \mathbf{GG}' W^2 J_{WW}^* \end{aligned} \tag{3.81}$$

The Lagrangian (known as the Kuhn-Tucker condition in economics [1]):

$$\begin{aligned} \gamma - \lambda + 2Da &\leqslant 0 \\ a'\gamma - \lambda a' 1 + 2a' D a &= 0 \end{aligned} \tag{3.82}$$

Hence, we solve for $< a, \lambda >$ as follows. The equilibrium interest rate is:

$$\begin{aligned} r &= \frac{\lambda^*}{W J_W} \\ &= a'\alpha - \frac{-J_{WW}}{J_W}\frac{\text{var}[W]}{W} - \sum_{i=1}^{k} \frac{-J_{WY_i}}{J_W}\frac{\text{cov}[W, Y_i]}{W} \end{aligned} \tag{3.83}$$

Substituting the above equilibrium interest rate back to (3.76): $\psi_b = 0$, we shall obtain the equilibrium expected return on contingent claims. $\psi_b = 0$ implies:

$$\begin{aligned} \beta &= r\underline{1} - (\mathbf{HG}'a + \mathbf{HH}'b) W J_{WW}/J_W - \mathbf{HS}' J_{WY}/J_W \\ &= a'\alpha\underline{1} + \left[-\frac{-J_{WW}}{J_W}\frac{\text{var}[W]}{W} - \sum_{i=1}^{k} \frac{-J_{WY_i}}{J_W}\frac{\text{cov}[W, Y_i]}{W} \right] \underline{1} \\ &\quad - (\mathbf{HG}'a + \mathbf{HH}'b) W J_{WW}/J_W - \mathbf{HS}' J_{WY}/J_W \end{aligned} \tag{3.84}$$

3.5 Reconciliation between Merton and Cox-Ingersoll-Ross

Several points merit a special mention. First, it is apparent that the basic assumptions of Merton and Cox-Ingersoll-Ross are exactly the same. Hence, there is no reason the results should be different. Indeed, we find that till (3.76), we do have an identical result to Merton for CIR. Yet, after (3.76), CIR start to deviate from Merton. CIR are interested in expressing the ICAPM in a format of derivative pricing (contingent claim pricing) as opposed to the format adopted traditionally by Sharpe and Merton. This leads to the next point.

In Merton, the derivation for general equilibrium is necessary (as in Sharpe) as otherwise, no equilibrium asset expected return can be derived. Furthermore,

[1]Interseted readers can refer to any microeconomic text for details.

Merton would like to show its convergence to Sharpes CAPM if there is only one state variable. In CIR, the equilibrium is assumed. At the beginning, CIR assume that the asset returns must follow a particular diffusion format. Then at the end, they show that indeed such an assumption is internally consistent. This is similar to finding the solution to a partial differential equation. A solution must be guessed first and then it is verified to be correct.

Thirdly, CIR are interested in expressing the ICAPM in a format of derivative pricing, which is useful for the derivation of various option pricing models. The most prominent examples are Vasicek's and CIR's term structure models. In their models, the underlying asset, which is the instantaneous short rate r, is not a traded asset and hence no-arbitrage pricing cannot be performed. As a result, the Black-Scholes type preference-free argument does not apply. And these models will include the preference parameter from the utility function. A similar situation takes place in random volatility stock option models. Given that the underlying volatility is not traded, the resulting option pricing model must include the preference parameter.

We should note that the preference structure must be set consistent with the diffusion processes, as demonstrated in Chapter 6. Literature regarding this issue is voluminous. Yet, this is a little beyond the scope of this book (after all, this book is about financial mathematics and not microeconomics) and I choose to leave it out.

Chapter 4

Continuous Time Math

This chapter draws heavily from Karlin and Taylor and Ludwig, and also Duffie. The basic stochastic calculus is taken from Karlin and Taylor. Ludwig is taken for his discussions on linear stochastic differential equations. Duffie is taken for his discussions on change of measure.

4.1 A Few Words on Measure Theory

The materials presented here in this chapter are not meant to be rigorous but designed to help readers who did not receive formal mathematical training acquire the necessary background to do the modeling work.

The measure theory is the foundation of the probability theory, although the origin of the probability theory was gambling. [1] [2] What we are familiar with

---

[1]This is a direct quote from Science Clarified (http://www.scienceclarified.com/Ph-Py/Probability-Theory.html): Probability theory was originally inspired by gambling problems. The earliest work on the subject was performed by Italian mathematician and physicist Girolamo Cardano (1501 – 1576). In his manual Liber de Ludo Aleae, Cardano discusses many of the basic concepts of probability complete with a systematic analysis of gambling problems. Unfortunately, Cardano's work had little effect on the development of probability because his manual did not appear in print until 1663 – and even then received little attention. In 1654, another gambler named Chevalier de Mr invented a system for gambling that he was convinced would make money. He decided to bet even money that he could roll at least one twelve in 24 rolls of two dice. However, when the Chevalier began losing money, he asked his mathematician friend Blaise Pascal (1623 – 1662) to analyze his gambling system. Pascal discovered that the Chevalier's system would lose about 51 percent of the time. Pascal became so interested in probability that he began studying more problems in this field. He discussed them with another famous mathematician, Pierre de Fermat (1601 – 1665) and, together they laid the foundation of probability theory.

[2]In this quote (from http://www.cc.gatech.edu/classes/cs6751_97_winter/Topics/stat-

from the basic calculus is the Riemann measure that is often used to compute an area or a volume. However, in stochastic calculus, such definition is deficient. We need a much more rigorous and sound definition of measure so that the area and volume can continue to be calculated. This measure is the Lebesgue measure (or Lebesgue-Stieltjes measure).

One of the well-known problems of the Riemann measure is its inability to calculate the area of the following:

$$f(x) = \begin{cases} 1 & x \text{ is rational} \\ 0 & x \text{ is irrational} \end{cases}$$

Since both rational numbers and irrational numbers are infinitely dense (*i.e.* there is always a rational/irrational number between any two rational/irrational numbers), it is impossible for the Riemann measure to define an area under this function.

The Lebesgue measure is the measure (as opposed to the Riemann measure) used in the stochastic calculus. In other words, with Lebesgue measure, we can use mathematical operations in the usual sense (like linearity, monotonicity, exchange of summation and integral, and exchange of limit and integral (known as the Lebesgue Monotonic Convergence Theorem)) without worrying about the fundamental measure theory. For example, under the Lebesgue measure the above function is well defined to be 0.

The derivation of the measure is not important. Yet we must check if the assumptions that guarantee (like positive measurable functions) the validity of the measure are satisfied. Luckily, all the finance problems satisfy the assumptions of the Lebesgue measure (e.g. all pricing functions in finance are all positive measurable.)

---

meas/probHist.html), we see the link between probability and measure: In 1812 Pierre de Laplace (1749-1827) introduced a host of new ideas and mathematical techniques in his book, Thorie Analytique des Probabilits. Before Laplace, probability theory was solely concerned with developing a mathematical analysis of games of chance. Laplace applied probabilistic ideas to many scientific and practical problems. The theory of errors, actuarial mathematics, and statistical mechanics are examples of some of the important applications of probability theory developed in the 19th century. One of the difficulties in developing a mathematical theory of probability has been to arrive at a definition of probability that is precise enough for use in mathematics, yet comprehensive enough to be applicable to a wide range of phenomena. The search for a widely acceptable definition took nearly three centuries and was marked by much controversy. The matter was finally resolved in the 20th century by treating probability theory on an axiomatic basis. In 1933 a monograph by a Russian mathematician A. Kolmogorov outlined an axiomatic approach that forms the basis for the modern theory. (Kolmogorov's monograph is available in English translation as Foundations of Probability Theory, Chelsea, New York, 1950.) Since then the ideas have been refined somewhat and probability theory is now part of a more general discipline known as measure theory.

4.2 Stochastic Processes

4.2.1 Probability Space

A population, Ω, heuristically, can be thought of as a collection of events (if the events are discrete). A simple example is the set of outcomes from throwing a dice. There are six possible outcomes of a dice, 1 through 6. Hence, $\Omega = \{1, 2, \cdots, 6\}$. In a general (still discrete) case, we write $\Omega = \{\omega_1, \omega_2, \cdots, \omega_n\}$ where there are n countable events. A population can be a set of uncountable events (such as the set of rational numbers). A population can be continuous (such as temperature).

A σ-algebra is a collection all subsets of a population. In the dice example, the σ-algebra is the following: $\{\phi, \{1\}, \cdots \{6\}, \{1, 2\}, \cdots \{5, 6\}, \cdots, \{1, 2, \cdots, 6\}\}$, which contains all the subsets of Ω. A filtration, $\mathfrak{F}$, is the smallest σ-algebra. In other words, a filtration contains all and yet non-repeating information of the population. Later when we introduce stochastic processes, the filtration is regarded as all the information up till the current time. For example, if we look at a stock (e.g. IBM) with a price $S(t)$ at time t. Then we write the filtration of IBM as $\mathfrak{F}_t$ to represent all the information regarding the stock IBM at time t. This filtration should include all the past information of the stock (i.e. $S(\tau)$ where $\tau < t$) and all the current information about IBM (e.g. financial statements, publicly announced news, etc.)

A probability measure, $\mathbb{P}$, is a function that maps a σ-algebra to a real number between 0 and 1, that is:

$$\mathbb{P} : \mathfrak{F} \to [0, 1].$$

In the dice example, we can define the probability measure over a fair dice as: $\mathbb{P}(1) = \mathbb{P}(2) = \cdots = 1/6$ or $\mathbb{P}(\{1, 2\}) = 1/3$ or $\mathbb{P}(\{2, 4, 6\}) = 1/2$ (that is the probability of having even numbers is half). Of course, we can also define an alternative measure $\mathbb{Q}$ that represents an unfair dice as: $\mathbb{Q}(1) = \mathbb{Q}(2) = 1/3$ and $\mathbb{Q}(3) = \cdots = \mathbb{Q}(6) = 1/12$.

A general case for a probability measure is:

- $\mathbb{P}(A) \in [0, 1]$ for any arbitrary $A \in \mathfrak{F}$, and
- $\int_\Omega \mathbb{P}(\omega) = 1$

Finally, a probability space is a triple: $(\Omega, \mathfrak{F}, \mathbb{P})$.

4.3 Random Variable

A random variable X is a measurable function from the population Ω to a real number. In other words, $X : \Omega \to R^1$. The expected value of the random variable is therefore defined as:

$$\mathbb{E}[X] = \int_{\Omega} X(\omega) d\mathbb{P}(\omega)$$

where $\omega \in \Omega$.

4.4 Stochastic Process

A stochastic process is best understood as a collection of random variables over time. More formally, we need to identify a probability space, $(\Omega, \mathcal{F}, \mathbb{P})$, and an arbitrary nonempty index I, such that a family $\{X(t); t \in I\}$ of R^d valued random variables can be called a stochastic process.

4.5 Wiener Process

$W(t)$ is a standard Wiener process if it obeys the following properties:

- $W(t+\tau) - W(t)$ has the same distribution as $W(s+\tau) - W(s)$ where $t \neq s$ and τ is arbitrary
- increments of $W(t)$ are i.i.d.
- $dW(t) \sim N(0, dt)$

Then, it follows that the auto-covariance is:

- $\mathbb{K}[W(t), W(s)] = \text{cov}[W(t), W(s)] = \min\{t, s\}$ which implies that the auto-correlation is $\rho[W(t), W(s)] = \sqrt{\frac{\min\{t,s\}}{\max\{t,s\}}}$.

4.6 Ito Calculus and Ito's Lemma (Tyler Series Expansion)

The major difference between Ito calculus and ordinary calculus is the notion of time. A stochastic process is nothing more than a collection of random variables over time. Let $X(t)$ be a stochastic process. In Ito calculus, the Wiener differential, $dW(t)$, has a size at the order of $\sqrt{dt}$; or $dW(t)^2$ is at the order of dt. As a result, different from the ordinary calculus, $dW(t)^2$ cannot be regarded as a higher order term and should not be ignored. In fact, the following property holds:

$$\mathbb{E}\left[\left(\int_0^t a(s)dW(s)\right)^2\right] \approx \int_0^t a(s)^2 ds \tag{4.1}$$

The above property allows us to manipulate various operations under Ito calculus.

Ito's Lemma. Let $X(t)$ be a stochastic process, and $C(X(t), t)$ be a function (contingent claim) of X and time. Then, the dynamics of the contingent claim, $dC(X(t), t)$, abbreviated as dC, can be written as:

$$dC = \frac{\partial C}{\partial X} dX(t) + \frac{1}{2}\frac{\partial^2 C}{\partial X^2}(dX)^2 + \frac{\partial C}{\partial t} dt + o(dt) \tag{4.2}$$

where $o(dt)$ is a collection of residual terms that are negligible in the order or dt. In other words, $\frac{o(dt)}{dt} \to 0$ as $dt \to 0$.

□

4.7 Stochastic Differential Equation (Ludwig's Chapter 6)

Let $X(t)$ be a stochastic process. It is often convenient (if it exists) to write its differential form as follows:

$$dX(t) = \mu(X(t), t)dt + \sigma(X(t), t)dW(t) \tag{4.3}$$

where $W(t)$ is a standard Wiener. μ and σ are called drift and diffusion (or volatility) respectively. If the stochastic differential equation (SDE) for a process exists, we can gain much mathematical convenience. For example, the Black-Scholes model is derived from the SDE of a geometric Brownian motion.

4.8 Stochastic Linear Differential Equation

If an SDE is linear (or can be transformed to be linear, such as log linear), the mathematics is much easier to deal with. We write a linear SDE as follows:

$$dX(t) = [A(t)X(t) + a(t)]dt + \left[\sum_{i=1}^{n} B_i(t)X(t) + b_i(t)\right] dW(t) \tag{4.4}$$

The above SDE has a unique solution provided that $A(t)$, $a(t)$, $B_i(t)$, and $b_i(t)$ are measurable and bounded.

While there is no closed-form solution for this general case, simple closed-form solutions do exist for special cases. We shall show two examples. The first one is constant diffusion as follows (i.e. $B_i(t) = 0$ and $b_i(t) = b(t)$):

$$dX(t) = [A(t)X(t) + a(t)]dt + b(t)dW(t) \tag{4.5}$$

We note that the diffusion now becomes independent of X. This simplification drastically reduces the mathematical complexity.

Theorem (Linear). The above SDE has the following solution:

$$X(t) = \Phi(t)\left(c + \int_0^t \Phi(s)^{-1}a(s)ds + \int_0^t \Phi(s)^{-1}b(s)dW(s)\right) \tag{4.6}$$

where

$$\Phi(t) = \exp\left(\int_0^t A(s)ds\right) \tag{4.7}$$

and $c = X(0)$ is a constant.

□

The proof of this theorem is left for an exercise.

Corollary. For the solution shown above, we also have:

$$\mathbb{E}[X(t)] = \Phi(t)\left(c + \int_0^t \Phi(s)^{-1}a(s)ds\right) \tag{4.8}$$

and also

$$\begin{aligned}\mathbb{K}[X(s), X(t)] &= \mathbb{E}[X(s) - \mathbb{E}[X(s)]]\mathbb{E}[X(t) - \mathbb{E}[X(t)]] \\ &= \Phi(s)\left(\int_0^{\min(s,t)} \left[\Phi(u)^{-1}b(u)\right]^2 du\right)\Phi(t)\end{aligned} \tag{4.9}$$

□

The second special case where a linear SDE has a closed-form solution is log linear as follows:

$$dX(t) = A(t)X(t)dt + B(t)X(t)dW(t) \tag{4.10}$$

Theorem (Log linear). The above SDE has the following solution:

$$X(t) = c\exp\left(\int_0^t A(u) - \frac{1}{2}B(u)^2du + \int_0^t B(u)dW(u)\right) \tag{4.11}$$

□

Again, the proof is left for an exercise.

There are other SDEs studied in the finance literature:

$$dX(t) = [A(t)X(t) + a(t)]dt + B(t)X(t)dW(t) \tag{4.12}$$

which is studied by Cox and Ross (1976), and

$$dX(t) = [A(t)X(t) + a(t)]dt + B(t)X(t)^\alpha dW(t) \tag{4.13}$$

where α is a constant. This process is studied by Cox (1976) and Cox, Ingersoll, and Ross (1985) and we shall examine them as we encounter them in the future. We should note that having a solution to the SDE or not is not crucial in solving a pricing problem. Often times, we can obtain closed-form solutions from the LaPlace transform, transition density, characteristic function, and so on. We shall see examples in later chapters.

4.9 Forward And Backward Equations (Karlin and Taylor, Chapter 15)

Let Z_j be i.i.d. random walk takes on value 1 (or Δ) or 1 (or Δ) with probabilities p and q respectively and $p + q = 1$, as demonstrated below:

$$Z_j = \begin{cases} \Delta & p \\ -\Delta & q \end{cases} \tag{4.14}$$

Hence,

$$\mathbb{E}[Z_j] = (p - q)\Delta\mathbb{E}[Z_j^2] = \Delta^2 \tag{4.15}$$

Let

$$X_n = X_0 + \sum_{j=1}^{n} Z_j \tag{4.16}$$

Hence (note that $\mathbb{E}[Z_i Z_j] = 0$)

$$\begin{aligned} \mathbb{E}[X_n - X_0] &= n\Delta(p-q) \\ \mathbb{E}[(X_n - X_0)^2] &= \mathbb{E}\left[\left(\sum_{j=1}^{n} Z_j\right)^2\right] \\ &= \sum_{j=1}^{n} \mathbb{E}[Z_j^2] \\ &= n\Delta^2 \end{aligned} \tag{4.17}$$

Note that in the binomial model of Cox-Ross-Rubinstein:[3]

$$\begin{aligned} p &= \frac{R-d}{u-d} \\ &= \frac{1+rh-(1-\sigma\sqrt{h})}{1+\sigma\sqrt{h}-(1-\sigma\sqrt{h})} \\ &= \frac{rh+\sigma\sqrt{h}}{2\sigma\sqrt{h}} \\ &= \frac{1}{2} + \frac{r}{2\sigma}\sqrt{h} \\ q = 1-p &= \frac{1}{2} - \frac{r}{2\sigma}\sqrt{h} \end{aligned} \tag{4.18}$$

and hence,

$$(p-q)^2 = \frac{r^2}{\sigma^2}h4pq = 1 - \frac{r^2}{\sigma^2}h \tag{4.19}$$

If we let $n = T/h$ and $\mathbb{E}[X_n - X_0] = \mu T$ and $\mathbb{E}[(X_n - X_0)^2] = \sigma^2 T$, then:

$$\mu = \frac{1}{h}(p-q)\Delta\sigma^2 = \frac{1}{h}\Delta^2 \tag{4.20}$$

and hence,

[3]Note that the binomial model can be constructed in multiple ways and the Cox-Ross-Rubinstein method is not unique.

$$\mathbb{E}[Z_j] = \mu h \mathbb{E}[Z_j^2] = \sigma^2 h \tag{4.21}$$

4.9.1 Forward Equation

We now turn to forward and backward equations. These equations are related to the numerical methods (finite difference methods) in Chapter 9. Readers are encouraged to compare the two materials. Define a "forward" random walk as Figure 4.1 shows:

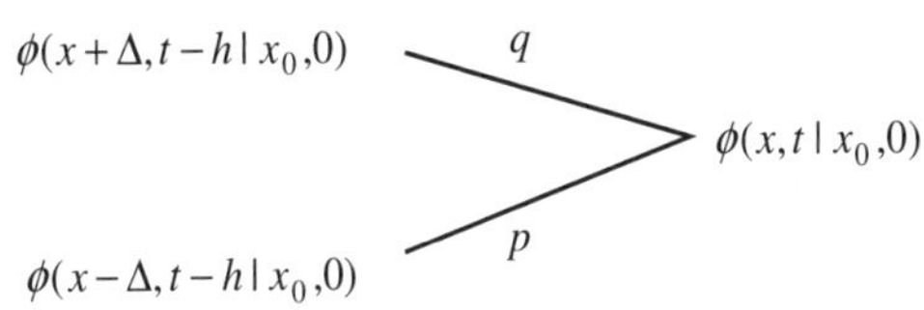

Figure 4.1: Forward Equation

and also define the following equation:

$$\phi(x,t|x_0,0) = \phi(x-\Delta, t-h|x_0,0)p + \phi(x+\Delta, t-h|x_0,0)q \tag{4.22}$$

By applying Taylor's series expansion to the first term of the righ-hand-side of the above equation, we get:

$$\begin{aligned}
&\phi(x-\Delta, t-h|x_0,0) = \phi(x,t|x_0,0) \\
&+\frac{\partial\phi(x-\Delta,t-h|x_0,0)}{\partial x}(-\Delta x) + \frac{1}{2}\frac{\partial^2\phi(x-\Delta,t-h|x_0,0)}{\partial x^2}(-\Delta x)^2 \\
&+\frac{\partial\phi(x-\Delta,t-h|x_0,0)}{\partial t}(-\Delta t)
\end{aligned} \tag{4.23}$$

Similarly, the expansion to the second term is:

$$\begin{aligned}
&\phi(x+\Delta, t-h|x_0,0) = \phi(x,t|x_0,0) \\
&+\frac{\partial\phi(x+\Delta,t-h|x_0,0)}{\partial x}\Delta x + \frac{1}{2}\frac{\partial^2\phi(x+\Delta,t-h|x_0,0)}{\partial x^2}(\Delta x)^2 \\
&+\frac{\partial\phi(x+\Delta,t-h|x_0,0)}{\partial t}(-\Delta t)
\end{aligned} \tag{4.24}$$

Substitute (4.23) and (4.24) back to (4.22), we have:

$$\phi(x,t|x_0,0) = \phi(x,t|x_0,0) - (p-q)\frac{\partial \phi}{\partial x}\Delta x - \frac{\partial \phi}{\partial t}\Delta t + \frac{1}{2}\frac{\partial^2 \phi}{\partial x^2}(\Delta x)^2 \tag{4.25}$$

When Δt is small (that is $\Delta t \to dt$ and $\Delta x \to dx$), we then can write the above equation as follows:

$$\frac{\partial \phi}{\partial t} = -\mu\frac{\partial \phi}{\partial x} + \frac{\sigma^2}{2}\frac{\partial^2 \phi}{\partial x^2} \tag{4.26}$$

This is a partial differential equation (PDE). The solution to this PDE is:

$$\phi = \frac{1}{\sqrt{2\pi\sigma^2 t}}\exp\left(-\frac{(x-x_0-\mu t)^2}{2\sigma^2 t}\right) \tag{4.27}$$

To verify this is indeed the right solution, we take the following partial derivatives:

$$\frac{\partial \phi}{\partial x} = \phi\frac{-(x-x_0-\mu t)}{\sigma^2 t} \tag{4.28}$$

$$\frac{\partial^2 \phi}{\partial x^2} = \phi\left(\frac{(x-x_0-\mu t)^2}{\sigma^4 t^2} + \frac{-1}{\sigma^2 t}\right) \tag{4.29}$$

$$\begin{aligned}\frac{\partial \phi}{\partial t} &= \frac{-1}{2\sqrt{2\pi\sigma^2 t^3}}\exp\left(-\frac{(x-x_0-\mu t)^2}{2\sigma^2 t}\right) + \frac{1}{\sqrt{2\pi\sigma^2 t}}\exp\left(-\frac{(x-x_0-\mu t)^2}{2\sigma^2 t}\right)\\ &\quad\times\left(-\frac{-2\mu(x-x_0-\mu t)2\sigma^2 t - (x-x_0-\mu t)^2 2\sigma^2}{4\sigma^4 t^2}\right)\\ &= \frac{-1}{2t}\phi + \phi\left(\frac{\mu(x-x_0-\mu t)}{\sigma^2 t} + \frac{(x-x_0-\mu t)^2}{2\sigma^2 t^2}\right)\\ &= \phi\left(\frac{-1}{2t} + \frac{\mu(x-x_0-\mu t)}{\sigma^2 t} + \frac{(x-x_0-\mu t)^2}{2\sigma^2 t^2}\right)\end{aligned} \tag{4.30}$$

It then follows that:

$$-\mu\frac{\partial \phi}{\partial x} = \mu\phi\frac{(x-x_0-\mu t)}{\sigma^2 t} \tag{4.31}$$

$$\frac{\sigma^2}{2}\frac{\partial^2 \phi}{\partial x^2} = \frac{\sigma^2}{2}\phi\left(\frac{(x-x_0-\mu t)^2}{\sigma^4 t^2} + \frac{-1}{\sigma^2 t}\right) = \phi\left(\frac{(x-x_0-\mu t)^2}{2\sigma^2 t^2} + \frac{-1}{2t}\right) \tag{4.32}$$

Hence the forward equation is satisfied. Note that this derivation is only heuristic. The more rigorous derivation can be found in Karlin and Taylor.

4.9.2 Backward Equation

Define a "backward" random walk as follows:

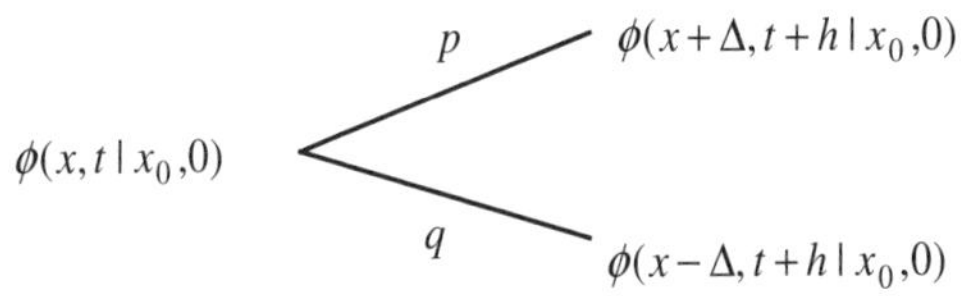

Figure 4.2: Backward Equation

and also define the equation:

$$\phi(x,t|x_0,0) = \phi(x+\Delta,t+h|x_0,0)p + \phi(x-\Delta,t+h|x_0,0)q \tag{4.33}$$

Apply Taylor's series expansion to the first term to get:

$$\begin{aligned}
&\phi(x+\Delta,t+h|x_0,0) = \phi(x,t|x_0,0)\\
&+\frac{\partial\phi(x+\Delta,t+h|x_0,0)}{\partial x}\Delta x + \frac{1}{2}\frac{\partial^2\phi(x+\Delta,t+h|x_0,0)}{\partial x^2}(\Delta x)^2\\
&+\frac{\partial\phi(x+\Delta,t+h|x_0,0)}{\partial t}(\Delta t)
\end{aligned} \tag{4.34}$$

and the second term to get:

$$\begin{aligned}
&\phi(x-\Delta,t+h|x_0,0) = \phi(x,t|x_0,0)\\
&+\frac{\partial\phi(x-\Delta,t+h|x_0,0)}{\partial x}(-\Delta x) + \frac{1}{2}\frac{\partial^2\phi(x-\Delta,t+h|x_0,0)}{\partial x^2}(-\Delta x)^2\\
&+\frac{\partial\phi(x-\Delta,t+h|x_0,0)}{\partial t}(\Delta t)
\end{aligned} \tag{4.35}$$

Substituting (4.34) and (4.35) back to (4.33), we have:

$$\phi(x,t|x_0,0) = \phi(x,t|x_0,0) + (p-q)\frac{\partial\phi}{\partial x}\Delta x + \frac{\partial\phi}{\partial t}\Delta t + \frac{1}{2}\frac{\partial^2\phi}{\partial x^2}(\Delta x)^2 \tag{4.36}$$

By letting $\Delta t \to dt$ and $\Delta x \to dx$, we end up with:

$$-\frac{\partial\phi}{\partial t} = \mu\frac{\partial\phi}{\partial x} + \frac{\sigma^2}{2}\frac{\partial^2\phi}{\partial x^2} \tag{4.37}$$

which is a PDE. The solution to this PDE is:

$$\phi = \frac{1}{\sqrt{2\pi\sigma^2 t}}\exp\left(-\frac{(x-x_0+\mu t)^2}{2\sigma^2 t}\right) \tag{4.38}$$

Again, to verify, we differentiate the solution with respect to x and t:

$$\frac{\partial\phi}{\partial x} = \phi\frac{-(x-x_0+\mu t)}{\sigma^2 t} \tag{4.39}$$

$$\frac{\partial^2\phi}{\partial x^2} = \phi\left(\frac{(x-x_0+\mu t)^2}{\sigma^4 t^2} + \frac{-1}{\sigma^2 t}\right) \tag{4.40}$$

$$\begin{aligned}\frac{\partial\phi}{\partial t} &= \frac{-1}{2\sqrt{2\pi\sigma^2 t^3}}\exp\left(-\frac{(x-x_0+\mu t)^2}{2\sigma^2 t}\right) + \frac{1}{\sqrt{2\pi\sigma^2 t}}\exp\left(-\frac{(x-x_0+\mu t)^2}{2\sigma^2 t}\right)\\ &\quad\times\left(-\frac{2\mu(x-x_0+\mu t)2\sigma^2 t - (x-x_0+\mu t)^2 2\sigma^2}{4\sigma^4 t^2}\right)\\ &= \frac{-1}{2t}\phi + \phi\left(-\frac{\mu(x-x_0+\mu t)}{\sigma^2 t} + \frac{(x-x_0+\mu t)^2}{2\sigma^2 t^2}\right)\\ &= \phi\left(\frac{-1}{2t} + \frac{-\mu(x-x_0+\mu t)}{\sigma^2 t} + \frac{(x-x_0+\mu t)^2}{2\sigma^2 t^2}\right)\end{aligned} \tag{4.41}$$

It then follows that:

$$\mu\frac{\partial\phi}{\partial x} = \phi\frac{-\mu(x-x_0+\mu t)}{\sigma^2 t} \tag{4.42}$$

$$\frac{\sigma^2}{2}\frac{\partial^2\phi}{\partial x^{2\partial}} = \frac{\sigma^2}{2}\phi\left(\frac{(x-x_0+\mu t)^2}{\sigma^4 t^2} + \frac{-1}{\sigma^2 t}\right) = \phi\left(\frac{(x-x_0+\mu t)^2}{2\sigma^2 t^2} + \frac{-1}{2t}\right) \tag{4.43}$$

Hence the backward equation is satisfied.

The standard Brownian motion (pure random walk) has a variance of t (instead of $\sigma^2 t$) and drift of 0 (instead of μt). Hence, the backward equation becomes:

$$\frac{\partial \phi}{\partial t} = \frac{1}{2}\frac{\partial \phi^2}{\partial x^2} \tag{4.44}$$

This is the famous heat equation that has the solution as:

$$\phi = \frac{1}{\sqrt{2\pi t}} \exp\left(-\frac{(x - x_0)^2}{2t} \right) \tag{4.45}$$

4.9.3 Kolmogorov Backward Equation (Karlin And Taylor)

The Kolmogorov backward equation is similar to the generic backward equation derived above. Yet, the solution to the Kolmogorov backward equation ties expicitly to a stochastic differential equation (which we have done for the generic backward equation previously). Hence, we begin by assuming a stochastic process:

$$dX(t) = \mu dt + \sigma dW(t) \tag{4.46}$$

where clearly,

$$\begin{cases} \mathbb{E}[dX(t)] = \mu dt \\ \mathbb{E}[dX(t)^2] = \sigma^2 dt \end{cases} \tag{4.47}$$

With the following time line,

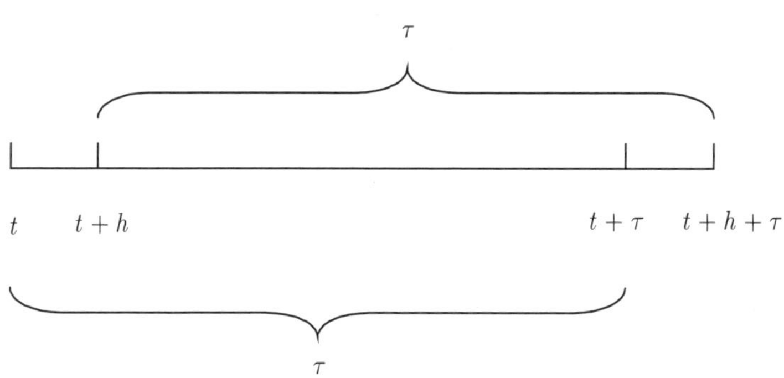

Figure 4.3: Time Line

we define u as a conditional expectation as a function of time to maturity and current value as follows:

$$u(\tau, X(t)) = \mathbb{E}[g(X(t+\tau))|X(t)] \tag{4.48}$$

where the first argument represents the time between now (t) and when the function is evaluated ($t+\tau$) and $g(X)$ is an arbitrary function of X. Then, $u(\tau, x)$ satisfies the previously defined PDE, known as the Kolmogorov Backward Equation:

$$\frac{\partial u}{\partial \tau} = \mu \frac{\partial u}{\partial x} + \frac{1}{2}\sigma^2 \frac{\partial^2 u}{\partial x^2} \tag{4.49}$$

The law of iterative expectations (law of total probability) states that:

$$\begin{aligned} u(h+\tau, X(t)) &= \mathbb{E}[g(X(t+h+\tau))|X(t)] \\ &= \mathbb{E}[\mathbb{E}[g(X(t+h+\tau))|X(t+h)]|X(t)] \\ &= \mathbb{E}[u(\tau, X(t+h))|X(t)] \end{aligned} \tag{4.50}$$

Applying Taylor's series expansion on $u(\tau, X(t+h))$, we obtain:

$$u(\tau, X(t+h)) = u(\tau, X(t)) + \frac{\partial u}{\partial x}(X(t+h) - X(t)) + \frac{1}{2}\frac{\partial^2 u}{\partial x^2}(X(t+h) - X(t))^2 \tag{4.51}$$

Let h be sufficiently small, i.e. $h \to dt$ (which implies $X(t+h) - X(t) \to dX$) and substitute (4.51) back to (4.50). Then we have:

$$\begin{aligned} u(\tau + dt, X(t)) &= u(\tau, X(t)) + \mathbb{E}\left[\frac{\partial u}{\partial x} dX + \frac{1}{2}\frac{\partial^2 u}{\partial x^2}(dX)^2 \middle| X(t) \right] \\ &= u(\tau, X(t)) + \frac{\partial u}{\partial x}\mu dt + \frac{1}{2}\frac{\partial^2 u}{\partial x^2}\sigma^2 dt \end{aligned} \tag{4.52}$$

Subtracting $u(\tau, X(t))$ and dividing by dt on both sides, we obtain:

$$\frac{\partial u}{\partial \tau} = \frac{\partial u}{\partial x}\mu + \frac{1}{2}\frac{\partial^2 u}{\partial x^2}\sigma^2 \tag{4.53}$$

A natural extension to the Kolmogorov backward equation is the Kolmogorov backward equation for the Kac functional. A functional is a real-valued function on a vector space of functions, that is $f : D \to R$ where D is a vector space and R is the real line. Here we are not discussing the Kac functional but only to recongize the Kolmogorov backward equation for the Kac functional, which is:

$$ku + \frac{\partial u}{\partial t} = \frac{\partial u}{\partial x}\mu + \frac{1}{2}\frac{\partial^2 u}{\partial x^2}\sigma^2 \tag{4.54}$$

and the solution is:

$$u(t, X(0)) = \mathbb{E}\left[\exp\left(-\int_0^t k(u)du\right) g(X(t)) \,\middle|\, X(0)\right] \tag{4.55}$$

Note that the Kolmogorov Backward Equation for the Kac functional differs from the Kolmogorov Backward Equation by just one term on the left hand side: ku. And in the solution, the function $g(X)$ is "discounted" at the rate k. The proof is similar and left for the readers to practice.

The results of the Kolmogorov Backward Equation and that for the Kac functional are particularly useful in derivatives pricing. Note that the function $g(X)$ is arbitrary. Hence, if we let X be the stock price S and $g(S) = \max\{S - K, 0\}$ where K is a constant strike price, then we can use the result of the Kolmogorov Backward Equation for the Kac functional to derive the Black-Scholes option model. This result does not need to solve the PDE.

Under the risk-neutral measure (which is briefly explored in Chapter (2) and will further detailed later in this chapter), (4.46) becomes:

$$dS(t) = rS(t)dt + \sigma S(t)d\hat{W}(t) \tag{4.56}$$

and (4.54) becomes:

$$ru + \frac{\partial u}{\partial t} = \frac{\partial u}{\partial S}rS + \frac{1}{2}\frac{\partial^2 u}{\partial S^2}\sigma^2 S^2 \tag{4.57}$$

which is the Black-Scholes PDE. The solution is:

$$u(t, S(0)) = \hat{\mathbb{E}}\left[\exp\left(-\int_0^t r(u)du\right) \max\{S(t) - K, 0\} \,\middle|\, S(0)\right] \tag{4.58}$$

In addition to the Black-Scholes model, there are a number of useful applications of the Kolmogorov backward equation. Given that the solution to the Kolmogorov backward equation is an expectation, we know that the following results must hold. First, the probability function must satisfy the Kolmogorov backward equation as the probability function is the expectation of an indicator function. For example, the in-the-money probability of a call option is (where S is the stock price and K is the strike price):

$$\Pr(S > K) = \hat{\mathbb{E}}[\mathbb{I}_{S>K}] \tag{4.59}$$

where $\mathbb{I}_A$ is an indicator function that takes 1 if event A is true and 0 otherwise. In the case of the Black-Scholes (see Chapter 5, Section IV-b), $N(d_1)$ and $N(d_2)$ are solutions to following Kolmogorov backward equations:

$$\frac{\partial u}{\partial t} = \frac{\partial u}{\partial S}(r + \sigma^2)S + \frac{1}{2}\frac{\partial^2 u}{\partial S^2}\sigma^2 S^2 \frac{\partial u}{\partial t} = \frac{\partial u}{\partial S} rS + \frac{1}{2}\frac{\partial^2 u}{\partial S^2}\sigma^2 S^2 \tag{4.60}$$

respectively. Readers can easily verify such result.

If the probability functions are solutions to the Kolomgorov backward equations, their derivatives, i.e. density functions, are also solutions to the same Kolmogorov backward equations. Taking derivative of (4.59) with respect to the strike price, K, yields:

$$\frac{d\Pr(S > K)}{dK} = \hat{\mathbb{E}}\left[\frac{d\mathbb{I}_{S>K}}{dK}\right] f(K) = \hat{\mathbb{E}}[\delta_K] \tag{4.61}$$

where $f(\cdot)$ is density function and δ_K is the Diracs delta function evaluated at point K. Hence, the density functions also satisfy the same Kolmogorov backward equations shown in (4.60).

Two more important functions that satisfy Kolmogorov backward equations are characteristic functions and LaPlace transform. In the Heston model introduced in Chapter 13 where the probability functions are not solvable, he finds the characteristic functions are and then probability functions are simply Fourier transforms of the characteristic functions. Details can be seen in Chapter 13. The characteristic function is:

$$\phi(u) = \hat{\mathbb{E}}\left[e^{-iu\ln S}\right] \tag{4.62}$$

where $i = \sqrt{-1}$ and S is the stock price and the Kolmogorov Backward Equation is:

$$\frac{\partial \phi}{\partial t} = \frac{\partial \phi}{\partial S} rS + \frac{1}{2}\frac{\partial^2 \phi}{\partial S^2}\sigma^2 S^2 \tag{4.63}$$

The characteristic function has a closed-form solutions in the Heston model under random volatility and in the Scott model under random volatility, random interest rates (of the Cox-Ingersoll-Ross type), and jumps. Similar to characteristic functions, LaPlace transforms also satisfy the same PDE. LaPlace transform is the following expectation:

$$\psi(u) = \hat{\mathbb{E}}\left[e^{uX}\right] \tag{4.64}$$

In the Cox-Ingersoll-Ross model of the term structure where $X = -\int_t^T r_u du$ where r is the instantaneous rate (see Chapter 7, Section III), $\psi(u)$ has a closed-form solution.

4.10 Martingale

The concept of martingales serves as an important foundation in quantitative finance. Martingales provide a convenient way to derive derivative prices via conditional expectations and help identify the correct drift adjustments necessary in identifying the right pricing measures to use.

4.10.1 The Basic

Definition (adapted process). A process $X(t)$ is adapted if for each t, $X(t)$ is a random variable with respect to $(\Omega, \mathcal{F})$.

□

Definition (martingale). An adapted process $X(t)$ is a martingale if $\mathbb{E}[X(s)|X(t)] = X(t)$ where $\mathbb{E}$ is an expectation defined under $(\Omega, \mathcal{F})$. (In other words, a process may be a martingale under one space but not the other.)

□

Definition (predictable process). An adapted process $X(t)$ is predictable with respect to $(\Omega, \mathcal{G})$ if $X(t)$ is adapted to $(\Omega, \mathcal{G})$ where $\mathcal{G}(t) = \mathcal{F}(t-1)$.

□

In other words, a process is predictable if it can be chosen a value based upon information given prior to time t.

Definition (stochastic integral). $\int_0^t \theta(\omega, s) dX(\omega, s)$ where $\omega \in \Omega$ is a random variable whose state space is a collection of path by path result of the Stieltjes integral.

□

Warning. Readers are to be warned that this definition of stochastic integral is not general enough to illustrate some economic principles. For such a purpose, readers are asked to refer to Duffies Dynamic Asset Pricing Theory for a formal definition.

□

Lemma. The trading profit is a martingale. In other words,
$\mathbb{E}\left[\sum_{j=1}^{n}\theta(T_j)[X(T_j)-X(T_{j-1})|X(t)]\right]=0.$

□

Proof. Given,

$$\begin{aligned}\mathbb{E}[X(T)|X(t)] &= X(t)\mathbb{E}\left[\sum_{j=1}^{n} X(T_j)-X(T_{j-1})|X(t)\right] \\ &= 0\end{aligned} \tag{4.65}$$

where $t=T_0<T_1<\cdots<T_n=T$.

Given that $\theta(t)$ is a predictive process, it is easy to show that:

$$\begin{aligned}&\mathbb{E}[\theta_1(X_1-X_0)|X(0)]+\mathbb{E}[\theta_2(X_2-X_1)|X(0)]+\cdots\\ &\qquad=\theta_1\mathbb{E}[(X_1-X_0)|X(0)]+\mathbb{E}[\mathbb{E}[\theta_2(X_2-X_1)|X(1)]|X(0)]+\cdots\\ &\qquad=0\end{aligned} \tag{4.66}$$

□

The lemma demonstrates that if an asset price process is a martingale, so is the trading gain.

Theorem. The trading profit in continuous time, $\mathbb{E}[\int_0^t\theta(t)X(t)]=0$, is a martingale.

□

4.10.2 Stock Price Process

Assume the stock price process follows the Geometric Brownian motion under the risk-neutral measure:

$$\frac{dS(t)}{S(t)}=rdt+\sigma d\hat{W}(t) \tag{4.67}$$

where both r is the risk-free rate and σ is volatility. Define $X(t) = e^{-rt}S(t)$. Then by Ito's lemma,

$$\begin{aligned} dX(t) &= \frac{\partial X}{\partial S}dS + \frac{1}{2}\frac{\partial^2 X}{\partial S^2}(dS)^2 + \frac{\partial X}{\partial t}dt \\ &= e^{-rt}[\mu S(t)dt + \sigma S(t)d\hat{W}(t)] + 0 + e^{-rt}S(t)(-r)dt \\ &= e^{-rt}\sigma S(t)d\hat{W}(t) \\ &= \sigma X(t)d\hat{W}(t) \end{aligned} \tag{4.68}$$

and clearly $X(t)$ is a martingale, since we have removed the drift from the $S(t)$ process. This implies $X(t) = \hat{\mathbb{E}}[X(T)|X(t)]$, or $e^{-rt}S(t) = e^{-rT}\hat{\mathbb{E}}[S(T)]$, or $S(t) = e^{-r(T-t)}\hat{\mathbb{E}}[S(T)]$, conditional on the information at time t.

In option pricing, we always need to deflate the asset price to achieve a martingale. The deflator is also known as the numraire. Here, the numeraire is e^{-rt} which is non-stochastic. If the numeraire is stochastic, then it involves change of measure that is discussed in Chapter 7.

4.10.3 Martingale Representation Theorem

The main result of the Martingale Representation Theorem (MRT) is to show that the Black-Scholes model is self-financing. That is, if the Black-Scholes model is the true model for the option, then there is a self-financing strategy consisting of the stock and bond that duplicates the option. In other words, the option is a redundant asset that can be "synthesized or duplicated by trading the stock and bond. We shall visit the MRT in the next chapter.

If $X(t)$ and $Y(t)$ are both martingales, then there exists a predictable process $\theta(t)$ such that:

$$Y(t) = Y(0) + \int_0^t \theta(s)dX(s) \tag{4.69}$$

which implies that $\mathbb{E}\left[\int_0^t \theta(s)dX(s)\right] = 0$.

The application of the Martingale Representation Theorem is to derive the Black-Scholes model with the self-financing trading strategy. A self-financing strategy is a replication strategy for an option. In other words, if the option is a redundant asset, then it must be able to be replicated with primary assets such as stock and bond.

To simplify expressions, we use time subscripts to represent time. Let $Y(t)$

be the option that is redundant, and hence can be replicated by stock and bond as follows:

$$\begin{aligned} Y(t) &= \phi(t)S(t) + \psi(t)B(t) \\ dY(t) &= \phi(t)dS(t) + \psi(t)dB(t) \end{aligned} \tag{4.70}$$

where $\phi(t)$ and $\psi(t)$ are hedge ratios. In the Black-Scholes case, the risk free rate is constant. So we can, for simplicity, write:

$$\begin{aligned} dS(t) &= rS(t)dt + \sigma S(t)d\hat{W} \\ B(t) &= e^{rt} \end{aligned} \tag{4.71}$$

Let $S(t)^* = S(t)/B(t) = e^{-rt}S(t)$. Then it is easy to show that:

$$\begin{aligned} dS(t)^* &= e^{-rt}dS(t) - rS(t)^* \\ &= e^{-rt}[rS(t)dt + \sigma S(t)d\hat{W}] - rS(t)^* \\ &= \sigma S(t)^* d\hat{W} \end{aligned} \tag{4.72}$$

which is a martingale. Similarly, let $Y(t)^* = Y(t)/B(t) = e^{-rt}Y(t)$. Then:

$$\begin{aligned} dY(t)^* &= e^{-rt}dY(t) - rY(t)^*dt \\ &= e^{-rt}(\phi(t)dS(t) + \psi(t)dB(t)) - rY(t)^*dt \\ &= e^{-rt}\left(\phi(t)rS(t)dt + \phi(t)\sigma S(t)d\hat{W} + \psi(t)re^{rt}dt\right) - rY(t)^*dt \\ &= \left(r\phi(t)S(t)^* + r\psi(t) - rY(t)^*\right)dt + \phi(t)\sigma S(t)^*d\hat{W} \\ &= \phi(t)\sigma S(t)^*d\hat{W} \\ &= \phi(t)dS(t)^* \end{aligned} \tag{4.73}$$

which is also a martingale. Hence, it must be true (by MRT) that:

$$Y(t)^* = Y(0) + \int_0^t \phi(s)dS(s)^* \tag{4.74}$$

(note that $Y(0)^* = Y(0)$) and as a result:

$$Y(t)^* = \hat{\mathbb{E}}[Y(T)^*|Y(t)] \tag{4.75}$$

4.11 Absolute Continuity (Duffie)

4.11.1 Definition

Consider two probability measures $\mathbb{P}$ and $\mathbb{Q}$ on a space $(\Omega, \mathfrak{F})$. For an event B, if $\mathbb{P}(B) = 0$ implies $\mathbb{Q}(B) = 0$, for any B in $\mathfrak{F}$, then $\mathbb{Q}$ is absolutely continuous with respective to $\mathbb{P}$. (That is every $\mathbb{P}$-negligible event is $\mathbb{Q}$-negligible.)

The above definition has an interesting and useful result: change of measure, for:

$$\mathbb{E}^{\mathbb{Q}}[x] = \int_{\Omega} x(\omega) d\mathbb{Q}(\omega) = \int_{\Omega} x(\omega)\eta(\omega) dP(\omega) = \mathbb{E}^{P}[\eta x] \tag{4.76}$$

The function h, often denoted $d\mathbb{Q}/d\mathbb{P}$, is called the Radon-Nykodym derivative (RND) of $\mathbb{Q}$ with respect to $\mathbb{P}$. It is not hard to see that:

$$\mathbb{E}^{\mathbb{P}}[\eta] = \int_{\Omega} \eta(\omega) d\mathbb{P}(\omega) = \int_{\Omega} d\mathbb{Q}(\omega) = 1 \tag{4.77}$$

4.11.2 Change of Measure (Duffie)

While absolute continuity defines two equivalent probability measures, the implementation of the change of measure requires non-trivial work. In continuous time, the change of measure of Brownian motions requires the Girsanov theorem. Define the Radon-Nykodim derivative as follows:

$$\eta(t) = \exp\left(-\int_0^t \theta(u) dW(u) - \frac{1}{2}\int_0^t \theta(u)^2 du\right) \tag{4.78}$$

Then the Girnasov theorem states the following theorem.

Theorem (Girsanov). If the following regularity condition is satisfied:

$$\mathbb{E}\left[\exp\left(\frac{1}{2}\int_0^T \theta(t)^2\right)\right] < \infty \tag{4.79}$$

and $W(t)$ is a standard Brownian motion, then:

$$X(t) = X(0) + \int_0^t \mu(s) ds + \int_0^t \sigma(s) dW(s) \tag{4.80}$$

can be written as:

$$X(t) = X(0) + \int_0^t \nu(s)ds + \int_0^t \sigma(s)d\hat{W}(s) \tag{4.81}$$

where

$$\hat{W}(t) = W(t) + \int_0^t \theta(s)ds\theta(t) = \frac{\mu(t) - \nu(t)}{\sigma(t)} \tag{4.82}$$

where $\hat{W}(t)$ is also a standard Brownian motion.

□

It is almost trivial to see the theorem in a differential form:

$$dX(t) = \mu(t)dt + \sigma(t)dW(t) = \nu(t)dt + \sigma(t)d\hat{W}(t) \tag{4.83}$$

where

$$d\hat{W}(t) = dW(t) + \theta(t)dt \tag{4.84}$$

In the remainder of the chapter, I provide two well-known examples. The first example is the Black-Scholes model and the second example is Jamshidian's forwrd measure. They will be seen again in later chapters.

Example 1: Black-Scholes. Define the stock price process as:

$$\frac{dS(t)}{S(t)} = \mu dt + \sigma dW(t) \tag{4.85}$$

which is a geometric Brownian motion with constant drift and diffusion. Define:

$$\theta = \frac{\mu - r}{\sigma} \tag{4.86}$$

as the per unit risk premium. And hence the RND is written as:

$$\eta = \exp\left(-\frac{\mu - r}{\sigma}W(t) - \frac{1}{2}\left(\frac{\mu - r}{\sigma}\right)^2 t\right) \tag{4.87}$$

It is easy to verify that $\mathbb{E}[\eta] = 1$. It is also trivial that $\frac{\mu - r}{\sigma}$ satisfies the regularity condition (since both μ and σ are constants). Hence, we can write:

$$d\hat{W}(t) = dW(t) + \frac{\mu - r}{\sigma} dt \tag{4.88}$$

and

$$\frac{dS(t)}{S(t)} = rdt + \sigma d\hat{W}(t) \tag{4.89}$$

This implies the following result:

$$\mathbb{E}[\eta(t)S(t)] = \mathbb{E}[\eta(t)]\hat{\mathbb{E}}[S(t)] = \hat{\mathbb{E}}[S(t)] \tag{4.90}$$

which can be easily verified as follows:

$$\begin{aligned}
\mathbb{E}[\eta(t)S(t)] &= \mathbb{E}\left[\exp\left(-\tfrac{\mu-r}{\sigma}W(t) - \tfrac{1}{2}\left[\tfrac{\mu-r}{\sigma}\right]^2 t\right) S_0 \exp\left((\mu - {}^1\!/_2\sigma^2)t + \sigma W(t)\right)\right] \\
&= S_0 e^{rt} \\
&= \hat{\mathbb{E}}[S(t)]
\end{aligned} \tag{4.91}$$

Often times, the trick is to find θ.

Example 2: Forward measure.

The call value under stochastic interest rates is usually written under the risk-neutral measure as follows (the condition in the expectation is ignored for convenience):

$$\begin{aligned}
C(t) &= \hat{\mathbb{E}}_t\left[\exp\left(-\int_t^T r(u)du\right)\max\{S(T) - K, 0\}\right] \\
&= P(t,T)\mathbb{E}_t^{\mathbb{F}(T)}[\max\{S(T) - K, 0\}]
\end{aligned} \tag{4.92}$$

where

$$P(t,T) = \hat{\mathbb{E}}_t\left[\exp\left(-\int_t^T r(u)du\right)\right] \tag{4.93}$$

is called \$1 zero coupon bond price. The second line is a result of change of measure:

$$\eta(t,T) = \frac{\exp\left(-\int_t^T r(u)du\right)}{\hat{\mathbb{E}}_t\left[\exp\left(-\int_t^T r(u)du\right)\right]} = \frac{\exp\left(-\int_t^T r(u)du\right)}{P(t,T)} \tag{4.94}$$

It is natural that $\hat{\mathbb{E}}[\eta(t,T)] = 1$. Now, we need to find θ. We first assume a similar process (geometric Brownian motion) for the bond price:

$$\frac{dP(t,T)}{P(t,T)} = rdt + \sigma_P(t)d\hat{W}_r(t) \tag{4.95}$$

We should note that the diffusion in this process is not a function of bond price. Taking log of the log normal stochastic process:

$$\begin{aligned} 0 &= \ln P(T,T) \\ &= \ln P(t,T) + \int_t^T r(u) - \frac{\sigma_P^2(u)}{2} du + \int_t^T \sigma_P(u)d\hat{W}_r(u) \frac{\exp\left(-\int_t^T r(u)\right)}{P(t,T)} \\ &= \exp\left(\int_t^T \sigma_P(u)d\hat{W}_r(u) - 1/2 \int_t^T \sigma_P^2(u)du\right) \end{aligned} \tag{4.96}$$

and hence the Girsanov transformation is defined over:

$$d\hat{W}_r(t) = dW_r^{\mathbb{F}(T)}(t) + \sigma_P(t)dt \tag{4.97}$$

Note that when the change of measure is performed on the bond price process, the stock price process will be affected due to the correlation between the two, $dW(t)$ and $dW_r(t)$, which is usually assumed as $dW(t)dW_r(t) = \rho dt$. Note that by Itos lemma,

$$\sigma_P(t) = \frac{1}{P(t,T)} \frac{\partial P(t,T)}{\partial r} \sigma(r,t) \tag{4.98}$$

We should note again that the LHS is independent of r or P while the RHS is a function of both r and P. Hence, to have the above result, the interest rate process and its resulting bond price solution need to be so "coincident" that the product results in a function only dependent of time, and not any state variable.

Chapter 5

Risk-Neutral Pricing

Risk-neutral pricing has become the standard of asset pricing today. Risk-neutral pricing is also known as no-arbitrage pricing. And no-arbitrage pricing obeys the law of one price which is the industry standard for marking to market and risk management. Risk-neutral pricing leverages upon the martingale theory.

5.1 Introduction

The risk-neutral pricing theory was first discovered by Cox and Ross (1976) where they pointed out that there is no risk preference in the Black-Scholes model. Prior to the Black-Scholes, economic pricing models are all based upon some utility theory where a risk parameter must be identified. Black and Scholes, via continuous rebalancing, miraculously removed the dependency on the utility. In other words, by continuous rebalancing, the option can be synthetically duplicated by stock and bond and hence all preferences should agree on one price – the no arbitrage price. This is known as the Martingale Representation Theory (or "self-financing") that is introduced in the previous chapter (Chapter 4). This finding was later formalized by a number of authors including Harrison and Kreps (1979), Harrison and Pliska (1981), and Duffie and Huang (1985). Basically, risk-neutral pricing states that under no arbitrage (later on we show that this is equivalent to complete markets), the current price of a security is a risk-neutral expectation of its future payoff, if properly deflated. In other words, after properly deflated, an asset price must follow a martingale.

5.2 A Binomial Example

The simple binomial model was first proposed by William Sharpe in his Investments textbook in which he demonstrates the basic no arbitrage idea in a very understandable fashion to his MBA students. Cox, Ross and Rubinstein in 1979 formalized it with a sound mathematical derivation. The pictorial representation of the binomial model can be shown as follows.

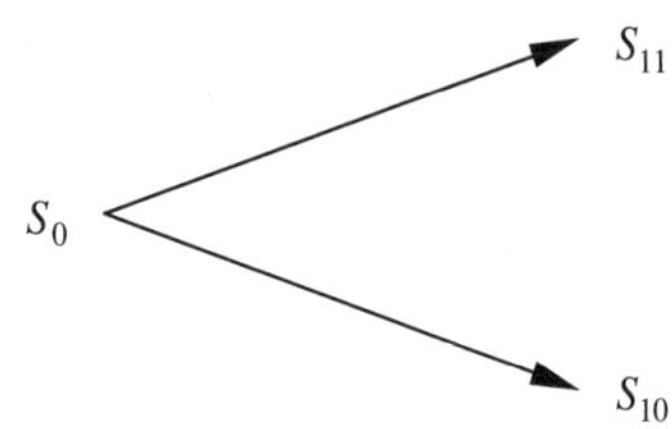

Figure 5.1: One-period Binomial Model for the Stock Price

The up probability is p and down probability is $1-p$. These up and down probabilities are known as subjective probabilities that reflect the real world distribution of the stock price. Taking ratios, $u = S_{11}/S_0$ and $d = S_{10}/S_0$ represent 1 plus rate of return in up and down states of economy respectively. In equilibrium, the current stock value must be worth the same as the expected future payoff, discounted at the proper risk-adjusted discount rate, k, as follows:

$$S_0 = \mathbb{E}[\tilde{S}_1]/k = \frac{pS_{11} + (1-p)S_{10}}{k} \tag{5.1}$$

but we don't know how to find k. The CAPM (see Chapter 3) will help but then we need to assume a quadratic utility function.

Suppose there is an option written on this stock. Then it must follow the same binomial model as follows:

Again, in equilibrium, we have:

$$\begin{aligned} C_0 &= \frac{\mathbb{E}[\tilde{C}_1]}{k_C} \\ &= \frac{pC_{11} + (1-p)C_{10}}{k_C} \end{aligned} \tag{5.2}$$

where k_C is the risk-adjusted discount rate for the option and certainly greater than

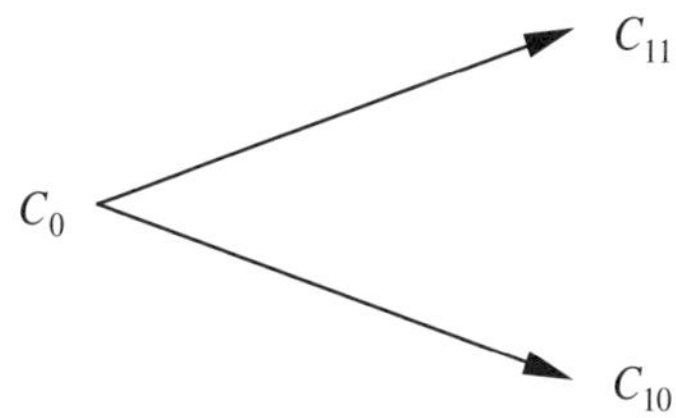

Figure 5.2: One-period Binomial Model for the Option Price

k because call is a lot more risky! We can certainly use the CAPM to calculate k_C but it is not easy. In fact, Black and Scholes demonstrated that in continuous time, the same CAPM formula applies to both stock and option.

The risk-neutral pricing saves the trouble. Imagine a person who is extremely INsensitive to risks. Then he would treat stock and call the same as the risk-free bond. His required return is therefore the risk free rate for all securities. For the stock,

$$S_0 = \frac{\hat{\mathbb{E}}[\tilde{S}_1]}{R} = \frac{\hat{p}S_{11} + (1-\hat{p})S_{10}}{R} \tag{5.3}$$

where $\hat{p}$ is how this risk-neutral person views the economy and $R = 1 + r_f$ is 1 plus the risk-free return. Now we can solve for $\hat{p}$ as follows.

$$\begin{aligned} \hat{p} &= \frac{RS_0 - S_{10}}{S_{11} - S_{10}} \\ &= \frac{RS_0 - dS_0}{uS_0 - dS_0} \\ &= \frac{R - d}{u - d} \end{aligned} \tag{5.4}$$

For the option,

$$\begin{aligned} C_0 &= \frac{\hat{\mathbb{E}}[\tilde{C}_1]}{R} \\ &= \frac{\hat{p}C_{11} + (1-\hat{p})C_{10}}{R} \end{aligned} \tag{5.5}$$

This risk-insensitive person does not have to really exist! Just by imagining him, we solve a very complex problem the easy way. The risk premium is represented by the difference between p and $\hat{p}$.

5.2.1 No arbitrage pricing

Following the same example,

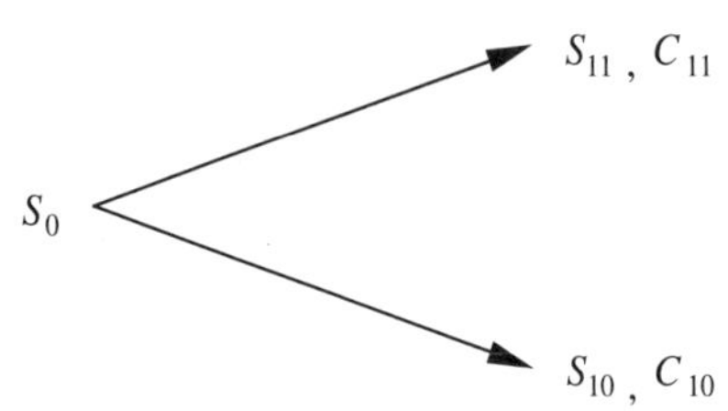

Figure 5.3: One-period Binomial Model for both Stock and Option Prices

we form a portfolio that consists of long 1 call and short h shares of stock where h is know as the "hedge ratio". Then portfolio values today and at maturity are:

- today $C_0 - hS_0$
- maturity $\begin{cases} C_{11} - hS_{11} \\ C_{10} - hS_{10} \end{cases}$

We can set the hedge ratio h so that the payoffs at maturity are equal. So h can be solved as:

$$\begin{aligned} C_{11} - hS_{11} &= C_{10} - hS_{10} \\ C_{11} - C_{10} &= h(S_{11} - S_{10}) \\ h &= \frac{C_{11} - C_{10}}{S_{11} - S_{10}} \end{aligned} \tag{5.6}$$

Substitute h back into the portfolio values today and at maturity to get:

- today $\frac{C_0 S_{11} - C_0 S_{10} - C_{11} S_0 + C_{10} S_0}{S_{11} - S_{10}}$
- maturity $\frac{C_{10} S_{11} - C_{11} S_{10}}{S_{11} - S_{10}}$

Since this is an investment with a certain return (note that the two maturity payoffs are now one), today's value must be the risk-free present value of the maturity value:

$$\frac{C_0 S_{11} - C_0 S_{10} - C_{11} S_0 + C_{10} S_0}{S_{11} - S_{10}} = \frac{1}{R} \frac{C_{10} S_{11} - C_{11} S_{10}}{S_{11} - S_{10}} \tag{5.7}$$

Let u be the up return and d be the down return (i.e. $S_{11} = S_0 u$, $S_{10} = S_0 d$). We rewrite:

$$\frac{C_0 u - C_0 d - C_{11} + C_{10}}{u - d} = \frac{1}{R} \frac{C_{10} u - C_{11} d}{u - d} \tag{5.8}$$

Re-arranging terms to solve for the current call price, we obtain:

$$\begin{aligned} C_0 &= \frac{1}{R}\left[\frac{C_{10} u - C_{11} d}{u - d} + \frac{R(C_{11} - C_{10})}{u - d}\right] \\ &= \frac{1}{R}\left[\frac{R - d}{u - d} C_{11} + \frac{u - R}{u - d} C_{10}\right] \\ &= \frac{1}{R}[\hat{p} C_{11} + (1 - \hat{p}) C_{10}] \end{aligned} \tag{5.9}$$

This result, although simple and straightforward, represents that the call option price is the risk-neutral expected value discounted at the risk-free rate. Note that $\hat{p} C_{11} + (1 - \hat{p}) C_{10}$ is the risk-neutral expectation and discounting it by the risk-free rate arrives at the call price. It should be interesting to note that

$$S_0 = \frac{1}{R}[\hat{p} S_{11} + (1 - \hat{p}) S_{10}] \tag{5.10}$$

that is, the risk-neutral pricing does not only apply to options, but also the underlying stocks.

From the above example, we can see that no-arbitrage pricing is equivalent to risk-neutral pricing. Under no-arbitrage, we arrive at the result that the option price is a discounted risk-neutral expected value of the payoff.

Example

In the following example, the initial stock price is set to be \$50, and the strike price is \$50. The risk-free rate is 5%. The future stock prices (up and down) are \$55 (i.e. 10% return) and \$45 (i.e. 10% return).

The hedge ratio can be computed as follows:

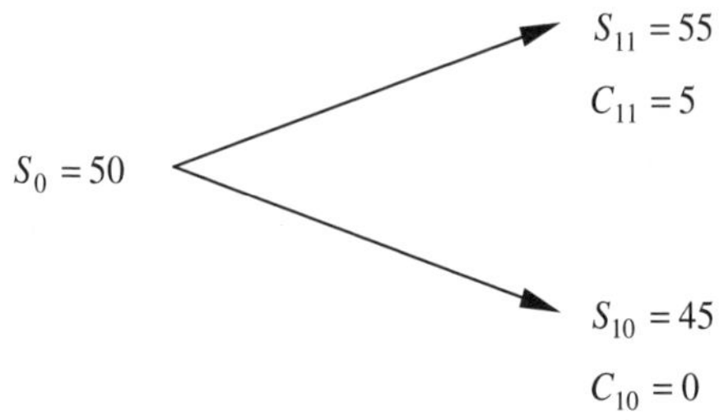

Figure 5.4: Numerical Example

$$h = \frac{5-0}{55-45} = \frac{1}{2}$$

and hence the portfolio value at maturity is as follows:

$$\begin{cases} C_{11} - hS_{11} = 5 - \dfrac{55}{2} = -22.5 \\ C_{10} - hS_{10} = 0 - \dfrac{45}{2} = -22.5 \end{cases}$$

Since the portfolio is risk-free (i.e. regardless of the uncertain future, the payoff is fixed at 22.5), the portfolio value today is just the payoff discounted at the risk-free rate:

$$\frac{-22.5}{1.05} = -21.43$$

This value must also equal the portfolio value today

$$C_0 - hS_0 = C_0 - \frac{50}{2} = C_0 - 25$$

As a result, we can solve for the call value today:

$$\begin{cases} C_0 - 25 = \dfrac{-22.5}{1+5\%} = -21.43 \\ C_0 = 3.57 \end{cases}$$

We can automate the above process and derive a pricing formula so we can quickly plug in inputs to arrive at the correct price, as shown below:

$$
\begin{aligned}
u &= \frac{S_{11}}{S_0} = 1.1 \\
d &= \frac{S_{10}}{S_0} = 0.9 \\
R &= 1 + 5\% = 1.05 \\
\hat{p} &= \frac{R-d}{u-d} = \frac{1.05-0.9}{1.1-0.9} = 0.75 \\
C_0 &= \frac{1}{1.05}(0.75 \times 5 + 0.25 \times 0) = \frac{3.75}{1.05} = 3.57
\end{aligned}
$$

5.2.2 Replication with Stock and Bond

Apparently, the possibility of risk-neutral pricing, that is the existence of the risk-neutral probabilities, $\hat{p}$ and $1 - \hat{p}$, relies on the fact that there are two states in the example. With two securities, stock and option, the market is complete and hence the risk-free portfolio is possible. In other words, stock and option replicate the payoff of the risk-free investment.

If the risk-free rate is tradable, that is, if there exists a zero-coupon bond that yields the risk-free rate of return, then stock and bond can also complete the market and option then becomes the redundant asset that can be replicated. Then, the option price can be derived accordingly. To see that, we put the payoffs of stock and bond in the following matrix:

$$
\begin{bmatrix} S_{11} & S_{10} \\ 1 & 1 \end{bmatrix}
$$

where the first row contains the stock payoffs in state 1 and state 0 respectively and the second row contains the bond payoff that is just principal (normalized to 1). The price of the risk-free zero-coupon bond is $\frac{1}{R}$.

By forming a portfolio of stock and bond, we can replicate the option payoff. In other words, set the portfolio weights as ϕ for stock and ψ for bond and then we can replicate the payoffs of the option:

$$
\begin{bmatrix} \phi & \psi \end{bmatrix} \begin{bmatrix} S_{11} & S_{10} \\ 1 & 1 \end{bmatrix} = \begin{bmatrix} C_{11} & C_{10} \end{bmatrix}
$$

As a result, we can easily solve for the portfolio weights:

$$\begin{aligned}\left[\begin{array}{cc}\phi & \psi\end{array}\right] &= \left[\begin{array}{cc}C_{11} & C_{10}\end{array}\right]\left[\begin{array}{cc}S_{11} & S_{10}\\ 1 & 1\end{array}\right]^{-1}\\ &= \frac{1}{S_{11}-S_{10}}\left[\begin{array}{cc}C_{11} & C_{10}\end{array}\right]\left[\begin{array}{cc}1 & -S_{10}\\ -1 & S_{11}\end{array}\right]\\ &= \left[\begin{array}{cc}\frac{C_{11}-C_{10}}{S_{11}-S_{10}} & \frac{-C_{11}S_{10}+C_{10}S_{11}}{S_{11}-S_{10}}\end{array}\right]\end{aligned}$$

The portfolio weight for stock is the similar one that we saw earlier, which is the delta of the option. The portfolio weight for bond is new. Knowing the portfolio weights, we can easily compute the option price today:

$$C_0 = \left[\begin{array}{cc}\phi & \psi\end{array}\right]\left[\begin{array}{c}S_0\\ \frac{1}{R}\end{array}\right] = \frac{C_{11}-C_{10}}{S_{11}-S_{10}}S_0 + \frac{-C_{11}S_{10}+C_{10}S_{11}}{S_{11}-S_{10}}\frac{1}{R}$$

Substituting $S_{11} = uS_0$ and $S_{10} = dS_0$ into the above formula for the call, we can simplify the above result and arrive at (5.9).

Example

Following the previous example, we can compute the portfolio weights as:

$$\left[\begin{array}{cc}\phi & \psi\end{array}\right] = \left[\begin{array}{cc}\frac{5-0}{55-45} & \frac{-225+0}{55-45}\end{array}\right] = \left[\begin{array}{cc}1/2 & -22.5\end{array}\right]$$

and the call value today is:

$$C_0 = \left[\begin{array}{cc}1/2 & -22.5\end{array}\right]\left[\begin{array}{c}50\\ \frac{1}{1.05}\end{array}\right] = 25 - 21.43 = 3.57$$

which is the same as the result of the previous section.

5.3 Continuous Time Limit

Apparently if the time interval is small, then the hedge ratio in (5.6) becomes:

$$h = \frac{\Delta C}{\Delta S} \tag{5.11}$$

Also, to tie the binomial model to the Black-Scholes model, we must specify the up and down as:

$$\begin{cases} u = e^{\sigma\sqrt{\Delta t}} \\ d = 1/u = e^{-\sigma\sqrt{\Delta t}} \end{cases} \tag{5.12}$$

Note that the expectation result, $\hat{\mathbb{E}}[S_1] = S_0 e^{r\Delta t}$, remains true because it is independent of what u and d are. The variance now becomes:

$$\begin{aligned} \widehat{\text{var}}[\ln S_1 - \ln S_0] &= \hat{\mathbb{E}}[(\ln S_1 - \ln S_0)^2] - \hat{\mathbb{E}}[\ln S_1 - \ln S_0]^2 \\ &= \hat{p}(\ln u)^2 + (1-\hat{p})(\ln d)^2 - \{\hat{p}\ln u + (1-\hat{p})\ln d\}^2 \\ &= \hat{p}\sigma^2\Delta t + (1-\hat{p})\sigma^2\Delta t - \{\hat{p}\sigma\sqrt{\Delta t} + (1-\hat{p})(-\sigma\sqrt{\Delta t})\}^2 \\ &= 4\hat{p}(1-\hat{p})\sigma^2\Delta t \end{aligned} \tag{5.13}$$

Note that $\hat{p} \to 1/2$ as $\Delta t \to 0$ (as an exercise). Then,

$$\widehat{\text{var}}[\ln S_1 - \ln S_0] \to \sigma^2\Delta t \tag{5.14}$$

in continuous time. We also note that:

$$\widehat{\text{var}}[\ln S_1 - \ln S_0] = \text{var}[\ln S_1 - \ln S_0] \tag{5.15}$$

This indicates that the risk-neutral variance is identical to the variance under the real measure. This is not surprising as change of measure does not affect the variance but only the drift. This demonstrates that the continuous time limit of the binomial model is the Black-Scholes model.

5.4 Black-Scholes Model

There are four ways to derive the Black-Scholes model. We have just seen how the binomial model can, in the continuous time limit (if the up and down are set properly), converge to the Black-Scholes model and now I shall present some alternative derivations. The original Black-Scholes derivation is to solve the partial differential equation (PDE). Then from the martingale theory, we can solve the Black-Scholes model using fundamental statistical tools. This is the most popular method used by a number of derivatives text books now. The last method is to use the Fourier transform. We shall see the PDE method and the martingale method in this chapter and the Fourier transform method in Chapter 13.

5.4.1 The Original – Journal of Political Economy, 1973

From Chapter 4, we continue to assume the stock price process as:

$$\frac{dS(t)}{S(t)} = \mu dt + \sigma dW(t) \tag{5.16}$$

where both and are constants. The Ito's lemma states that the dynamics of an option must satisfy the following process:

$$\begin{aligned} dC &= C_S dS + \frac{1}{2} C_{SS} (dS)^2 + C_t dt \\ &= \left[C_S \mu S + \frac{1}{2} C_{SS} \sigma^2 S^2 + C_t \right] dt + C_S \sigma S dW(t) \end{aligned} \tag{5.17}$$

where subscripts here represent partial derivatives, i.e. $C_S = \frac{\partial C}{\partial S}, C_{SS} = \frac{\partial^2 C}{\partial S^2}$, and $C_t = \frac{\partial C}{\partial t}$. Time index is drop for easy exposition.

As in the binomial model, we form a risk-free portfolio by heging the option with h units of stock, that is $V = C - hS$. To render the portfolio risk-free, it must be that the portfolio contains no risk, i.e. no dW. In order to do that, by comparing (5.16) and (5.17), we must set the hedge ratio as $h = C_S$. Then,

$$\begin{aligned} dV &= dC - hdS \\ &= \left[\frac{1}{2} C_{SS} \sigma^2 S^2 + C_t \right] dt \end{aligned} \tag{5.18}$$

which is deterministic. Deterministic investments should yield the risk-free return and hence we can write equation (5.18) as:

$$\begin{aligned} &\left[\frac{1}{2} C_{SS} \sigma^2 S^2 + C_t \right] dt = rV dt = r(C - C_S S) dt \\ &\frac{1}{2} C_{SS} \sigma^2 S^2 + C_t + r C_S S = rC \end{aligned} \tag{5.19}$$

which is the famous Black-Scholes PDE. Solving this PDE with the following boundary conditions:

$$\begin{cases} C(\infty, t) = \infty \\ C(0, t) = 0 \\ C(S(T), T) = \max\{S(T) - K, 0\} \end{cases} \tag{5.20}$$

yields the famous Black-Scholes model. To see how the PDE can be transformed into the famous heat equation and solved, please refer to the Appendix at the end of this Chapter.

We would like to do the same thing in the risk-neutral space instead of the real space. We now define the stock price process under the risk-neutral measure using the result in Chapter 4 with the following change of measure: $d\hat{W} = dW + \frac{\mu - r}{\sigma}dt$. The stock then earns the risk-free rate as follows:

$$dS = rSdt + \sigma Sd\hat{W} \tag{5.21}$$

As a result,

$$\begin{aligned} dC &= C_S dS + \frac{1}{2}C_{SS}(dS)^2 + C_t dt \\ &= \left[C_S rS + \frac{1}{2}C_{SS}\sigma^2 S^2 + C_t\right] dt + C_S \sigma Sd\hat{W} \\ &= rCdt + C_S\sigma Sd\hat{W} \end{aligned} \tag{5.22}$$

It is clear that both stock and option now earn the risk-free rate under the risk-neutral measure. In fact, all assets under the risk-neutral measure earn the risk-free rate. We have shown in Chapter 4 via the Martingale Representation Theorem that normalized stock and option prices are martingales under the risk-neutral measure. This makes the risk-neutral measure special and unique. Hence, the risk-neutral measure is also called the equivalent martingale measure.

5.4.2 Martingale Representation Theorem

In Chapter 4, we derive the MRT in general. We know that the MRT is to prove how an option is a redundant asset and can be replicated by the stock and bond – so called self-financing. Here we re-visit the MRT with the Black-Scholes model. From (5.19), we know that

$$\tfrac{1}{2}C_{SS}\sigma^2 S^2 + C_t + rC_S S = rC \tag{5.23}$$

From (5.17), we also know that

$$dC = (C_S\mu S + \tfrac{1}{2}C_{SS}\sigma^2 S^2 + C_t)dt + C_S\sigma SdW(t) \tag{5.24}$$

Substituting (5.19) back into (5.17), we get:

$$\begin{aligned}\frac{dC}{C} &= rdt + C_S\frac{S}{C}(\mu - r)dt + C_S\frac{S}{C}\sigma dW(t) \\ &= (1-\eta)rdt + \eta\frac{dS}{S}\end{aligned} \tag{5.25}$$

where $\eta = C_S S/C$ as the elasticity of the option with respect to the stock. (5.25) states that the return of the option is a weighted average of the stock return and the risk-free return. This is precisely the MRT in a different format. (5.25) states that the return of the option is represented by the return of the risk-free rate and the return of the stock. In other words, options are redundant.

5.4.3 Alternative Derivation

There are a few ways to derive the Black-Scholes model. The one shown here is the most popular and intuitive. Under the risk-neutral pricing theory, any security must have a price as the risk-neutral expectation of its future payoff discounted at the risk-free rate. In the case of a call option, we have:

$$\begin{aligned}C(t) &= e^{-r(T-t)}\hat{\mathbb{E}}_t[\max\{S(T)-K,0\}] \\ &= e^{-r(T-t)}\hat{\mathbb{E}}_t[(S(T)-K)\mathbb{I}_{\{S(T)>K\}}] \\ &= e^{-r(T-t)}\hat{\mathbb{E}}_t[S(T)\mathbb{I}_{\{S(T)>K\}}] - e^{-r(T-t)}K\hat{\mathbb{E}}_t[\mathbb{I}_{\{S(T)>K\}}]\end{aligned} \tag{5.26}$$

Note that the second expectation in the last line of equation (5.26) is a straightforward probability that can be derived in closed-form as:

$$\begin{aligned}\hat{\mathbb{E}}_t[\mathbb{I}_{\{S(T)>K\}}] &= \int_K^\infty \hat{\varphi}(S)dS \\ &= \int_{\frac{\ln K - m}{v}}^\infty \hat{\varphi}(Z)dZ \\ &= 1 - N\left(\frac{\ln K - m}{v}\right) \\ &= N\left(\frac{m - \ln K}{v}\right)\end{aligned} \tag{5.27}$$

where

$$
\begin{aligned}
Z &= \frac{\ln S - m}{v} \\
m &= \ln S(t) + r(T-t) - \frac{1}{2}\sigma^2(T-t) \\
v &= \sigma\sqrt{T-t}
\end{aligned} \tag{5.28}
$$

The first expectation in the last line of equation (5.26) requires further change of measure. From Chapter 4, we know that:

$$
\begin{aligned}
\hat{\mathbb{E}}_t[S(T)\mathbb{I}_{\{S(T)>K\}}] &= \hat{\mathbb{E}}_t[S(T)]\hat{\mathbb{E}}_t^{(S)}[\mathbb{I}_{\{S(T)>K\}}] \\
&= S(t)e^{r(T-t)}\hat{\mathbb{E}}_t^{(S)}[\mathbb{I}_{\{S(T)>K\}}]
\end{aligned} \tag{5.29}
$$

where $\hat{\mathbb{E}}_t^{(S)}[\cdot]$ represents the expectation where the stock price serves the numeraire. The discussion of numeraire is given in the next section. The first two terms in the above equation are the standard result of risk-neutral expectation (i.e. all assets under the risk-neutral measure yield the risk-free rate). The second term is obtained by the following change of measure. The RND of the change of measure is:

$$
\eta(t) = \frac{S(T)}{\hat{E}_t[S(T)]} = \frac{S(T)}{e^{r(T-t)}S(t)} \tag{5.30}
$$

Note that:

$$
\begin{aligned}
S(T) &= S(t)\exp\left(\int_t^T (r - 1/2\sigma^2)du + \int_t^T \sigma d\hat{W}(u)\right) \\
&= S(t)\exp\left((r - 1/2\sigma^2)(T-t) + \sigma r - 1/2\sigma^2 d\hat{W}(u)\right)
\end{aligned} \tag{5.31}
$$

Hence,

$$
\begin{aligned}
\eta(t) &= \frac{S(T)}{e^{r(T-t)}S(t)} \\
&= \exp\left(\int_t^T \sigma d\hat{W}(u) - \int_t^T 1/2\sigma^2 du\right)
\end{aligned} \tag{5.32}
$$

It is clear that $\theta(t) = -\sigma$ in the Girsanov:

$$
d\hat{W}^{(S)}(t) = d\hat{W}(t) + \sigma dt \tag{5.33}
$$

Hence, in equation (5.26), the first expectation $\bar{\mathbb{E}}_t[\mathbb{I}_{\{S(T)>K\}}]$ can be derived as:

$$
\begin{aligned}
\bar{\mathbb{E}}_t[\mathbb{I}_{\{S(T)>K\}}] &= \int_K^{\infty} \hat{\hat{\varphi}}(S)dS \\
&= \int_{\frac{\ln K-\hat{m}}{v}}^{\infty} \hat{\hat{\varphi}}(Z)dZ \\
&= 1 - N\left(\frac{\ln K - \hat{m}}{v}\right) \\
&= N\left(\frac{\hat{m} - \ln K}{v}\right)
\end{aligned}
\tag{5.34}
$$

where

$$
\begin{aligned}
\hat{m} &= \ln S(t) + r(T-t) - {}^1\!/\!_2\sigma^2(T-t) + \sigma^2(T-t) \\
&= \ln S(t) + r(T-t) + {}^1\!/\!_2\sigma^2(T-t)
\end{aligned}
\tag{5.35}
$$

which is parallel to equation (5.27). Now we can write the final Black-Scholes model as:

$$
C(t) = S(t)N(d_1) - e^{-r(T-t)}KN(d_2) \tag{5.36}
$$

where

$$
\begin{aligned}
d_1 &= \frac{\ln S(t) - \ln K + r(T-t) + {}^1\!/\!_2\sigma^2(T-t)}{\sigma\sqrt{T-t}} \\
d_2 &= \frac{\ln S(t) - \ln K + r(T-t) - {}^1\!/\!_2\sigma^2(T-t)}{\sigma\sqrt{T-t}}
\end{aligned}
$$

Later on, this model is extended to include Gaussian interest rates (Merton (1973), Rabinovitch (1989)). See Chapter 7.

5.5 Concept of Numeraire

The alternative derivation of the Black-Scholes model in the previous section relates to the concept of numraire. The concept of numeraire is that a security price after normalizing by the numeraire is a martingale. Simply speaking, if:[1]

[1]The expectations here carry a time subscript to represent conditional expectations given the information till time t.

$$\frac{X(t)}{A(t)} = \mathbb{E}_t^{(A)}\left[\frac{X(T)}{A(T)}\right] \tag{5.37}$$

then A is known as the numeraire and $\mathbb{E}_t^{(A)}$ is the expectation defined over the probability measure that is adjusted by the numeraire. In the previous chapter, we mentioned the money market as the numeraire under the risk-neutral measure, that is to let $A(T) = e^{r(T-t)}$, then (5.37) becomes:

$$\begin{aligned} \frac{X(t)}{1} &= \hat{\mathbb{E}}_t\left[\frac{X(T)}{e^{r(T-t)}}\right] \\ X(t) &= e^{-r(T-t)}\hat{\mathbb{E}}_t[X(T)] \end{aligned} \tag{5.38}$$

The alternative derivation of the Black-Scholes model in the previous section uses the (reciprocal of) stock price as the numeraire. Let $A(T) = e^{r(T-t)}/S(T)$. Then it follows that $A(t) = 1/S(t)$ and hence $A(T)/A(t)$ is identical to (5.30) as the RND. Another useful application of the concept of numeraire is the stochastic discount factor in microeconomics. In microeconomics, the following is true:

$$X(t) = \mathbb{E}_t[M(t,T)X(T)] \tag{5.39}$$

where $M(t,T)$ is known as the Marginal Rate of Substitution (MRS) or stochastic discount factor. Note that the above expectation is under the real measure, not the risk-neutral measure. Briefly speaking, the MRS of an individual trades off his/er current consumption with his/er future consumption. As a result, the MRS depends on a utility function and its degree of risk aversion.

This traditional and standard microeconomic theory can be interpreted using the concept of numeraire. Effectively, $M(t,T) = 1/A(T)$ is the (reciprocal of) numeraire. Note that $M(t,t) = 1$ and (5.39) is identical to (5.37). From the state price theory introduced in Chapter 2, it is true that:

$$\mathbb{E}_t[M(t,T)] = P(t,T) \tag{5.40}$$

that is, the real expectation of the MRS is the risk-free discount factor. It follows naturally that:

$$\begin{aligned} X(t) &= \mathbb{E}_t[M(t,T)]\hat{\mathbb{E}}_t[X(T)] \\ &= P(t,T)\hat{\mathbb{E}}_t[X(T)] \end{aligned} \tag{5.41}$$

Rewrite (5.39) as:

$$X(t) = P(t,T)\mathbb{E}_t\left[\frac{M(t,T)}{P(t,T)}X(T)\right] \tag{5.42}$$

and it is apparent that $M(t,T)/P(t,T)$ is an RND which takes the form of:

$$\eta(t) = \exp\left(-\int_0^t \theta(u)dW(u) - {}^1\!/_2\int_0^t \theta^2(u)du\right) \tag{5.43}$$

Applying the Girsanov theorem, we have:

$$d\hat{W}(t) = dW(t) + \theta(t)dt \tag{5.44}$$

This new measure defined under $\hat{W}(t)$ must be the same measure that is used to compute the expectation in (5.41), which is the risk-neutral measure. As a result, the drift adjustment $\theta(t)$ is the risk premium that can be in general stochastic; or can be as simple as $(\mu - r)/\sigma$ in the Black-Scholes case. The MRS then can be written as:

$$M(t,T) = \exp\left(\ln P(t,T) - \int_t^T \theta(u)dW(u) - {}^1\!/_2\int_t^T \theta^2(u)du\right) \tag{5.45}$$

which implies that if $\theta(t)$ is deterministic then the MRS follows a log normal process. It also implies that even if the interest rate is deterministic or constant (i.e., $\ln P(t,T) = -r(T-t)$), the MRS can still be stochastic, simply because the risk premium can be stochastic. In fact, many recent researchers follow this approach and model the risk premium directly.

The last application of the numeraire is the forward measure where the random risk-free discount factor is the numeraire.

5.6 Appendix

5.6.1 Solving Black-Scholes PDE

Recall the Black-Scholes PDE as:

$$\frac{1}{2}C_{SS}\sigma^2S^2 + C_t + rC_SS = rC \tag{5.46}$$

with the boundary condition (only the t boundary condition) $C(S(T),T) = \max\{S(T)-K, 0\}$ where T is maturity time and K is the strike price. First we need to perform the following change of variable (equation (9) of the Black-Scholes paper on page 643):

$$C(S,t) = e^{-r(T-t)} f(S^*, t^*) \tag{5.47}$$

where

$$\begin{aligned} S^* &= \frac{2}{\sigma^2}(r - {}^1\!/\!{}_2\sigma^2)\left[\ln\frac{S}{K} + (r - {}^1\!/\!{}_2\sigma^2)(T-t)\right] \\ t^* &= \frac{2}{\sigma^2}(r - {}^1\!/\!{}_2\sigma^2)^2(T-t) \end{aligned} \tag{5.48}$$

Then the following results hold:

$$\begin{aligned} C_S &= e^{-r(T-t)}\frac{\partial f}{\partial S^*}\frac{\partial S^*}{\partial S} \\ &= e^{-r(T-t)}\frac{\partial f}{\partial S^*}\frac{2}{\sigma^2}(r - {}^1\!/\!{}_2\sigma^2)\frac{1}{S} \\ C_{SS} &= e^{-r(T-t)}\frac{\partial^2 f}{(\partial S^*)^2}\frac{2}{\sigma^2}(r - {}^1\!/\!{}_2\sigma^2)\frac{1}{S} + e^{-r(T-t)}\frac{\partial f}{\partial S^*}\frac{2}{\sigma^2}(r - {}^1\!/\!{}_2\sigma^2)\frac{-1}{S^2} \\ C_t &= re^{-r(T-t)} f + e^{-r(T-t)}\left(\frac{\partial f}{\partial S^*}\frac{\partial S^*}{\partial t} + \frac{\partial f}{\partial t^*}\frac{\partial t^*}{\partial t}\right) \\ &= re^{-r(T-t)} f - e^{-r(T-t)}\frac{\partial f}{\partial S^*}\frac{2}{\sigma^2}(r - {}^1\!/\!{}_2\sigma^2)^2 - e^{-r(T-t)}\frac{\partial f}{\partial t^*}\frac{2}{\sigma^2}(r - {}^1\!/\!{}_2\sigma^2)^2 \end{aligned} \tag{5.49}$$

Substitute these results back to the PDE and we get the famous heat equation (note that the three f_{S^*} terms cancel out):

$$f_{S^*S^*} = f_{t^*} \tag{5.50}$$

that has been well studied. Depending on the boundary conditions, the heat equation has different solutions. To arrive at the boundary condition of the Black-Scholes model for the heat equation, we first reverse the change of variable as follows:

$$\begin{aligned} S &= K \exp\left\{\frac{{}^1\!/\!{}_2\sigma^2 S^*}{r - {}^1\!/\!{}_2\sigma^2} - \frac{{}^1\!/\!{}_2\sigma^2 t^*}{r - {}^1\!/\!{}_2\sigma^2}\right\} \\ t &= T - t^*\frac{{}^1\!/\!{}_2\sigma^2}{(r - {}^1\!/\!{}_2\sigma^2)^2} \end{aligned} \tag{5.51}$$

and then it follows that the boundary condition becomes:

$$\max(S(T) - K, 0\} = \max\left\{K\left[\exp\left\{\frac{1/2\sigma^2 S^*}{r - 1/2\sigma^2}\right\} - 1\right], 0\right\} \tag{5.52}$$

The heat equation subject to the above boundary condition has a well-known solution as follows:

$$f(S^*, t^*) = \frac{1}{\sqrt{2\pi}} \int_{-S^*/\sqrt{2t^*}}^{\infty} K\left[\exp\left\{\frac{1/2\sigma^2(S^* + x\sqrt{2t^*})}{r - \frac{1}{2}\sigma^2}\right\} - 1\right] e^{-1/2x^2} dx \tag{5.53}$$

Subtituting the results of S^* and t^*, we obtain the Black-Scholes model.

5.6.2 Exchange Option Formula

Let the joint process of X and Y under the risk-neutral process be:

$$\begin{bmatrix} d\ln X \\ d\ln Y \end{bmatrix} = \begin{bmatrix} r - 1/2\sigma_X^2 \\ r - 1/2\sigma_Y^2 \end{bmatrix} dt + \begin{bmatrix} \sigma_X & 0 \\ 0 & \sigma_Y \end{bmatrix} \begin{bmatrix} \sqrt{1-\rho^2} & \rho \\ 0 & 1 \end{bmatrix} \begin{bmatrix} d\hat{W}_X(t) \\ d\hat{W}_Y(t) \end{bmatrix} \tag{5.54}$$

and the call option valuation is as follows:

$$\begin{aligned} C_t &= e^{-r(T-t)}\hat{\mathbb{E}}_t[\max X(T) - Y(T), 0\}] \\ &= e^{-r(T-t)}\hat{\mathbb{E}}_t[Y \max\{X(T)/Y(T) - 1, 0\}] \\ &= e^{-r(T-t)}\hat{\mathbb{E}}_t[Y]\bar{\mathbb{E}}_t[\max\{Z(T) - 1, 0\}] \\ &= Y(t)\bar{\mathbb{E}}_t[\max\{Z(T) - 1, 0\}] \end{aligned} \tag{5.55}$$

The RND is:

$$\eta = \frac{Y}{\hat{\mathbb{E}}_t[Y]} = \exp\left(\int_t^T \sigma_Y d\hat{W}_Y(u) - \frac{1}{2}\int_t^T \sigma_Y^2 du\right) \tag{5.56}$$

and hence the Girnosav Theorem implies the change of measure to be $\bar{dW}_Y = d\hat{W}_Y - \sigma_Y dt$. Under the new measure, (5.54) becomes:

$$
\begin{aligned}
\begin{bmatrix} d\ln X \\ d\ln Y \end{bmatrix} &= \begin{bmatrix} r - {}^1\!/\!_2\sigma_X^2 \\ r - {}^1\!/\!_2\sigma_Y^2 \end{bmatrix} dt + \begin{bmatrix} \sigma_X & 0 \\ 0 & \sigma_Y \end{bmatrix} \begin{bmatrix} \sqrt{1-\rho^2} & \rho \\ 0 & 1 \end{bmatrix} \begin{bmatrix} d\hat{W}_X(t) \\ d\hat{W}_Y(t) + \sigma_Y dt \end{bmatrix} \\
&= \begin{bmatrix} r - {}^1\!/\!_2\sigma_X^2 + \rho\sigma_X\sigma_Y \\ r + {}^1\!/\!_2\sigma_Y^2 \end{bmatrix} dt + \begin{bmatrix} \sigma_X & 0 \\ 0 & \sigma_Y \end{bmatrix} \begin{bmatrix} \sqrt{1-\rho^2} & \rho \\ 0 & 1 \end{bmatrix} \begin{bmatrix} d\hat{W}_X(t) \\ d\hat{W}_Y(t) \end{bmatrix}
\end{aligned} \tag{5.57}
$$

where

$$
\begin{aligned}
&\ln X(T) \sim (\ln X(t) + \left(r - \tfrac{1}{2}\sigma_X^2 + \rho\sigma_X\sigma_Y\right)(T-t), \sigma_X^2(T-t)\big) \\
&\ln Y(T) \sim \hat{n}(\ln Y(t) + \left(r + \tfrac{1}{2}\sigma_Y^2\right)(T-t), \sigma_Y^2(T-t)\big) \\
&\ln X(T)/Y(T) \sim \hat{n}(m_X - m_Y, V_X + V_Y - 2\text{cov}[\ln X, \ln Y])
\end{aligned} \tag{5.58}
$$

and

$$
\begin{aligned}
m_X - m_Y &= m_Z \\
&= \ln X(t) - \ln Y(t) - ({}^1\!/\!_2\sigma_X^2 - \rho\sigma_X\sigma_Y + {}^1\!/\!_2\sigma_Y^2)(T-t) \\
V_X + V_Y - 2\,\text{cov}[\ln X, \ln Y] &= V_Z \\
&= (\sigma_X^2 + \sigma_Y^2 - 2\rho\sigma_X\sigma_Y)(T-t)
\end{aligned} \tag{5.59}
$$

Hence,

$$
\begin{aligned}
C_t &= e^{-r(T-t)} e^{r(T-t)} Y(t) \left[\frac{X(t)}{Y(t)} N(d_1) - N(d_2) \right] \\
&= X(t)N(d_1) - Y(t)N(d_2)
\end{aligned} \tag{5.60}
$$

where

$$
\begin{aligned}
d_2 &= \frac{m_Z - 0}{\sqrt{V_Z}} \\
&= \frac{\ln \frac{X(t)}{Y(t)} - {}^1\!/\!_2(\sigma_X^2 + \sigma_Y^2 - 2\rho\sigma_X\sigma_Y)(T-t)}{\sqrt{\sigma_X^2 + \sigma_Y^2 - 2\rho\sigma_X\sigma_Y}}
\end{aligned} \tag{5.61}
$$

5.6.3 FX Option Formula

The derivation of the FX option formula is identical. See Chapter 11.

Chapter 6

Term Structure of Interset Rates

This chapter is heavily drawn from Ingersoll's Theory of Financial Decision Making and Chen's Understanding and Managing Interest Rate Risk. In this chapter, I talk about the theory in general. The term structure models here are all based upon the pure expectation hypothesis (The two other popular hypotheses are liquidity premium and preferred habitat.)

6.1 A Note on Notation

In this chapter, we introduce the two most important term structure models in the literature – the Vasicek model (1977) and the Cox-Ingersoll-Ross model (original working paper in 1978 and published version in 1985). We adopt a trick in Ingersoll to derive the risk premiums for the two models. We focus on the equilibrium argument in this chapter and leave the closed-form solutions to the two models for Chapter 7.

Note that in this chapter, we simplify the notation in order to help readers grasp the basic meanings as quickly and as efficiently as possible. For example, $\hat{\alpha}$ and $\hat{\mu}$ represent the risk-adjusted mean reversion speed and reversion level respectively in both Cox-Ingersoll-Ross and Vasicek models, although they are different quantities in the two models. In the CIR model, $\hat{\alpha} = \alpha + \lambda$ and $\hat{\mu} = \frac{\alpha\mu}{\alpha+\lambda}$ where λ is the market price of risk under the square-root dynamics and in the Vasicek model, $\hat{\alpha} = \alpha$ and $\hat{\mu} = \mu + \frac{\sigma\lambda}{\alpha}$ where λ is the market price of risk under the Gaussian dynamics. The two λs are different values.

6.2 Partial Differential Equation

6.2.1 Cox-Ingersoll-Ross Model

The Cox-Ingersoll-Ross model is a square-root process for the instantaneous short rate:

$$dr = \alpha(\mu - r)dt + \sigma\sqrt{r}dW \tag{6.1}$$

Let $P(t,T)$ represent the zero-coupon bond price at time t of \$1 paid at time T. Ito's lemma states that:

$$\begin{aligned} dP &= \frac{\partial P}{\partial r}dr + \frac{1}{2}\frac{\partial^2 P}{\partial r^2}(dr)^2 + \frac{\partial P}{\partial t}dt \\ &= \left[\frac{\partial P}{\partial r}\alpha(\mu - r) + \frac{1}{2}\frac{\partial^2 P}{\partial r^2}\sigma^2 r + \frac{\partial P}{\partial t}\right]dt + \frac{\partial P}{\partial r}\sigma\sqrt{r}dW \end{aligned} \tag{6.2}$$

Form a portfolio with two bonds (underlying r is not traded):

$$V = P(t,T_1) + hP(t,T_2) \tag{6.3}$$

where $P_1 = P(t,T_1)$ and $P_2 = P(t,T_2)$. Then,

$$\begin{aligned} dV &= dP_1 + hdP_2 \\ &= \left\{\left[\frac{\partial P_1}{\partial r}\alpha(\mu - r) + \frac{1}{2}\frac{\partial^2 P_1}{\partial r^2}\sigma^2 r + \frac{\partial P_1}{\partial t}\right]\right. \\ &\quad \left. + h\left[\frac{\partial P_2}{\partial r}\alpha(\mu - r) + \frac{1}{2}\frac{\partial^2 P_2}{\partial r^2}\sigma^2 r + \frac{\partial P_2}{\partial t}\right]\right\}dt \\ &\quad + \left[\frac{\partial P_1}{\partial r} + h\frac{\partial P_2}{\partial r}\right]\sigma\sqrt{r}dW \end{aligned} \tag{6.4}$$

To eliminate uncertainty in order to arrive at the risk-free portfolio, we set the hedge ratio as:

$$h = -\frac{\partial P_1/\partial r}{\partial P_2/\partial r} \tag{6.5}$$

As a result, V is a risk-free portfolio and its return must be the risk-free rate as follows:

$$\begin{aligned} dV &= \left\{ \left[\frac{\partial P_1}{\partial r}\alpha(\mu - r) + \frac{1}{2}\frac{\partial^2 P_1}{\partial r^2}\sigma^2 r + \frac{\partial P_1}{\partial t} \right] \right. \\ &\quad \left. - \frac{\partial P_1/\partial r}{\partial P_2/\partial r}\left[\frac{\partial P_2}{\partial r}\alpha(\mu - r) + \frac{1}{2}\frac{\partial^2 P_2}{\partial r^2}\sigma^2 r + \frac{\partial P_2}{\partial t} \right] \right\} dt \\ &= rV\,dt \\ &= r\left[P_1 - -\frac{\partial P_1/\partial r}{\partial P_2/\partial r} P_2 \right] \end{aligned} \tag{6.6}$$

Rearranging terms, we get the PDE as follows:

$$\begin{aligned} &\left[\frac{\partial P_1}{\partial r}\alpha(\mu - r) + \frac{1}{2}\frac{\partial^2 P_1}{\partial r^2}\sigma^2 r + \frac{\partial P_1}{\partial t} - rP_1 \right] \\ -\frac{\partial P_1/\partial r}{\partial P_2/\partial r}&\left[\frac{\partial P_2}{\partial r}\alpha(\mu - r) + \frac{1}{2}\frac{\partial^2 P_2}{\partial r^2}\sigma^2 r + \frac{\partial P_2}{\partial t} - rP_2 \right] = 0 \end{aligned} \tag{6.7}$$

Note that the solution to:

$$\frac{\partial P}{\partial r}\alpha(\mu - r) + \frac{1}{2}\frac{\partial^2 P}{\partial r^2}\sigma^2 r + \frac{\partial P}{\partial t} - rP = 0 \tag{6.8}$$

is also a solution to our PDE of (6.7). Furthermore, note that the solution to:

$$\frac{\partial P}{\partial r}\alpha(\mu - r) + \frac{1}{2}\frac{\partial^2 P}{\partial r^2}\sigma^2 r + \frac{\partial P}{\partial t} - rP - \frac{\partial P}{\partial r}b = 0 \tag{6.9}$$

is also a solution to our PDE of (6.7). Hence, the PDE of (6.7) therefore has no unique solution, since b is arbitrary. As a result, we need another constraint to help identify a unique solution, and that is the ICAPM introduced in Chapter 3. The ICAPM provides the following result:

$$\mathbb{E}\left[\frac{dP}{P}\right] = \left[r + \frac{1}{P}\frac{\partial P}{\partial r}\lambda r \right] dt \tag{6.10}$$

where λ is an arbitrary constant. Equate this with the drift of dP to get:

$$\begin{gathered} \frac{\partial P}{\partial r}\alpha(\mu - r) + \frac{1}{2}\frac{\partial^2 P}{\partial r^2}\sigma^2 r + \frac{\partial P}{\partial t} = rP + \frac{\partial P}{\partial r}\lambda r, \\ \text{or} \\ \frac{\partial P}{\partial r}(\alpha\mu - (\alpha + \lambda)r) + \frac{1}{2}\frac{\partial^2 P}{\partial r^2}\sigma^2 r + \frac{\partial P}{\partial t} - rP = 0 \end{gathered} \tag{6.11}$$

This is the PDE that has a unique solution. The boundary condition is $P(T,T) = 1$. Then we get the CIR solution (1985).

6.2.2 Vasicek Model

The Vasicek model (1977) assumes the following instantaneous short rate process:

$$dr = \alpha(\mu - r)dt + \sigma dW \tag{6.12}$$

Note the ICAPM for the bond return under the Vasicek model (which is different from equation (6.10), is:

$$\mathbb{E}\left[\frac{dP}{P}\right] = \left[r + \frac{1}{P}\frac{\partial P}{\partial r}\lambda\sigma\right] dt \tag{6.13}$$

which gives the following PDE for the Vasicek model:

$$\frac{\partial P}{\partial r}(\alpha(\mu - r) - \lambda\sigma) + \frac{1}{2}\frac{\partial^2 P}{\partial r^2}\sigma^2 + \frac{\partial P}{\partial t} - rP = 0 \tag{6.14}$$

As in the CIR model, this PDE has a unique solution with the boundary condition $P(T,T) = 1$.

6.3 Risk Premia in Term Structure Models

We now turn to the justification of the risk premium: $(1/P)(\partial P/\partial r)\lambda r$ for the CIR model and $(1/P)(\partial P/\partial r)\lambda$ for the Vasicek model. We shall first establish some equilibrium results. Let a generic state variable, x, follow an Ito process:

$$dx(t) = \mu(x,t)dt + \sigma(x,t)dW(t) \tag{6.15}$$

Then a contingent claim written on x has a diffusion equation, by Ito's lemma as:

$$\begin{aligned} dF &= \left[\frac{1}{2}\frac{\partial^2 F}{\partial x^2}\sigma(x,t)^2 + \frac{\partial F}{\partial t} + \frac{\partial F}{\partial x}\mu(x,t)\right] dt + \frac{\partial F}{\partial x}\sigma(x,t)dW(t) \\ &= \mu_F(x,t)F dt + \sigma_F(x,t)F dW(t) \end{aligned} \tag{6.16}$$

which can be re-written as:

$$\begin{aligned}\frac{dF}{F} &= \mu_F(x,t)dt + \sigma_F(x,t)dW(t) \\ &= \mu_F(x,t)dt + \frac{1}{F}\frac{\partial F}{\partial x}\sigma(x,t)dW(t)\end{aligned} \tag{6.17}$$

No arbitrage results in the following PDE by definition:

$$\frac{1}{2}\frac{\partial^2 F}{\partial x^2}\sigma(x,t)^2 + \frac{\partial F}{\partial t} + \frac{\partial F}{\partial x}\mu(x,t) - \mu_F(x,t)F = 0 \tag{6.18}$$

If there is another claim, G, also written on x, then we have:

$$\begin{aligned}\frac{dG}{G} &= \mu_G(x,t)dt + \sigma_G(x,t)dW(t) \\ &= \mu_G(x,t)dt + \frac{1}{G}\frac{\partial G}{\partial x}\sigma(x,t)dW(t)\end{aligned} \tag{6.19}$$

and

$$\frac{1}{2}\frac{\partial^2 G}{\partial x^2}\sigma(x,t)^2 + \frac{\partial G}{\partial t} + \frac{\partial G}{\partial x}\mu(x,t) - \mu_G(x,t)G = 0 \tag{6.20}$$

Consider a portfolio of, F, G and the risk-free asset (B) with the respective weights so that $V = w_F F + w_G G + w_r B$ and $w_F + w_G + w_r = 1$. Then the portfolio return is:

$$\begin{aligned}\mu_V &= w_F\mu_F + w_G\mu_G + w_r r \\ &= w_F\mu_F + w_G\mu_G + (1 - w_F - w_G)r \\ &= r + w_F(\mu_F - r) + w_G(\mu_G - r)\end{aligned} \tag{6.21}$$

where r is the risk-free rate, and

$$\sigma_V = w_F\sigma_F + w_G\sigma_G \tag{6.22}$$

No arbitrage condition states that if $\sigma_V = 0$ (no risk), then $\mu_V = r$ (return must be the risk-free rate). Therefore, the following famous "no arbitrage" result, first given by Merton, must hold (see the Appendix for details):

$$\frac{\mu_F - r}{\sigma_F} = \frac{\mu_G - r}{\sigma_G} = \frac{\mu_V - r}{\sigma_V} \equiv a(x,t) \tag{6.23}$$

where $a(x,t)$ is known as the price of risk. Substitute what represent σ_F, σ_G, and σ_V back to (6.23) to arrive at:

$$\frac{\mu_F - r}{\frac{1}{F}\frac{\partial F}{\partial r}} = \frac{\mu_G - r}{\frac{1}{G}\frac{\partial G}{\partial r}} = \frac{\mu_V - r}{\frac{1}{V}\frac{\partial V}{\partial r}} = \sigma(x,t)a(x,t) = \lambda(x,t) \tag{6.24}$$

In our case, $\sigma(x,t) \equiv \sigma_0$ is a constant. And we can postulate $a(x,t) = a_0 + a_1 x$ which is independent of t. Hence, $\lambda(x,t) = a_0\sigma_0 + a_1\sigma_0 x = \lambda_0 + \lambda_1 x$ (in the CIR model, $\lambda_0 = 0$, and in the Vasicek model $\lambda_1 = 0$). Now, we need to show that the proposition of the linear premium does not violate arbitrage. Following this formulation, we let P be the zero-coupon bond which is a contingent claim on x, and let x be the CIR interest rate r. Then we have:

$$\mu_P - r = \frac{1}{P}\frac{\partial P}{\partial r}(\lambda_0 + \lambda_1 r) \tag{6.25}$$

Similar to equations (6.18) and (6.20) and , the PDE for P is:

$$\begin{gathered}\frac{1}{2}\frac{\partial^2 P}{\partial r^2}\sigma^2 r + \frac{\partial P}{\partial t} + \frac{\partial P}{\partial r}(\alpha(\mu - r) - \lambda_0 - \lambda_1 r) = rP \\ \text{or} \\ \frac{1}{2}\frac{\partial^2 P}{\partial r^2}\sigma^2 r + \frac{\partial P}{\partial t} + \frac{\partial P}{\partial r}\hat{\alpha}(\hat{\mu} - r) = rP\end{gathered} \tag{6.26}$$

where

$$\hat{\alpha} = \alpha + \lambda_1$$
$$\hat{\mu} = \frac{\alpha\mu - \lambda_0}{\hat{\alpha}}$$

There is a solution to this PDE:

$$P(t,T) = A(t,T)e^{-rB(t,T)} \tag{6.27}$$

Taking the partial derivative of the bond price with respect to r, we get:

$$\frac{\partial P}{\partial r} = -PB \tag{6.28}$$

and equation (6.24) becomes:

$$\begin{aligned} \mu_P - r &= (-B)(\lambda_0 + \lambda_1 r) \\ \sigma_P &= \sigma\sqrt{r}B \end{aligned} \tag{6.29}$$

When $r \to 0$, $\mu_P \to \lambda_0 B$, and $\sigma_P \to 0$. So there is no volatility; but we have a locally risk-free bond that makes a non-zero return. Hence the only way to avoid arbitrage is that $\lambda_0 = 0$.

Now let x be the Vasicek interest rate r. Then we have:

$$\begin{gathered} \frac{1}{2}\frac{\partial^2 P}{\partial r^2}\sigma^2 + \frac{\partial P}{\partial t} + \frac{\partial P}{\partial r}(\alpha(\mu - r) - \lambda_0 - \lambda_1 r) = rP \\ \text{or} \\ \frac{1}{2}\frac{\partial^2 P}{\partial r^2}\sigma^2 + \frac{\partial P}{\partial t} + \frac{\partial P}{\partial r}\hat{\alpha}(\hat{\mu} - r) = rP \end{gathered} \tag{6.30}$$

There is a solution to this PDE:

$$P(t,T) = e^{-rF(t,T) - G(t,T)} \tag{6.31}$$

and hence:

$$\frac{\partial P}{\partial r} = -PF \tag{6.32}$$

and equation (6.24) becomes:

$$\begin{cases} \mu_P - r = (-B)(\lambda_0 + \lambda_1 r) \\ \sigma_P = \sigma\sqrt{r}B \end{cases} \tag{6.33}$$

Note that the volatility now is independent of r. As a result, the risk premium, $\mu_P - r$, must also be independent of r, which implies $\lambda_1 = 0$. Due to different forms of the risk premium, the solutions to the Vasicek model and to the CIR model are different. In the next chapter, we derive explicit closed-form solutions to both models.

With the above derived market prices of risk of both models, we can now write down the risk-neutral processes of both models. Note that equation (9.20) is the Kolmogorov Back Equation with a drift term equal to $\alpha(\mu - r) - \lambda_0 - \lambda_1 r$. Hence the short rate process under the risk-neutral measure for the CIR model is:

$$\begin{aligned} dr &= [\alpha(\mu - r) - \lambda_1 r]dt + \sigma\sqrt{r}d\hat{W} \\ &= \hat{\alpha}(\hat{\mu} - r)dt + \sigma\sqrt{r}d\hat{W} \end{aligned} \tag{6.34}$$

where

$$\begin{aligned} \hat{\alpha} &= \alpha + \lambda_1 \\ \hat{\mu} &= \frac{\alpha\mu}{\hat{\alpha}} \end{aligned}$$

and short rate process for the Vasicek model is:

$$\begin{aligned} dr &= [\alpha(\mu - r) - \lambda_0]dt + \sigma d\hat{W} \\ &= \hat{\alpha}(\hat{\mu} - r)dt + \sigma d\hat{W} \end{aligned} \tag{6.35}$$

where

$$\begin{aligned} \hat{\alpha} &= \alpha \\ \hat{\mu} &= \frac{\alpha\mu + \lambda_0}{\hat{\alpha}} \end{aligned}$$

In each of the models, the change of measure is derived accordingly as follows:

$$\begin{cases} \text{CIR} & dW = d\hat{W} - \frac{\lambda_1}{\sigma}\sqrt{r}dt \\ \text{Vasicek} & dW = d\hat{W} - \lambda_0\sqrt{r}dt \end{cases} \tag{6.36}$$

6.4 Risk Neutral Pricing

An alternative method to obtain the model for the zero coupon bond (discount factor) is to employ the Kolomogorov Backward Equation for the Kac Functional. In Chapter 4, we demonstrate that the solution to the PDE can be solved by computing the risk-neutral expectation of the discounted payoff. As a result, the same solution (CIR model or Vasicek model) can be found by solving the following expectation:

$$P(t,T) = \hat{\mathbb{E}}_t\left[\exp\left(-\int_t^T r(u)du\right)\right] \tag{6.37}$$

In Chapter 7, we shall leverage on this result and find risk-neutral expectations for various pricing models.

6.5 Appendix

Vectors $X_1, X_2, \cdots, X_n$ are said to be linearly dependent iff there exist scalars, $c_1, c_2, \cdots, c_n$, not all zero, such that:

$$c_1 \begin{bmatrix} X_{11} \\ X_{21} \\ \vdots \\ X_{n1} \end{bmatrix} + c_2 \begin{bmatrix} X_{11} \\ X_{21} \\ \vdots \\ X_{n1} \end{bmatrix} + \cdots + c_n \begin{bmatrix} X_{11} \\ X_{21} \\ \vdots \\ X_{n1} \end{bmatrix} = \begin{bmatrix} 0 \\ 0 \\ \vdots \\ 0 \end{bmatrix} \tag{6.38}$$

or

$$\begin{bmatrix} X_{11} & X_{12} & \cdots & X_{1n} \\ X_{21} & X_{22} & \cdots & X_{2n} \\ \vdots & \vdots & \ddots & \vdots \\ X_{n1} & X_{n2} & \cdots & X_{nn} \end{bmatrix} \begin{bmatrix} c_1 \\ c_2 \\ \vdots \\ c_n \end{bmatrix} = \begin{bmatrix} 0 \\ 0 \\ \vdots \\ 0 \end{bmatrix} \tag{6.39}$$

If no such scalars exist, the vectors are said to be linearly dependent. In order for this matrix equation to have a non-trivial solution, the determinant must be 0. In other words, the vectors are linear dependent if:

$$\begin{vmatrix} X_{11} & X_{12} & \cdots & X_{1n} \\ X_{21} & X_{22} & \cdots & X_{2n} \\ \vdots & \vdots & \ddots & \vdots \\ X_{n1} & X_{n2} & \cdots & X_{nn} \end{vmatrix} = 0 \tag{6.40}$$

In our case where

$$\begin{bmatrix} \mu_F - r & \mu_G - r \\ \sigma_F & \sigma_G \end{bmatrix} \begin{bmatrix} w_F \\ w_G \end{bmatrix} = \begin{bmatrix} 0 \\ 0 \end{bmatrix} \tag{6.41}$$

and then it must be true that:

$$\begin{vmatrix} \mu_F - r & \mu_G - r \\ \sigma_F & \sigma_G \end{vmatrix} = 0 \tag{6.42}$$

which immediately implies:

$$\frac{\mu_F - r}{\sigma_F} = \frac{\mu_G - r}{\sigma_G} \tag{6.43}$$

It is then straightforward to show that:

$$\frac{\mu_F - r}{\sigma_F} = \frac{\mu_G - r}{\sigma_G} = \frac{\mu_V - r}{\sigma_V} \tag{6.44}$$

Chapter 7

Cox-Ingersoll-Ross and Vasicek Models

Now we can talk about explicit solutions to some term structure models. Specifically we need to first go over the two most important one-factor models.

7.1 Forward Measure (Jamshidian, 1987, 1989)

Forward measure is introduced by Jamshidian (1987) whose first paper is never published. He applies the change of measure technique introduced in Chapter 4 to derive a forward measure under which expectations generate forward prices. Let us have a general interest rate process:

$$dr(t) = \hat{\mu}(r,t)dt + \sigma(r,t)d\hat{W}(t) \tag{7.1}$$

A zero coupon bond price (T-maturity, \$1 face value) is regarded as an interest rate contingent claim. Hence, Ito's lemma applies:

$$\begin{aligned} dP(t,T) &= P_r(t,T)dr + \frac{1}{2}P_{rr}(t,T)(dr)^2 + P_t(t,T)dt \\ &= \left[P_r(t,T)\hat{\mu}(r,t) + \frac{1}{2}P_{rr}(t,T)\sigma(r,t)^2 + P_t(t,T)\right]dt + P_r(t,T)\sigma(r,t)d\hat{W}(t) \end{aligned} \tag{7.2}$$

Usually we write the above equation in a return format as follows:

$$\begin{aligned}\frac{dP(t,T)}{P(t,T)} &= \frac{1}{P(t,T)}\left[P_r(t,T)\hat{\mu}(r,t) + \frac{1}{2}P_{rr}(t,T)\sigma(r,t)^2 + P_t(t,T)\right]dt \\ &\quad + \frac{P_r(t,T)}{P(t,T)}\sigma(r,t)d\hat{W}(t) \\ &= rdt + \sigma_P(r,t,T)d\hat{W}(t)\end{aligned} \tag{7.3}$$

Note that the second equality holds by definition that under the risk-neutral measure all assets must earn the risk-free return. Later on, we shall be more specific about what interest rate process we assume. Typically, it is either Vasicek or Cox-Ingersoll-Ross.

To compute any discounted present value (or price of a European contract) of a payoff X under the random interest rate r, we have:

$$\begin{aligned}\hat{\mathbb{E}}_t\left[\exp\left(-\int_t^T r(u)du\right)X(T)\right] &= \hat{\mathbb{E}}_t\left[\exp\left(-\int_t^T r(u)du\right)\right]\tilde{\mathbb{E}}_t^{(T)}[X(T)] \\ &= P(t,T)\tilde{\mathbb{E}}_t^{(T)}[X(T)\end{aligned} \tag{7.4}$$

where $P(t,T) = \hat{\mathbb{E}}_t\left[\exp\left(-\int_t^T r(u)du\right)\right]$ is the result to be found in Chpater 6, equation (6.37). The martingale result (that the deflated asset price under the risk-neutral measure must follow a martingale) implies that the LHS of equation (7.4) must equal $X(t)$, and hence:

$$X(t) = P(t,T)\tilde{\mathbb{E}}_t^{(T)}[X(T)] \tag{7.5}$$

Note that the forward measure is maturity dependent. Clearly, the RND is also maturity-dependent (usually, the RND is just $\eta(t)$):

$$\eta(t,T) = \frac{\exp\left(-\int_t^T r(u)du\right)}{P(t,T)} \tag{7.6}$$

Similarly, for $t < T < s$, we have:

$$\begin{aligned}P(t,s) &= \hat{\mathbb{E}}_t\left[\exp\left(-\int_t^T r(u)du\right)P(T,s)\right] \\ &= P(t,T)\tilde{\mathbb{E}}_t^{(T)}[P(T,s)]\end{aligned} \tag{7.7}$$

where the first equality is a result of law of iterative expectations; and naturally,

$$\begin{aligned}\tilde{\mathbb{E}}_t^{(T)}[P(T,s)] &= \frac{P(t,s)}{P(t,T)} \\ &= \Psi(t,T,s)\end{aligned} \tag{7.8}$$

which is the forward price. Equation (7.8) confirms the no-arbitrage for the forward price: the forward price of a zero coupon bond is the ratio of two zero coupon bonds.

Now we have to find the "Girsanov term", θ, for the RND. Applying Ito's lemma on the log bond price at maturity, i.e. $\ln P(T,T) = \ln(1) = 0$,

$$\begin{aligned}0 &= \ln P(T,T) \\ &= \ln P(t,T) + \int_t^T \frac{1}{P(u,T)}\left[P_u(u,T)du + P_r(u,T)dr + \frac{1}{2}P_{rr}(u,T)(dr)^2\right] \\ &\quad - \int_t^T \frac{1}{2}\left[\frac{\sigma(r,u)P_r(u,T)}{P(u,T)}\right]^2 du \\ &= \ln P(t,T) + \int_t^T \frac{1}{P(u,T)}\left[P_u(u,T)du + P_r(u,T)\hat{\mu}(r,u) + \frac{1}{2}P_{rr}(u,T)\sigma(r,u)^2\right] du \\ &\quad + \int_t^T \frac{1}{P(u,T)}P_r(u,T)\sigma(r,u)d\hat{W}(u) - \int_t^T \frac{1}{2}\left[\frac{\sigma(r,u)P_r(u,T)}{P(u,T)}\right]^2 du \\ &= \ln P(t,T) + \int_t^T r(u)du \\ &\quad + \int_t^T \frac{P_r(u,T)\sigma(r,u)}{P(u,T)}d\hat{W}(u) - \int_t^T \frac{1}{2}\left[\frac{\sigma(r,u)P_r(u,T)}{P(u,T)}\right]^2 du\end{aligned} \tag{7.9}$$

Letting

$$\theta(t,T) = -\frac{\sigma(r,t)P_r(t,T)}{P(t,T)} \tag{7.10}$$

and moving the first two terms to the left, we obtain:

$$-\int_t^T r(u)du - \ln P(t,T) = \int_t^T -\theta(u,T)d\hat{W}(u) - \int_t^T \frac{1}{2}\theta(u,T)^2 du \tag{7.11}$$

which can be interpreted as the RND in the following:

$$
\begin{aligned}
\eta(t,T) &= \frac{\exp\left(-\int_t^T r(u)du\right)}{P(t,T)} \\
&= \exp\left(\int_t^T -\theta(u,T)d\hat{W}(u) - \int_t^T \frac{1}{2}\theta(u,T)^2 du\right)
\end{aligned}
\tag{7.12}
$$

This implies the Girsonav transformation of the following:

$$
\begin{aligned}
\tilde{W}^{(T)}(t) &= \hat{W}(t) + \int_t^T \theta(u)dt \\
&= \hat{W}(t) - \int_t^T \sigma(r,u)\frac{P_r(u,T)}{P(u,T)}du
\end{aligned}
\tag{7.13}
$$

The interest rate process under the forward measure henceforth becomes:

$$
dr(t) = \left[\hat{\mu}(r,t) + \sigma(r,t)^2 \frac{P_r(t,T)}{P(t,T)}\right] dt + \sigma(r,t)d\tilde{W}^{(T)}(t) \tag{7.14}
$$

Note that the forward measure is quite general. It does not depend on any specific assumption on the interest rate process.

7.2 The Vasicek model

The Vasicek model assumes a Gaussian, mean-reverting (Ornstein-Uhlenbeck) process for the instantaneous short rate:

$$
\begin{aligned}
dr &= \alpha(\mu - r)dt + \sigma dW \\
&= \alpha(\hat{\mu} - r)dt + \sigma d\hat{W}
\end{aligned}
\tag{7.15}
$$

where, as shown in the previous chapter that $\hat{\mu} = \mu - \frac{\sigma\lambda}{\alpha}$. Applying the result of linear SDE in Chapter 4 with $\Phi(s-t) = e^{-\alpha(s-t)}$ for any $s > t$, we get:

$$
\begin{aligned}
r(s) &= e^{-\alpha(s-t)}\left(r(t) + \alpha\mu\int_t^s e^{\alpha(u-t)}du + \sigma\int_t^s e^{\alpha(u-t)}dW(u)\right) \\
&= e^{-\alpha(s-t)}r(t) + \mu\left(1 - e^{-\alpha(s-t)}\right) + e^{-\alpha(s-t)}\sigma\int_0^t e^{\alpha(u-t)}dW(u)
\end{aligned}
\tag{7.16}
$$

which results in:

$$\mathbb{E}_t[r(s)] = e^{-\alpha(s-t)}r(t) + \mu\left(1 - e^{-\alpha(s-t)}\right) \tag{7.17}$$

and

$$\begin{aligned}
\mathbb{V}_t[r(s)] &= e^{-2\alpha(s-t)}\sigma^2 V\left[\int_t^s e^{\alpha(u-t)}dW(u)\right] \\
&= e^{-2\alpha(s-t)}\sigma^2 \int_t^s e^{2\alpha(u-t)}du \\
&= e^{-2\alpha(s-t)}\sigma^2 \tfrac{1}{2\alpha}\left(e^{2\alpha(s-t)} - 1\right) \\
&= \tfrac{\sigma^2}{2\alpha}\left(1 - e^{-2\alpha(s-t)}\right)
\end{aligned} \tag{7.18}$$

In the risk-neutral case, we have:

$$\hat{\mathbb{E}}_t[r(s)] = e^{-\alpha(s-t)}r(t) + \hat{\mu}\left(1 - e^{-\alpha(s-t)}\right) \tag{7.19}$$

and

$$\hat{\mathbb{V}}_t[r(s)] = \mathbb{V}[r(s)] = \tfrac{\sigma^2}{2\alpha}\left(1 - e^{-2\alpha(s-t)}\right) \tag{7.20}$$

Using the closed-form result of the Vasicek model shown in the next section in equation (7.30), we have the Girsanov's theorem under the Vasicek model as follows:

$$\theta(t,T) = -\frac{\sigma(r,t)P_r(t,T)}{P(t,T)} = \sigma F(t,T) \tag{7.21}$$

which implies the forward-adjusted process:

$$\begin{aligned}
dr &= \alpha(\mu - r)dt + \sigma dW \\
&= (\alpha\mu - \sigma\lambda - \alpha r)dt + \sigma d\hat{W} \\
&= (\alpha\mu - \sigma\lambda - \sigma^2 F(t,T) - \alpha r)dt + \sigma d\tilde{W}^{(T)}
\end{aligned} \tag{7.22}$$

Applying the result of (7.16), we can write the solution as follows:

$$
\begin{aligned}
r(T) &= e^{-\alpha(T-t)} \left[r(t) + \int_t^T e^{\alpha(u-t)} \left(\alpha\hat{\mu} + F(u,T)\sigma^2 \right) du + \int_t^T e^{\alpha(u-t)} \sigma d\tilde{W}^{(T)}(u) \right] \\
&= e^{-\alpha(T-t)} \left[r(t) + \alpha\hat{\mu} \int_t^T e^{\alpha(u-t)} du + \sigma^2 \int_t^T e^{\alpha(u-t)} F(u,T) du \right. \\
&\qquad \left. + \sigma \int_t^T e^{\alpha(u-t)} d\tilde{W}^{(T)}(u) \right]
\end{aligned} \tag{7.23}
$$

Substitute $\frac{1-e^{-\alpha(T-u)}}{\alpha}$ for $F(u,T)$ and re-write the equation as:

$$
\begin{aligned}
r(T) = e^{-\alpha(T-t)} &\left[r_t + \hat{\mu} \left(e^{\alpha(T-t)} - 1 \right) + \frac{\sigma^2}{\alpha^2} \left(\left(e^{\alpha(T-t)} - 1 \right) \right. \right. \\
&\left. \left. - \frac{1}{2} \left(e^{\alpha(T-t)} - e^{-\alpha(T-t)} \right) \right) + \sigma \int_t^T e^{\alpha(u-t)} d\tilde{W}^{(T)}(u) \right]
\end{aligned} \tag{7.24}
$$

Re-arranging and cleaning up terms leads to:

$$
\begin{aligned}
r(T) &= e^{-\alpha(T-t)} r(t) + \hat{\mu} \left(1 - e^{-\alpha(T-t)} \right) + \frac{\sigma^2}{\alpha^2} \left(1 - e^{-\alpha(T-t)} \right) \\
&\quad - \frac{\sigma^2}{2\alpha^2} \left(1 - e^{-2\alpha(T-t)} \right) + \sigma \int_t^T e^{-\alpha(T-u)} d\tilde{W}^{(T)}(u) \\
&= e^{-\alpha(T-t)} r(t) + \hat{\mu} \left(1 - e^{-\alpha(T-t)} \right) + \frac{\sigma^2}{2\alpha^2} \left(1 - e^{-\alpha(T-t)} \right)^2 \\
&\quad + \sigma \int_t^T e^{-\alpha(T-u)} d\tilde{W}^{(T)}(u) \\
&= e^{-\alpha(T-t)} r(t) + \hat{\mu} \left(1 - e^{-\alpha(T-t)} \right) + \frac{\sigma^2}{2} F(t,T)^2 + \sigma \int_t^T e^{-\alpha(T-u)} d\tilde{W}^{(T)}(u)
\end{aligned} \tag{7.25}
$$

Hence,

$$
\begin{aligned}
\tilde{\mathbb{E}}_t^{(T)}[r(T)] &= e^{-\alpha(T-t)} r_t + \hat{\mu} \left(1 - e^{-\alpha(T-t)} \right) - \frac{\sigma^2}{2} F(t,T)^2 \\
\tilde{\mathbb{V}}_t^{(T)}[r(T)] &= \sigma^2 \int_t^T e^{-2\alpha(T-u)} du = \frac{\sigma^2}{2\alpha} \left(1 - e^{-2\alpha(T-t)} \right)
\end{aligned} \tag{7.26}
$$

Recall that under the risk-neutral measure, the mean and variance are shown in equations (7.19) and (7.20). The variance is the same as one under the risk-neutral measure (which is the same as the one under the real measure) and the mean is adjusted by a $\frac{\sigma^2}{2}F(t,T)^2$ term.

7.2.1 The bond model

Let R be the accumulation of r over time:

$$R(t,T) = \int_t^T r(s)ds \tag{7.27}$$

Then R is normal with mean and variance as follows:

$$\begin{aligned}\hat{\mathbb{E}}_t[R(t,T)] &= \int_t^T \hat{\mathbb{E}}[r(s)]ds \\ &= \int_t^T \left[r(t)e^{-\alpha(s-t)} - \hat{\mu}\left(1-e^{-\alpha(s-t)}\right)\right] ds \\ &= r(t)\left(\tfrac{1-e^{-\alpha(T-t)}}{\alpha}\right) + \hat{\mu}\left[T-t-\left(\tfrac{1-e^{-\alpha(T-t)}}{\alpha}\right)\right]\end{aligned} \tag{7.28}$$

and

$$\begin{aligned}\hat{\mathbb{V}}_t[R(t,T)] &= \text{cov}[R,R] \\ &= \int_t^T \int_t^s 2\,\text{cov}[r(u),r(s)]duds \quad (\forall u < s) \\ &= 2\int_t^T \int_t^s \tfrac{\sigma^2}{2\alpha}e^{-\alpha(s+u-2t)}\left(e^{2\alpha(u-t)}-1\right)duds \\ &= \tfrac{\sigma^2}{\alpha^2}\left(T-t-\tfrac{e^{-2\alpha(T-t)}}{2\alpha}+2\tfrac{e^{-\alpha(T-t)}}{\alpha}-\tfrac{3}{2\alpha}\right)\end{aligned} \tag{7.29}$$

The Vasicek model for the discount factor (zero coupon bond price):

$$\begin{aligned}P(t,T) &= \hat{\mathbb{E}}_t\left[\exp\left(-\int_t^T r(s)ds\right)\right] \\ &= \hat{\mathbb{E}}_t\left[e^{-R(t,T)}\right] \\ &= e^{-\hat{\mathbb{E}}[R(t,T)]+{}^1\!/\!_2 V[R(t,T)]} \\ &= e^{-r(t)F(t,T)-G(t,T)}\end{aligned} \tag{7.30}$$

where

$$F(t,T) = \frac{1 - e^{-\alpha(T-t)}}{\alpha}$$

$$G(t,T) = \left(\mu - \frac{\sigma^2}{2\alpha^2} - \frac{\sigma\lambda}{\alpha}\right)(T - t - F(t,T)) + \frac{\sigma^2 F(t,T)^2}{4\alpha}$$

Note that both F and G are functions of time to maturity and not calendar time, satisfying that the model is stationary.

Example

Inputs	
α	0.2456
μ	0.0648
σ	0.0289
λ	-0.2718
$r(t)$	0.0600

and the output yield curve is given below:

Term Structure					
$T-t$	$P(t,T)$	$y(T-t)$	$T-t$	$P(t,T)$	$y(T-t)$
1	0.9380	0.0641	16	0.2639	0.0833
2	0.8740	0.0673	17	0.2413	0.0836
3	0.8106	0.07	18	0.2207	0.0840
4	0.7491	0.0722	19	0.2017	0.0842
5	0.6905	0.0741	20	0.1844	0.0845
6	0.6352	0.0756	21	0.1686	0.0848
7	0.5835	0.077	22	0.1541	0.085
8	0.5354	0.0781	23	0.1409	0.0852
9	0.4908	0.0791	24	0.1288	0.0854
10	0.4497	0.0799	25	0.1177	0.0856
11	0.4118	0.0807	26	0.1076	0.0857
12	0.3769	0.0813	27	0.0984	0.0859
13	0.3449	0.0819	28	0.0899	0.086
14	0.3155	0.0824	29	0.0822	0.0862
15	0.2886	0.0829	30	0.0751	0.0863

Note that the bond price is not bounded by 1 since the interest rate can be negative, which is a bad property of the Vasicek model.

7.2.2 The option formula (Jamshidian – 1989 Journal of Finance)

There are a number of different ways to derive the option formula (for the zero coupon bond). The easiest way is to leverage on the fact that (for $t < T < s$) $\ln P(T,s) = -r(T)F(T,s) - G(T,s)$ is normally distributed, just like the Black-Scholes. And hence, we can find the mean and variance of $\ln P(T,s)$ as follows:

$$\begin{aligned} m_P &= \tilde{\mathbb{E}}_t^{(T)}[\ln P(T,s)] \\ &= -\tilde{\mathbb{E}}_t^{(T)}[r(T)]F(T,s) - G(T,s) \\ &= -F(T,s)\left[e^{-\alpha(T-t)}r(t) + \mu\left(1 - e^{-\alpha(T-t)}\right) + \tfrac{1}{2}\sigma^2 F(t,T)^2\right] - G(T,s) \\ &= \ln\frac{P(t,s)}{P(t,T)} - \frac{1}{2}V_P \end{aligned} \tag{7.31}$$

and

$$\begin{aligned} V_P &= \tilde{\mathbb{V}}_t^{(T)}[\ln P(T,s)] \\ &= \tilde{\mathbb{V}}_t^{(T)}[r(T)]F(T,s)^2 \\ &= \frac{1 - e^{-2\alpha(T-t)}}{2\alpha}\sigma^2 F(T,s)^2 \end{aligned} \tag{7.32}$$

The objective is to find $\Pr(\ln P(T,s) > \ln K)$ under the forward measure. Recall in the Black-Scholes model that (equation (5.27) in Chapter 5):

$$\Pr(\ln S_T > \ln K) = N\left(\frac{m - \ln K}{V}\right)$$

under the risk neutral measure where $m = \hat{\mathbb{E}}_t[\ln S_T] = \ln S_t + \left(r - \frac{1}{2}\sigma^2\right)(T-t) = \ln[S_t/e^{-r(T-t)}] - V/2$ and $V = \text{var}[\ln S_T] = \sigma^2(T-t)$.

Hence, it must be true that for bond option that:

$$\Pr(\ln P(T,s) > \ln K) = N\left(\frac{m_P - \ln K}{V_P}\right) \tag{7.33}$$

where m_P and V_P are given by (7.31) and (7.32) respectively. Recall from equation (7.8) that $\tilde{\mathbb{E}}_t^{(T)}[P(T,s)] = \frac{P(t,s)}{P(t,T)}$ and at the same time,

$$
\begin{aligned}
\tilde{\mathbb{E}}_t^{(T)}[P(T,s)] &= \tilde{\mathbb{E}}_t^{(T)}[e^{\ln P(T,s)}] \\
&= \exp\left(\tilde{\mathbb{E}}_t^{(T)}[\ln P(T,s)] + \tfrac{1}{2}\tilde{\mathbb{V}}_t^{(T)}[\ln P(T,s)]\right)
\end{aligned} \tag{7.34}
$$

So, $m_P = \tilde{\mathbb{E}}_t^{(T)}[\ln P(T,s)] = \ln \frac{P(t,s)}{P(t,T)} - \frac{1}{2}V_P$. Clearly this is not true under the risk-neutral measure, since $\hat{\mathbb{E}}_t[P(T,s)] \neq \frac{P(t,s)}{P(t,T)}$.

Now we turn to the option model. The Vasicek bond option pricing model is,

$$
\begin{aligned}
C(t) &= \hat{\mathbb{E}}_t\left[\exp\left(-\int_t^T r(u)du\right)\max\{P(T,s)-K,0\}\right] \\
&= P(t,T)\tilde{\mathbb{E}}_t^{(T)}[(P(T,s)-K)\mathbb{I}_{P>K}] \\
&= P(t,T)\tilde{\mathbb{E}}_t^{(T)}[P(T,s)\mathbb{I}_{P>K}] - P(t,T)K\tilde{\mathbb{E}}_t^{(T)}[\mathbb{I}_{P>K}]
\end{aligned} \tag{7.35}
$$

The second probability is derived above, that is $\tilde{\mathbb{E}}_t^{(T)}[\mathbb{I}_{P>K}] = N\left(\frac{m_P - \ln K}{\sqrt{V_P}}\right)$. The first probability is derived as follows,

$$
\begin{aligned}
\tilde{\mathbb{E}}_t^{(T)}[P(T,s)\mathbb{I}_{P>K}] &= \tilde{\mathbb{E}}_t^{(T)}[P(T,s)]\bar{E}_t^{(T)}[\mathbb{I}_{P>K}] \\
&= \frac{P(t,s)}{P(t,T)}\bar{E}_t^{(T)}[\mathbb{I}_{P>K}]
\end{aligned} \tag{7.36}
$$

Note that the RND is:

$$
\eta = \frac{P(T,s)}{\tilde{\mathbb{E}}_t^{(T)}[P(T,s)]} \tag{7.37}
$$

where $\ln \tilde{\mathbb{E}}_t^{(T)}[P(T,s)] = \tilde{\mathbb{E}}_t^{(T)}[\ln P(T,s)] + \frac{1}{2}\tilde{\mathbb{V}}_t^{(T)}[\ln P(T,s)]$. Also note that since $\ln P(T,s)$ is normal, it must take the following form:

$$
\ln P(T,s) = m_P + \int_t^T x(u)d\tilde{W}^{(T)}(u) \tag{7.38}
$$

which leads to the following alternative form of mean and variance of $\ln P(T,s)$:

$$
\begin{aligned}
\tilde{\mathbb{E}}_t^{(T)}[\ln P(T,s)] &= m_P \\
\tilde{\mathbb{V}}_t^{(T)}[\ln P(T,s)] &= V_P = \int_t^T x(u)^2 du
\end{aligned}
$$

We know that $m_P = \ln \frac{P(t,s)}{P(t,T)} - \frac{1}{2}V_P$. Also we know that $\ln \tilde{\mathbb{E}}_t^{(T)}[P(T,s)] = \ln \frac{P(t,s)}{P(t,T)}$. Hence,

$$\begin{aligned}
\frac{P(T,s)}{\tilde{\mathbb{E}}_t^{(T)}[P(T,s)]} &= \frac{\exp\left\{m_P + \int_t^T x(u)d\tilde{W}^{(T)}(u)\right\}}{\exp\left\{\ln \frac{P(t,s)}{P(t,T)}\right\}} \\
&= \exp\left(-1/2 V_P + \int_t^T x(u)d\tilde{W}^{(T)}(u)\right) \\
&= \exp\left(-1/2 \int_t^T x(u)^2 du + \int_t^T x(u)d\tilde{W}^{(T)}(u)\right)
\end{aligned} \tag{7.39}$$

Hence, $\theta(t) = -x(t)$ for the Girsanovs transformation. That is:

$$d\bar{W}(t)^{(T)} = d\tilde{W}(t)^{(T)} - x(t)dt \tag{7.40}$$

As a result,

$$\begin{aligned}
\ln P(T,s) &= m_P + \int_t^T x(u)d\tilde{W}^{(T)}(u) \\
&= m_P + \int_t^T x(u)^2 du + \int_t^T x(u)d\bar{W}^{(T)}(u)
\end{aligned} \tag{7.41}$$

Hence,

$$\begin{aligned}
\bar{m}_P &= \bar{E}_t^{(T)}[\ln P(T,s)] \\
&= m_P + V_P \\
&= \ln \frac{P(t,s)}{P(t,T)} + 1/2 V_P
\end{aligned} \tag{7.42}$$

Now we write (7.35) as follows:

$$\begin{aligned}
C(t) &= P(t,T)\tilde{\mathbb{E}}_t^{(T)}[P(T,s)\mathbb{I}_{P>K}] - P(t,T)K\tilde{\mathbb{E}}_t^{(T)}[\mathbb{I}_{P>K}] \\
&= P(t,T)\left\{\frac{P(t,s)}{P(t,T)}N\left(\frac{\bar{m}_P - \ln K}{\sqrt{V_P}}\right) - KN\left(\frac{m_P - \ln K}{\sqrt{V_P}}\right)\right\} \\
&= P(t,s)N\left(\frac{\bar{m}_P - \ln K}{\sqrt{V_P}}\right) - P(t,T)KN\left(\frac{m_P - \ln K}{\sqrt{V_P}}\right)
\end{aligned} \tag{7.43}$$

where $\bar{m}_P$, m_P, and V_P are defined in (7.42), (7.31), and (7.32) respectively.

So we have the Black-Scholes result under the Vasicek model. Note that this result holds only under Gaussian. Also note that (7.38) can be written so because of Gaussian.

Example

Given the same parameter values, a 6-month option on a 5-year bond with a strike of 0.6 is \$0.1088 out of \$1 face value.

7.2.3 The futures model (Carr – 1988 WP and Chen – 1992 JFQA)

The futures price is straightforward. It is just the moment generating function of the interest rate under the risk neutral measure.

$$\begin{aligned}\Phi(t,T,s) &= \hat{\mathbb{E}}[P(T,s)] \\ &= \hat{\mathbb{E}}\left[e^{-r(T)F(T,s)-G(T,s)}\right] \\ &= e^{-G(T,s)}e^{-F(T,s)E[r(T)]+0.5F(T,s)^2V[r(T)]} \\ &= e^{-r(t)X(t,T,s)-Y(t,T,s)}\end{aligned} \tag{7.44}$$

where

$$\begin{aligned}X(t,T,s) &= F(t,s)-F(t,T) \\ Y(t,T,s) &= \left(\mu-\tfrac{\lambda\sigma}{\alpha}-\tfrac{\sigma^2}{2\alpha^2}\right)[s-T-X(t,T,s)] \\ &\quad -\tfrac{\sigma^2}{2\alpha^2}\left[X(t,T,s)-\tfrac{\alpha}{2}X(t,T,s)^2-F(T,s)\right]\end{aligned} \tag{7.45}$$

Example

Taking the same parameter values, the futures price of a 1 year futures on a 5 year zero bond is: \$73.55 out of \$100 face value.

7.2.4 The futures option model (Chen – 1992 JFQA

The futures option model deserves a special attention. This is because the forward-adjusted expectation of a futures price does not lead to another futures price or a forward price.

$$
\begin{aligned}
C(t,T,s,w) &= \hat{\mathbb{E}}_t\left[\exp\left(\int_t^T r(u)du\right)\max\{\Phi(T,s,w)-K,0\}\right] \\
&= P(t,T)\tilde{\mathbb{E}}_t^{(T)}[\max\{\Phi(T,s,w)-K,0\}] \\
&= H(t)N(d_H) - P(t,T)KN(d_H - \sqrt{V_H})
\end{aligned}
\tag{7.46}
$$

where

$$
\begin{aligned}
H(t) &= P(t,T)\Phi(t,s,w)\exp\left(\tfrac{\sigma^2}{2}F(t,T)^2X(T,s,w)\right) \\
d_H &= \frac{1}{\sqrt{V_H}}\left(\ln\frac{H(t)}{P(t,T)K} + \frac{V_H}{2}\right) \\
V_H &= \mathrm{var}[\ln H(T)] = X(T,s,w)^2\frac{\sigma^2\left(1-e^{-2\alpha(T-t)}\right)}{2\alpha}
\end{aligned}
$$

Note that $H(t) = \hat{\mathbb{E}}_t\left[\exp\left(-\int_t^T r(u)du\right)\Phi(T,s,w)\right]$ which is an asset that pays the futures price at time T. This is the underlying asset that the futures option is written on.

7.3 Cox-Ingersoll-Ross model

Critics of the Vasicek family of models argue that nominal interest rates cannot fall below zero or arbitrage will take place. The Cox-Ingersoll-Ross model, on the other hand, is free frmo such a problem. The CIR model is a mean-reverting squared-root process:

$$
\begin{aligned}
dr &= \alpha(\mu - r)dt + \sigma\sqrt{r}dW \\
&= \hat{\alpha}(\hat{\mu} - r)dt + \sigma\sqrt{r}d\hat{W}
\end{aligned}
\tag{7.47}
$$

where, as shown in Chapter 6, $\hat{\alpha} = \alpha + \lambda$ and $\hat{\mu} = \frac{\alpha\mu}{\alpha+\lambda}$ are the risk neutral parameters. This is not an SDE and therefore must be studied differently. While there is no solution to $r(t)$, Feller worked out the transition density of this distribution:

$$
f(r(s)|r(t)) = ce^{-c(r(s)+\xi)}\left(\frac{r(s)}{\xi}\right)^{d/2} I_d(2c\sqrt{r(s)\xi}) \tag{7.48}
$$

where

$$
\begin{aligned}
c &= \frac{2\alpha}{\sigma^2\left(1-e^{-\alpha(s-t)}\right)} \\
\xi &= r(t)e^{-\alpha(s-t)} \\
d &= \frac{2\alpha\mu}{\sigma^2} - 1
\end{aligned}
$$

and I_d is the modified Bessel function of the first kind of order d, which is a scaled non-central chi-squared distribution. Let $x = 2cr$, we have a non-central chi-squared density function as follows:

$$
f(x(s)|x(t)) = \frac{1}{2}e^{-1/2(x(s)+\Lambda)}\left(\frac{x(s)}{\Lambda}\right)^{d/2} I_d(\sqrt{\Lambda x(s)}) \tag{7.49}
$$

where $\Lambda = 2c\xi$ is the degrees of non-centrality. The degrees of freedom is $\upsilon = 2d + 2 = \frac{4\alpha\mu}{\sigma^2}$. The mean and variance of a non-central chi-squared distribution are, respectively, $\upsilon + \Lambda$ and $2(\upsilon + 2\Lambda)$, which lead to:

$$
\begin{aligned}
\mathbb{E}[x(s)|x(t)] &= \upsilon + \Lambda \\
&= \tfrac{4\alpha\mu}{\sigma^2} + 2cr(t)e^{-\alpha(s-t)}
\end{aligned} \tag{7.50}
$$

and

$$
\begin{aligned}
\mathbb{V}[x(s)|x(t)] &= 2(\upsilon + 2\Lambda) \\
&= 2\left(\tfrac{4\alpha\mu}{\sigma^2} + 4cr(t)e^{-\alpha(s-t)}\right)
\end{aligned} \tag{7.51}
$$

Replacing x with $2cr$, we arrive at the mean and variance of the interest rate for the CIR model:

$$
\begin{aligned}
\mathbb{E}[r(s)|r(t)] &= \frac{1}{2c}\mathbb{E}[x(s)|x(t)] \\
&= r(t)e^{-\alpha(s-t)} + \mu\left(1-e^{-\alpha(s-t)}\right)
\end{aligned} \tag{7.52}
$$

and

$$
\begin{aligned}
\mathbb{V}[r(s)|r(t)] &= \frac{1}{4c^2}\mathbb{V}[x(s)|x(t)] \\
&= \frac{\sigma^2}{2\alpha}\left(1-e^{-\alpha(s-t)}\right)^2 \mu + r(t)\left(e^{-2\alpha(s-t)} - e^{-\alpha(s-t)}\right)
\end{aligned} \tag{7.53}
$$

Under the risk neutral measure, $\hat{x} = 2\hat{c}r$ is a non-central chi-square variable with $2\hat{d}+2 = \frac{4\alpha\mu}{\sigma^2}$ degrees of freedom and $\hat{\Lambda} = 2\hat{c}\hat{\xi}$ non-centrality where:

$$\begin{aligned}
\hat{c} &= \frac{2\hat{\alpha}}{\sigma^2\left(1-e^{-\hat{\alpha}(s-t)}\right)} \\
\hat{\xi} &= r(t)e^{-\hat{\alpha}(s-t)} \\
\hat{d} &= \frac{2\hat{\alpha}\hat{\mu}}{\sigma^2} - 1 = \frac{2\alpha\mu}{\sigma^2} - 1 \\
\hat{\alpha} &= \alpha + \lambda
\end{aligned} \tag{7.54}$$

Under the forward adjusted measure, $\tilde{x} = 2(\phi+\psi)r$ is a non-central chi-square variable with $\frac{4\alpha\mu}{\sigma^2}$ degrees of freedom and $\tilde{\Lambda} = \frac{2\phi^2 r(t)e^{\gamma(s-t)}}{\phi+\psi}$ degrees of non-centrality where:

$$\begin{aligned}
\phi &= \frac{2\gamma}{\sigma^2\left(1-e^{-\gamma(s-t)}\right)} \\
\psi &= \frac{\hat{\alpha}+\gamma}{\sigma^2} \\
\gamma &= \sqrt{\hat{\alpha}^2 + 2\sigma^2}
\end{aligned} \tag{7.55}$$

The best summary of the three measures under the CIR model is summarized in the following table.

Various Measures	
Real process	$2cr(s) \to \chi^2_{\text{nc}}\left(\frac{4\alpha\mu}{\sigma^2}, 2cr(t)e^{-\alpha(s-t)}\right)$
Risk-neutral process	$2\hat{c}r(s) \to \chi^2_{\text{nc}}\left(\frac{4\alpha\mu}{\sigma^2}, 2\hat{c}r(t)e^{-(\alpha+\lambda)(s-t)}\right)$
Forward-adjusted process	$2(\phi+\psi)r(s) \to \chi^2_{\text{nc}}\left(\frac{4\alpha\mu}{\sigma^2}, \frac{2\phi^2 r(t)e^{\gamma(s-t)}}{\phi+\psi}\right)$

Now we proceed to value the zero coupon bond. $\exp\left(-\int_t^T r(s)ds\right)$ is not Gaussian. Hence:

$$P(t,T) = \hat{\mathbb{E}}_t\left[\exp\left(-\int_t^T r(s)ds\right)\right] \tag{7.56}$$

cannot be solved easily. We must solve a LaPlace transform problem. We guess the solution to be:

$$P(t,T) = A(t,T)e^{-r(t)B(t,T)} \tag{7.57}$$

Taking partials to get:

$$\begin{aligned} P_r &= -PB \\ P_{rr} &= PB^2 \\ P_t &= A_t e^{-rB} + Ae^{-rB}(-r)B_t \end{aligned} \tag{7.58}$$

where subscripts represent partial derivatives. Substituting the above back to the PDE, we obtain:

$$\begin{gathered} \frac{1}{2}\sigma^2 rPB^2 + \hat{\alpha}(\hat{\mu} - r)(-PB) + A_t e^{-rB} + Ae^{-rB}(-r)B_t = rP \\ \text{or} \\ \frac{1}{2}\sigma^2 rB^2 + \hat{\alpha}(\hat{\mu} - r)(-B) + \frac{A_t}{A} + (-r)B_t = r \end{gathered} \tag{7.59}$$

Collecting r terms to get an ODE for B:

$$\frac{1}{2}\sigma^2 B^2 + \hat{\alpha}B - 1 = B_t \tag{7.60}$$

subject to the boundary condition ($P(t,t) = 1$) that $B(t,t) = 0$ for all t and leaving the remaining terms:

$$\begin{gathered} \frac{A_t}{A} = \hat{\alpha}\hat{\mu}B \\ \text{or} \\ \frac{d\ln A}{dt} = \hat{\alpha}\hat{\mu}B \end{gathered} \tag{7.61}$$

which leads to:

$$A = \exp\left(\hat{\alpha}\hat{\mu}\int B dt\right) \tag{7.62}$$

subject to the boundary condition that $A(t,t) = 1$. The solutions to B and A are given as:

$$
\begin{aligned}
A(t,T) &= \left[\frac{2\gamma e^{(\alpha+\lambda+\gamma)(T-t)/2}}{(\alpha+\lambda+\gamma)\left(e^{\gamma(T-t)}-1\right)+2\gamma}\right]^{2\alpha\mu/\sigma^2} \\
B(t,T) &= \frac{2\left(e^{\gamma(T-t)}-1\right)}{(\alpha+\lambda+\gamma)\left(e^{\gamma(T-t)}-1\right)+2\gamma} \\
\gamma &= \sqrt{(\alpha+\lambda)^2+2\sigma^2}
\end{aligned}
\tag{7.63}
$$

Example

Inputs	
α	0.2456
μ	0.0648
σ	0.14998
λ	-0.129
$r(t)$	0.06

The resulting term structure by the CIR model is given as follows.

Term Structure					
$T-t$	$P(t,T)$	$y(T-t)$	$T-t$	$P(t,T)$	$y(T-t)$
1	0.9379	0.0641	16	0.2641	0.0832
2	0.8738	0.0675	17	0.2417	0.0835
3	0.8099	0.0703	18	0.2212	0.0838
4	0.7481	0.0726	19	0.2025	0.0841
5	0.6891	0.0745	20	0.1853	0.0843
6	0.6336	0.0761	21	0.1696	0.0845
7	0.5818	0.0774	22	0.1552	0.0847
8	0.5337	0.0785	23	0.142	0.0849
9	0.4893	0.0794	24	0.1299	0.085
10	0.4483	0.0802	25	0.1189	0.0852
11	0.4107	0.0809	26	0.1088	0.0853
12	0.3761	0.0815	27	0.0996	0.0854
13	0.3443	0.082	28	0.0911	0.0856
14	0.3152	0.0825	29	0.0834	0.0857
15	0.2886	0.0829	30	0.0763	0.0858

Note that the bond price is bounded by $A(t,T)$ (which is bounded by 1) when the interest rate is 0.

7.3.1 The option formula

From Section I, we have:

$$\theta(t,T) = -\frac{\sigma(r,t)P_r(t,T)}{P(t,T)} = \sigma\sqrt{r}B(t,T) \tag{7.64}$$

The the forward adjusted process is:

$$\begin{aligned} dr &= \alpha(\mu - r)dt + \sigma\sqrt{r}dW \\ &= (\alpha\mu - \lambda r - \alpha r)dt + \sigma\sqrt{r}d\hat{W} \\ &= (\alpha\mu - \lambda r - \sigma^2 rB(t,T) - \alpha r)dt + \sigma\sqrt{r}d\tilde{W} \end{aligned} \tag{7.65}$$

The option pricing model for a zero coupon bond can now be derived as:

$$\begin{aligned} C(t,T,s) &= \hat{\mathbb{E}}_t\left[\exp\left(\int_t^T r(u)du\right)\max\{P(T,s)-K,0\}\right] \\ &= P(t,T)\tilde{\mathbb{E}}_t^{(T)}[\max\{P(T,s)-K,0\}] \\ &= P(t,T)\int_K^{A(T,s)}[P(T,s)-K]f(P(T,s))d\widetilde{P(T,s)} \\ &= P(t,T)\int_0^{r*}\left[A(T,s)e^{-r(T)B(T,s)} - A(T,s)e^{-r^*B(T,s)}\right]f(r(T)|r(t))d\widetilde{r(T)} \end{aligned} \tag{7.66}$$

where the last line is the change of variable from the bond price, P, to the interest rate, r and therefore $r^* = \ln(A(T,s)/K)/B(T,s)$. Since the distribution of P is unknown, we must go back to the distribution of r. Recall under the forward measure:

$$2(\phi+\psi)r(s) \to \chi^2_{\text{nc}}\left(\tfrac{4\alpha\mu}{\sigma^2}, \tfrac{2\phi^2 r(t)e^{\gamma(s-t)}}{\phi+\psi}\right) \tag{7.67}$$

Hence, the pricing formula is:[1]

$$\begin{aligned} C(t,T,s) = P(t,s)\chi^2_{\text{nc}}&\left(2(\phi+\psi+B(T,s))r^*; \tfrac{4\alpha\mu}{\sigma^2}, \tfrac{2\phi^2 e^{\gamma(T-t)}r_t}{\phi+\psi+B(T,s)}\right) \\ -P(t,T)K\chi^2_{\text{nc}}&\left(2(\phi+\psi)r^*; \tfrac{4\alpha\mu}{\sigma^2}, \tfrac{2\phi^2 e^{\gamma(T-t)}r_t}{\phi+\psi}\right) \end{aligned} \tag{7.68}$$

[1]The first probability is tedious to derive but it is a straightforward change of variable. The non-centrality parameter is $\bar{\Lambda} = \frac{2\phi^2 e^{\gamma(T-t)}r_t}{\phi+\psi+B(T,s)}$

where

$$\phi = \frac{2\gamma}{\sigma^2\left(e^{\gamma(T-t)}-1\right)}$$
$$\psi = \frac{\alpha+\lambda+\gamma}{\sigma^2}$$
$$r^* = \frac{\ln\frac{A(T,s)}{K}}{B(T,s)}$$

Example

Given the same parameter values, a 6-month option on a 5-year bond with a strike of 0.6 is \$0.1085 out of \$1 face value.

7.3.2 The Futures model

The moment generation function of the noncentral Chi-square $\mathbb{E}[e^{tx}]$ is (where Λ is degrees of non-centrality and v is degrees of freedom):

$$\frac{\exp\left(\frac{\Lambda t}{1-2t}\right)}{(1-2t)^{v/2}} \tag{7.69}$$

Hence,

$$\begin{aligned}\Phi(t,T,s) &= \hat{\mathbb{E}}_t[P(T,s)] = \hat{\mathbb{E}}_t\left[A(T,s)e^{-r(T)B(T,s)}\right]\\ &= \hat{\mathbb{E}}_t\left[A(T,s)e^{-x(T)\frac{1}{2\hat{c}}B(T,s)}\right]\\ &= A(T,s)\frac{\exp\left(\frac{-B(T,s)r(t)e^{-\hat{\alpha}(T-t)}}{1+B(T,s)/\hat{c}}\right)}{\left(1+\frac{B(T,s)}{\hat{c}}\right)^{2\alpha\mu/\sigma^2}}\\ &= C(t,T,s)e^{-r(t)D(t,T,s)}\end{aligned} \tag{7.70}$$

where

$$
\begin{aligned}
C(t,T,s) &= A(T,s)\left(\tfrac{\hat{c}}{\hat{c}+B(T,s)}\right)^{2\alpha\mu/\sigma^2} \\
D(t,T,s) &= B(T,s)\tfrac{\hat{c}e^{-(\alpha+\lambda)(T-t)}}{\hat{c}+B(T,s)} \\
\hat{c} &= \tfrac{2(\alpha+\lambda)}{\sigma^2\left(1-e^{-(\alpha+\lambda)(T-t)}\right)}
\end{aligned}
$$

Example

Taking the same parameter values, the futures price of a 1 year futures on a 5 year zero bond is: \$73.33 out of \$100 face value.

7.3.3 The Futures Option Model

$$
\begin{aligned}
C(t,T,s,w) = {} & H(t)\chi^2_{\text{nc}}\left(2(\phi+\psi+D(T,s,w))r^*; \tfrac{4\alpha\mu}{\sigma^2}, \Lambda^*\right) \\
& -P(t,T)K\chi^2_{\text{nc}}\left(2(\phi+\psi)r^*; \tfrac{4\alpha\mu}{\sigma^2}, \tilde{\Lambda}\right)
\end{aligned} \tag{7.71}
$$

where

$$
\begin{aligned}
H(t) &= P(t,T)C(T,s,w)\exp\left(\frac{\frac{-D(T,s,w)\tilde{\Lambda}}{2(\phi+\psi)}}{1+\frac{D(T,s,w)}{\phi+\psi}}\right)\left(1+\frac{D(T,s,w)}{\phi+\psi}\right)^{-2\alpha\mu/\sigma^2} \\
\tilde{\Lambda} &= \tfrac{2\phi^2 e^{\gamma(T-t)} r_t}{\phi+\psi} \\
\Lambda^* &= \tfrac{2\phi^2 e^{\gamma(T-t)} r_t}{\phi+\psi+D(T,s,w)}
\end{aligned} \tag{7.72}
$$

Note that $C(T,s,w)$ and $D(T,s,w)$ are given in the futures model.

7.4 Coupon bearing Option formula

7.4.1 Coupon bond and coupon bond futures

Risk free coupon bonds and coupon bond futures are simply weighted sums of zero coupon bonds and zero coupon bond futures, because the expectation operation is linear.

$$\begin{aligned} B &= c\sum_{j=1}^{n} P(t,T_j) + P(t,T_n) \\ \Upsilon &= c\sum_{j=1}^{n} \Phi(t,T_f,T_j) + \Phi(t,T_f,T_n) \end{aligned} \tag{7.73}$$

where for convenience $T_f < T_1$ to avoid coupon adjustment.

7.4.2 Coupon Bond Option

Coupon bond options usually do not have closed form solutions as effectively we are modeling options on portfolios (note that a coupon bond is a portfolio of zero coupon bonds). However, Farshid Jamshidian showed that under single factor affine models, given that the state variable (instantaneous short rate) is log linear in bond price, Jensen's equality disappears and coupon bond options can be expressed as a portfolio of zero coupon bond options.

$$\begin{aligned} C(t,T_c,\underline{T}) &= \hat{\mathbb{E}}_t\left[\exp\left(-\int_t^{T_c} r(u)du\right)\max\{B-K,0\}\right] \\ &= \hat{\mathbb{E}}_t\left[\exp\left(-\int_t^{T_c} r(u)du\right)\max\left\{\sum_{j=1}^{n} c_jP(t,T_j)-K,0\right\}\right] \quad T_c<T_1 \\ &= \hat{\mathbb{E}}_t\left[\exp\left(-\int_t^{T_c} r(u)du\right)\max\left\{\sum_{j=1}^{n} c_jP(t,T_j)-\sum_{j=1}^{n} c_jK_j,0\right\}\right] \\ &= \hat{\mathbb{E}}_t\left[\exp\left(-\int_t^{T_c} r(u)du\right)\sum_{j=1}^{n} c_j\max\{P(t,T_j)-K_j,0\}\right] \\ &= \sum_{j=1}^{n} c_j\hat{\mathbb{E}}_t\left[\exp\left(-\int_t^{T_c} r(u)du\right)\max\{P(t,T_j)-K_j,0\}\right] \\ &= \sum_{j=1}^{n} c_jC(t,T_c,T_j) \end{aligned} \tag{7.74}$$

due to the fact that selective exercise has no value here, where

$$P(t,T_j) = A(t,T_j)e^{-r(t)B(t,T_j)}K_j = A(t,T_j)e^{-\bar{r}B(t,T_j)} \tag{7.75}$$

Note that if for any j, $P(t,T_j) > K_j$, it must be that $r(t) < \bar{r}$ due to monotonicity , then $P(t,T_j) > K_j$ for all j, and hence selective exercise has no value. Note that this is not true for two factor models but it is still a good approximation. Since coupon, $A(t,T_j)$ and $B(t,T_j)$ are known, we can search for $\bar{r}$ with $\sum_{j=1}^{n} c_jK_j = K$.

7.4.3 Coupon Bond Futures Option

Coupon bond futures options can be derived in the same way:

$$\begin{aligned} C(t,T_c,\underline{T}) &= \hat{\mathbb{E}}_t\left[\exp\left(-\int_t^{T_c} r(u)du\right)\max\{\Upsilon - K, 0\}\right] \\ &= \sum_{j=1}^{n} c_j C(t,T_c,T_f,T_j) \end{aligned} \tag{7.76}$$

for $T_c < T_f < T_1$ and $K_j = C(t,T_f,T_j)e^{-\bar{r}D(t,T_f,T_j)}$.

7.5 Other Term Structure Models

There have been a huge number of term structure models in the literature. The following table roughly summarized (by no means exhausive but hopefully representative). There is no clear way to classify interest models. Roughly we use the following matrix:

Term Structure Models		
	Factor models (also called Stru. / Eqm. models)	No-arbitrage models (also called Reduced-form models)
Single-factor	Vasicek (1977) Cox-Ingersoll-Ross (1985) Dothan (1978) Ball-Torous (1983) Duffie-Kan (1996)	Ho-Lee (1986) Black-Derman-Toy (1989) Black-Karasinzinsky (1990) Bloomberg model (flat vol) Heath-Jarrow-Morton (1992)
Multi-factor	Cox-Ingersoll-Ross (1985) Langetieg (1980) Brennan-Schwartz (1978) Longstaff and Schwartz (1992) Duffie-Pan-Singleton (2000)	Hull-White (1990) Heath-Jarrow-Morton (1992) Scott (1997) Chen-Yang (2002)
Others	Quadratic Constantinides (1992) Leippold-Wu (2002)	Random field Goldstein (2000)

The no-arbitrage models are discussed in Chapter 12, yet only Ho-Lee, Black-Derman-Toy, and Heath-Jarrow-Morton models are covered. So far in this chapter, only Vasicek and CIR models are discussed. In this section, we try to bring in some practical flavor, given that Vasicek and CIR models while provide tremendous insight into term structure theories, are too limited to be used in the real world.

The Dothan model is similar to the Vasicek and CIR models but assumes a lognormally distributed short rate. It has been shown that this model could be unstable on the long end of the yield curve.

Ball and Torous model bonds with a Brownian Bridge process as the BB process is still Gaussian and yet the end of the variable can be pinned at any given value (such as face value of the bond). However, as Cheng (1991) pointed out, this model is not arbitrage-free.

One factor models bring tremendous insight to the term structure theories and yet they are not useful in reality. Empirists have found that one factor models are too limited in explaining the curvature and dynamics and the term structure. Hence, researchers have developed multi-factor models. The first such proposal was by Cox, Ingersoll, and Ross. In their paper where the seminal one factor model was introduced, they proposed the following multi-factor framework: $r = \sum_{i=1}^{n} y_j$ where r is the the short rate and it is a sum of multiple independent factors y_j. However, no explicit solution is given by CIR.

Langetieg, using the CIR suggestion, derived an explicit solution to the Vasicek model. This is straighforward as the sum of Gaussian variables is still Gaussian. This is difficult to do in the CIR model as the sum of non-central chi-square variables (unless they are independent) is not a non-central chi-square variable. The mean reverting processes used in Vasicek and CIR models are AR(1) processes and hence they are not independent.

Brennan and Schwartz explicitly specify a separate long rate to build their two factor model. This approach is better than the CIR suggestion in that these factors are observable (as opposed to unobservable y_j). The model has no closed form solution and needs to be solved numerically (e.g. finite difference method). However, since term rates cannot be modeled directly by seperate processes as they may conflict each other, the Brennan-Schwartz model is not arbitrage-free.

The Duffie-Kan model broadened the scope and established the general solution of the affine family of term structure models. The Duffie-Pan-Singleton model extends the Duffie-Kan model to include credit risk. We shall discuss them separately.

Outside of the affine family, the solutions become unmanageable. There are two proposals in the literature. One is the quadratic approach where $r = \sum_{j=1}^{n} y_j^2$. This approach is plausible as the short rate will not be negative and each factor y_j can be normally distributed. Constantinides first proposed this idea and then was followed by other researchers, such as Leippold and Wu. The problem is that the solution is expressed in terms of y_j which are non-observable. Empirists have found that such models can explain the dynamics of the term structure better.

Lastly, Goldstein adopted the Random Field theory in physics to model the term structure. In physics, a field is a physical quantity associated to each point of spacetime. A field may be thought of as extending throughout the whole of space. The Random Field theory has given the term structure model infinite degrees of freedom so it is rich in nature. Yet the model is not practical and has not been tested empirically.

7.6 Appendix

7.6.1 Solving PDE

$$\begin{aligned} &Ay^2 + By + C + \tfrac{dy}{dx} = 0 \\ &\left(Ay^2 + By + C\right) dx + dy = 0 \end{aligned} \tag{7.77}$$

Multiplying the equation by (known as integrating factor) the following

$$\frac{1}{Ay^2 + By + C} \tag{7.78}$$

we get

$$-dx = \frac{dy}{Ay^2 + By + C} \tag{7.79}$$

Re-expressing the quadric term as

$$Ay^2 + By + C = A\left[\left(y + \tfrac{B}{2A}\right)^2 - \left(\tfrac{\sqrt{B^2-4AC}}{2A}\right)^2\right] \tag{7.80}$$

Plug back and solve for x to get (use special integration functions, see calculus book)

$$-x = \left[\frac{1}{A}\right]\left[\frac{1}{2\left(\frac{\sqrt{B^2-4AC}}{2A}\right)}\right] \ln\left[\frac{\frac{\sqrt{B^2-4AC}}{2A} + \left(y + \frac{B}{2A}\right)}{\frac{\sqrt{B^2-4AC}}{2A} - \left(y + \frac{B}{2A}\right)}\right] + c \tag{7.81}$$

Rearrange terms to get

$$y = \frac{\sqrt{B^2 - 4AC}}{2A} \frac{e^{(x+c)\sqrt{B^2-4AC}} - 1}{e^{(x+c)\sqrt{B^2-4AC}} + 1} - \frac{B}{2A} \tag{7.82}$$

Note that the boundary condition is $y(T) = 0$ and hence we can solve for the constant c as

$$\begin{aligned} y(T) = 0 &= \frac{\sqrt{B^2 - 4AC}}{2A} \frac{e^{(T+c)\sqrt{B^2-4AC}} - 1}{e^{(T+c)\sqrt{B^2-4AC}} + 1} - \frac{B}{2A} \\ c &= \frac{1}{\sqrt{B^2 - 4AC}} \ln \left[\frac{\sqrt{B^2 - 4AC} + B}{\sqrt{B^2 - 4AC} - B} \right] - T \end{aligned} \tag{7.83}$$

Finally, the solution ($x = t$) can be found by substituting the expression for c back

$$y = \frac{\sqrt{B^2 - 4AC}}{2A} \frac{\left(\frac{\sqrt{B^2-4AC}+B}{\sqrt{B^2-4AC}-B}\right) e^{-\sqrt{B^2-4AC}(T-t)} - 1}{\left(\frac{\sqrt{B^2-4AC}+B}{\sqrt{B^2-4AC}-B}\right) e^{-\sqrt{B^2-4AC}(T-t)} + 1} - \frac{B}{2A} \tag{7.84}$$

Note that this is identical to CIRs result when $C = 1$:

$$y = \frac{2\left(e^{\sqrt{B^2-4A}(T-t)} - 1\right)}{\left(\sqrt{B^2 - 4A} - B\right)\left(e^{\sqrt{B^2-4A}(T-t)} - 1\right) + 2\sqrt{B^2 - 4A}} \tag{7.85}$$

7.6.2 Equity Option Pricing Under Random Interest Rates (Merton 1973)

In this appendix, we discuss how random interest rates can affect equity option prices. Merton (1973) was the first person to derive a model for equity option prices under normally distributed interest rates. Rabinovitch (1989) made it more concrete with the Vaicek model where interest rates are normally distributed. In Chapter 12, we also show how the Ho-Lee model (an no-arbitrage interest rate model) can be incorporated in the Merton model.

The following joint process is a combination of Black-Scholes and Vasicek models:

$$\begin{bmatrix} \frac{dS}{S} \\ dr \end{bmatrix} = \begin{bmatrix} r \\ \alpha(\theta - r) \end{bmatrix} dt + \begin{bmatrix} \sigma & 0 \\ 0 & \delta \end{bmatrix} \times \begin{bmatrix} \sqrt{1-\rho^2} & \rho \\ 0 & 1 \end{bmatrix} \times \begin{bmatrix} d\hat{W}_S \\ d\hat{W}_r \end{bmatrix} \tag{7.86}$$

This is the model that falls in the Merton model and explicitly solved by Rabinovitch (1989). Using the forward measure of this Chapter, we have the following process for the interest rate:

$$\begin{aligned} dr(t) &= \alpha(\theta - r(t))dt + \delta d\hat{W}_r(t) \\ &= \left[\alpha(\theta - r(t)) + \delta^2 \frac{P_r(t,T)}{P(t,T)}\right] dt + \delta d\tilde{W}_r^{(T)}(t) \end{aligned} \tag{7.87}$$

and the joint process becomes:

$$\begin{bmatrix} \frac{dS(t)}{S(t)} \\ dr(t) \end{bmatrix} = \begin{bmatrix} r(t) + \rho\sigma\delta \frac{P_r(t,T)}{P(t,T)} \\ \alpha(\mu - r(t)) + \delta^2 \frac{P_r(t,T)}{P(t,T)} \end{bmatrix} dt + \begin{bmatrix} \sigma & 0 \\ 0 & \delta \end{bmatrix} \times \begin{bmatrix} \sqrt{1-\rho^2} & \rho \\ 0 & 1 \end{bmatrix} \times \begin{bmatrix} d\hat{W}_S(t) \\ d\tilde{W}_r^{(T)}(t) \end{bmatrix} \tag{7.88}$$

Given that $r(t)$ is a normal process, $S(t)$ remains log normal and hence has the following solution:

$$S(T) = S(t) \exp\left(\int_t^T \left[r(u) + \rho\sigma\delta \frac{P_r(u,T)}{P(u,T)}\right] du + \int_t^T \sigma d\tilde{W}^{(T)}(u) - \frac{1}{2}\int_t^T \sigma^2 du\right) \tag{7.89}$$

where

$$d\tilde{W}^{(T)}(t) = \sqrt{1-\rho^2} d\hat{W}_S(t) + \rho d\tilde{W}_r^{(T)}(t)$$

From before,

$$\begin{aligned} 0 &= \ln P(t,T) + \int_t^T r(u)du + \int_t^T \delta \frac{P_r(u,T)}{P(u,T)} d\hat{W}_r(u) - \int_t^T \frac{1}{2}\left[\delta \frac{P_r(u,T)}{P(u,T)}\right]^2 du \\ &= \ln P(t,T) + \int_t^T r(u)du + \int_t^T \delta \frac{P_r(u,T)}{P(u,T)} d\tilde{W}_r(u) + \int_t^T \frac{1}{2}\left[\delta \frac{P_r(u,T)}{P(u,T)}\right]^2 du \end{aligned} \tag{7.90}$$

and hence,

$$\int_t^T r(u)du = -\ln P(t,T) - \int_t^T \delta \frac{P_r(u,T)}{P(u,T)} d\tilde{W}_r(u) - \int_t^T \frac{1}{2}\left[\delta \frac{P_r(u,T)}{P(u,T)}\right]^2 du \tag{7.91}$$

Substitute this back to the stock equation:

$$S(T) = \frac{S(t)}{P(t,T)} \exp\left(\left(\int_t^T \left[-\frac{1}{2}\left[\delta \frac{P_r(u,T)}{P(u,T)} \right]^2 + \rho\sigma\delta \frac{P_r(u,T)}{P(u,T)} - \frac{1}{2}\sigma^2 \right] du \right.\right.$$
$$\left. - \int_t^T \delta \frac{P_r(u,T)}{P(u,T)} d\tilde{W}_r^{(T)}(u) + \int_t^T \sigma d\tilde{W}^{(T)}(u) \right) \tag{7.92}$$

Note that the expectation under the forward measure $\sim$:

$$\begin{aligned} &\tilde{\mathbb{E}}_t^{(T)} \left[\left\{ \int_t^T -\delta \frac{P_r(u,T)}{P(u,T)} d\tilde{W}_r^{(T)}(u) + \int_t^T \sigma d\tilde{W}^{(T)}(u) \right\}^2 \right] \\ &= \int_t^T \left\{ \left[\delta \frac{P_r(u,T)}{P(u,T)} \right]^2 + \sigma^2 - 2\rho\sigma\delta \frac{P_r(u,T)}{P(u,T)} \right\} du \end{aligned} \tag{7.93}$$

and the covariance term (last term) is obtained because:

$$\begin{aligned} d\tilde{W}^{(T)}(t) d\tilde{W}_r^{(T)}(t) &= \left(\sqrt{1-\rho^2} d\hat{W}_S(t) + \rho d\tilde{W}_r^{(T)}(t) \right) d\tilde{W}_r^{(T)}(t) \\ &= \rho dt \end{aligned} \tag{7.94}$$

Hence,

$$\tilde{\mathbb{E}}_t^{(T)}[S(T)] = \frac{S(t)}{P(t,T)} \tag{7.95}$$

is a forward price. Note that this solution does not need normality assumption. We obtain this result with no specific assumption of the interest rate. Now we turn to the option pricing problem. Using the call option as an example, we have the:

$$\hat{\mathbb{E}}_t \left[\exp\left(-\int_t^T r(u) du \right) \max\{S(T) - K, 0\} \right] = P(t,T) \tilde{\mathbb{E}}_t^{(T)} [\max\{S(T) - K, 0\} \tag{7.96}$$

The change of measure to the forward measure is same as above. Recall that $S(T)$ is still log normal under the forward measure. Hence, the BS result applies except that we need to work out mean and variance of $\ln S(T)$ under the forward measure. From above, we know that:

$$\begin{aligned}\ln S(T) &= \ln S(t) - \ln P(t,T) \\ &+ \int_t^T \left[-\frac{1}{2}\left[\delta \frac{P_r(u,T)}{P(u,T)}\right]^2 + \rho\sigma\delta \frac{P_r(u,T)}{P(u,T)} - \frac{1}{2}\sigma^2 \right] du \\ &- \int_t^T \delta \frac{P_r(u,T)}{P(u,T)} d\tilde{W}_r^{(T)}(u) + \int_t^T \sigma d\tilde{W}^{(T)}(u)\end{aligned} \tag{7.97}$$

Hence,

$$\begin{aligned}\tilde{\mathbb{E}}_t^{(T)}[\ln S(T)] &= \ln S(t) - \ln P(t,T) + \frac{1}{2}V(t,T) \\ \tilde{\mathbb{V}}_t^{(T)}[\ln S(T)] &= V(t,T)\end{aligned} \tag{7.98}$$

where

$$V(t,T) = \int_t^T \left\{ \left[\delta \frac{P_r(u,T)}{P(u,T)}\right]^2 - 2\rho\sigma\delta \frac{P_r(u,T)}{P(u,T)} + \sigma^2 \right\} du \tag{7.99}$$

So far this is the Merton result (1973). Merton did not specify which bond pricing model for $P(u,T)$. But the interest rate model specified in here is the Vasicek model that has a closed form solution to $P(u,T)$ as shown in (7.30). As a result:

$$\begin{aligned}\frac{P_r(u,T)}{P(u,T)} &= -F(u,T) \\ &= -\frac{1 - e^{-\alpha(T-u)}}{\alpha}\end{aligned} \tag{7.100}$$

Then, equation (7.99) can be solved in closed form as:

$$\begin{aligned}V(t,T) &= \int_t^T \left\{ \frac{\delta^2}{\alpha^2}\left(1 - 2e^{-\alpha(T-u)} + e^{-2\alpha(T-u)}\right) + 2\frac{\rho\sigma\delta}{\alpha}\left(1 - e^{-\alpha(T-u)}\right) + \sigma^2 \right\} du \\ &= \frac{\delta^2}{\alpha^2}\left[T - t - \frac{2}{\alpha}\left(1 - e^{-\alpha(T-t)}\right) + \frac{1}{2\alpha}\left(1 - e^{-2\alpha(T-t)}\right)\right] \\ &+ 2\frac{\rho\sigma\delta}{\alpha}\left[T - t - \frac{1}{\alpha}\left(1 - e^{-\alpha(T-t)}\right)\right] + \sigma^2(T-t)\end{aligned} \tag{7.101}$$

From the standard BS result, we have the final solution to the call price as follows:

$$C(t) = S(t)N(d_1) - P(t,T)KN(d_2) \tag{7.102}$$

where

$$\begin{aligned} d_1 &= \frac{\ln S(t) - \ln P(t,T) + V(t,T)/2}{\sqrt{V(t,T)}} \\ d_2 &= d_1 - V(t,T) \end{aligned} \tag{7.103}$$

This is the Rabinovitch result (1989) in JFQA. As we can see, this is similar to the Black-Scholes model with the substitution of the volatility.

Chapter 8

Term Structure Estimation

This chapter is based upon Chen-Scott (Journal of Fixed Income 1993 article) and Chen-Yang (Journal of Housing Markets 1993). More complete and robust estimation procedures have been proposed, such as Kalman filter and GMM but they are not discussed here.

8.1 Introduction

Estimation of a model is crucial in using the model properly. Many reduced form term structure models (or also known as no-arbitrage models) take the current term structure as given and build in hypothetical parameter values. This is good for marking to market and risk management but not so useful in prediction and exploring trading opportunities. Many modelers recognize the importance of historical estimates of the parameters and the implications one can draw from these estimates.

While complex term structure models are very difficult to estimate (e.g. quadratic term structure models) and almost always one must resort to non-parametric methods such as Generalized Method of Moments (GMM), affine models can be easily estimated via the standard Maximum Likelihood Estimation (MLE).

8.2 Maximum Likelihood Estimation

Lets first look at the Vasicek model where the underlying interest rate is normally distributed. In the Vasicek model, the state variable (instantaneous short rate) follow the Ornstein-Uhlenbeck process under the real measure:

$$dr = \alpha(\mu - r)dt + \sigma dW \tag{8.1}$$

whose transition density is Gaussian:

$$N(m_t, V) = N\left(\mu\left(1 - e^{-\alpha h}\right) + e^{-\alpha h} r_t, \frac{1-e^{-2\alpha h}}{2\alpha}\sigma^2\right) \tag{8.2}$$

over the time interval h. To write it explicitly where $h = s - t$:

$$\begin{aligned} f(r_s|r_t) &= \frac{1}{\sqrt{2\pi V}} e^{-\frac{(r_s-m_t)^2}{2V}} \\ \ln f(r_s|r_t) &= -\tfrac{1}{2}\ln 2\pi - \tfrac{1}{2}\ln V - \tfrac{(r_s-m_t)^2}{2V} \end{aligned} \tag{8.3}$$

The joint distribution of r over time, i.e. $< r_0, r_1, r_2, \cdots, r_T >$ is:

$$\begin{aligned} f(r_1, r_2, \cdots, r_T) &= f(r_T|r_{T-1}) f(r_{T-1}|r_{T-2}) \cdots f(r_2|r_1) f(r_1|r_0) f(r_0) \\ &= \frac{1}{(2\pi V)^{1/2 T}} e^{-\sum_{j=1}^{T} \frac{(r_j-m_{j-1})^2}{2V}} f(r_0) \\ &= \frac{1}{(2\pi V)^{1/2 T}} e^{-\sum_{j=1}^{T} \frac{(r_j-m_{j-1})^2}{2V}} r_0 \end{aligned} \tag{8.4}$$

due to the Markov property. Note that $f(r_0)$ should be the asymptotic distribution but for convenience, usually it is replaced with the current interest rate (from line 2 to line 3 in the above equation).

In the market, we do not observe r but we observe the price, which is defined as follows:

$$\begin{aligned} P_{t,T} &= \hat{\mathbb{E}}_t \left[\exp\left(-\int_t^T r_s ds\right)\right] \\ &= e^{-r_t F_{T-t} - G_{T-t}} \end{aligned} \tag{8.5}$$

where

$$\begin{aligned} G_{T-t} &= (T - t - B_{T-t})\left(\mu + \tfrac{\sigma^2}{2\alpha^2} - \tfrac{\sigma\lambda}{\alpha}\right) + \tfrac{\sigma^2 F_{T-t}^2}{4\alpha} \\ F_{T-t} &= \tfrac{1-e^{-\alpha(T-t)}}{\alpha} \end{aligned}$$

and λ is price of risk. Inverting the result, we can write the state variable in terms of price:

$$r_t = -\frac{\ln P_{t,T} + G_{T-t}}{F_{T-t}} \tag{8.6}$$

Substituting equation (8.6) back to equation (8.4) and performing change of variable, we get the likelihood function:

$$\begin{aligned}
\mathcal{L} &= f(r_1, r_2, \cdots, r_T) \\
&= f\left(-\tfrac{\ln P_{1,2}+G_h}{F_h}, -\tfrac{\ln P_{2,3}+G_h}{F_h}, \cdots, -\tfrac{\ln P_{T-1,T}+G_h}{F_h}\right) \begin{vmatrix} \frac{\partial r_1}{\partial P_{1,2}} & & 0 \\ & \ddots & \\ 0 & & \frac{\partial r_T}{\partial P_{T-1,T}} \end{vmatrix} \\
&= \frac{1}{2\pi V^{1/2 T}} \exp\left\{-\sum_{j=1}^{T} \frac{1}{2V}\left(-\tfrac{\ln P_{j-1,j}+G_h}{F_h} - m_{j-1}\right)^2\right\}\left(-\tfrac{\ln P_{0,1}+G_h}{F_h}\right)\frac{1}{F_h^T}
\end{aligned} \tag{8.7}$$

where $1/F_h^T$ is the Jacobian. This is the likelihood function. The log likelihood is:

$$\ln \mathcal{L} = -1/2 T \ln[2\pi V] - \sum_{j=1}^{T} \frac{1}{2V}\left(-\tfrac{\ln P_{j-1,j}+G_h}{F_h} - m_{j-1}\right)^2 + \ln[-\tfrac{\ln P_{0,1}+G_h}{F_h}] - T \ln F_h \tag{8.8}$$

The first order conditions are:

$$\begin{aligned}
\frac{\partial \ln \mathcal{L}}{\partial \alpha} &= \sum_{i=1}^{T} \frac{1}{f}\frac{\partial f}{\partial r}\frac{\partial r}{\partial \alpha} - T\frac{1}{F}\frac{\partial F}{\partial \alpha} \\
&= \sum_{i=1}^{T} \sqrt{2\pi V} e^{\frac{(r-m)^2}{2V}} \left[\frac{1}{\sqrt{2\pi V}} e^{-\frac{(r-m)^2}{2V}} \frac{2(r-m)\left(1-e^{-\alpha h}\right)}{V}\right]\left[\frac{\frac{\partial G}{\partial \alpha}F + G\frac{\partial B}{\partial \alpha}}{F^2}\right] \\
&\quad - T\frac{1}{F}\frac{\partial F}{\partial \alpha} \\
&= \sum_{i=1}^{T} \left[\frac{r}{V}\right]\left[\frac{\frac{\partial G}{\partial \alpha}F + G\frac{\partial F}{\partial \alpha}}{F^2}\right] - T\frac{1}{F}\frac{\partial F}{\partial \alpha} = 0,
\end{aligned} \tag{8.9}$$

$$\begin{aligned}
\frac{\partial \ln \mathcal{L}}{\partial \mu} &= \sum_{i=1}^{T} \frac{1}{f}\frac{\partial f}{\partial x}\frac{\partial r}{\partial \mu} \\
&= \sum_{i=1}^{T} \left[\frac{r}{V}\right]\left[\frac{\frac{\partial G}{\partial \mu}}{F}\right] = 0,
\end{aligned} \tag{8.10}$$

$$\frac{\partial \ln \mathcal{L}}{\partial \sigma} = \sum\nolimits_{i=1}^{T} \left[\frac{r}{V}\right] \left[\frac{\frac{\partial G}{\partial \sigma}}{F}\right] = 0, \tag{8.11}$$

and

$$\frac{\partial \ln \mathcal{L}}{\partial \lambda} = \sum\nolimits_{i=1}^{T} \left[\frac{r}{V}\right] \left[\frac{\frac{\partial G}{\partial \lambda}}{F}\right] = 0 \tag{8.12}$$

where

$$\begin{cases} \dfrac{\partial G}{\partial \alpha} = \dfrac{\partial F}{\partial \alpha}\left(\mu + \frac{\sigma^2}{2\alpha^2} - \frac{\sigma\lambda}{\alpha}\right) + F\left(-\frac{\sigma^2}{\alpha^3} + \frac{\sigma\lambda}{\alpha^2}\right) - \frac{\sigma^2 F^2}{4\alpha^2} \\ \dfrac{\partial F}{\partial \alpha} = \dfrac{T-t}{\alpha} e^{-\alpha(T-t)} - \dfrac{1-e^{-\alpha(T-t)}}{\alpha^2} \end{cases} \tag{8.13}$$

$$\begin{cases} \dfrac{\partial G}{\partial \mu} = F \\ \dfrac{\partial F}{\partial \mu} = 0 \end{cases} \tag{8.14}$$

$$\begin{cases} \dfrac{\partial G}{\partial \sigma} = F\left(-\frac{\sigma}{\alpha^2} + \frac{\lambda}{\alpha^2}\right) - \frac{\sigma F^2}{2\alpha^2} \\ \dfrac{\partial F}{\partial \sigma} = 0 \end{cases} \tag{8.15}$$

$$\begin{cases} \dfrac{\partial G}{\partial \lambda} = \dfrac{\sigma F}{\alpha^2} \\ \dfrac{\partial F}{\partial \lambda} = 0 \end{cases} \tag{8.16}$$

Use a single bond price series (it needs to be zero coupon bond series), we can then solve the simultaneous equation system (which contains non-linear equations) for the parameters α, μ, σ, and λ.

The theory of Fisher information states that the sampling distribution of a maximum likelihood estimator is asymptotically unbiased and also asymptotically normal with its variance-covariance matrix obtained from the inverse Fisher information matrix which is defined as:

$$I_{jk} = \mathbb{E}\left[\frac{\partial \ln \mathcal{L}}{\partial \theta_j}\frac{\partial \ln \mathcal{L}}{\partial \theta_k}\right] \tag{8.17}$$

where θ_j represents α, μ, σ, or λ. In the case of the Vasicek model, the joint distribution is normal:

$$\begin{bmatrix} r_1 \\ r_2 \\ \vdots \\ r_T \end{bmatrix} [\underline{m}, \mathbb{K}] \tag{8.18}$$

where each element of the variance-covariance matrix $\mathbb{K}$ can be computed as follows:

$$\mathbb{K}_{s,t} = \text{cov}[r(s), r(t)] = \frac{\sigma^2}{2\alpha}\left(e^{-\alpha(s-t)} - e^{-\alpha(s+t)}\right) \tag{8.19}$$

for $s > t$. As a result, the information matrix can be written as:

$$I_{jk} = \frac{\partial \underline{m}}{\partial \theta_j}\mathbb{K}^{-1}\frac{\partial \underline{m}}{\partial \theta_k} + \frac{1}{2}\text{tr}\left(\mathbb{K}^{-1}\frac{\partial \underline{m}}{\partial \theta_j}\mathbb{K}^{-1}\frac{\partial \underline{m}}{\partial \theta_k}\right) \tag{8.20}$$

where the partial derivatives of each element in the mean vector, $\underline{m}$, with respective to the parameters are:

$$\begin{aligned} \frac{\partial m_t}{\partial \alpha} &= (\mu - r_t)(s-t)e^{-\alpha(s-t)} \\ \frac{\partial m_t}{\partial \mu} &= 0 \\ \frac{\partial m_t}{\partial \sigma} &= 0 \end{aligned} \tag{8.21}$$

and

$$\begin{aligned} \frac{\partial \mathbb{K}_{s,t}}{\partial \alpha} &= \sigma^2 \frac{2\alpha(s-t)e^{-2\alpha(s-t)} - \left(1 - e^{-2\alpha(s-t)}\right)}{2\alpha^2} \\ \frac{\partial \mathbb{K}_{s,t}}{\mu} &= 0 \\ \frac{\partial \mathbb{K}_{s,t}}{\partial \sigma} &= 2\sigma\frac{1 - e^{-2\alpha(s-t)}}{2\alpha} \end{aligned} \tag{8.22}$$

8.3 Regression

8.3.1 The Vasicek Model

The Vasicek model is a one-factor mean reverting Gaussian model with the following short rate dynamics

$$dr = \alpha(\mu - r)dt + \sigma dW \tag{8.23}$$

under the real measure. Recall the result of the stochastic linear differential equation introduced in Chapter 3. The mean and variance of r are given as follows:

$$\mathbb{E}[r_s|r_t] = \left(1 - e^{-\alpha(s-t)}\right)\mu + e^{-\alpha(s-t)}r_t \tag{8.24}$$

and

$$\mathbb{V}[r_s|r_t] = \sigma^2 \frac{1 - e^{-2\alpha(s-t)}}{2\alpha} \tag{8.25}$$

We also know from Chapter 3, r is normally distributed. Hence, we can write an AR(1) process in discrete time for r:

$$r_t = \left(1 - e^{-\alpha h}\right)\mu + e^{-\alpha h}r_{t-h} + e_t \tag{8.26}$$

where h is time interval and e_t has mean 0 and variance$\frac{1-e^{-2\alpha h}}{2\alpha}\sigma^2$. This essentially an OLS with normally distributed error terms:

$$r_t = a + br_{t-h} + e_t \tag{8.27}$$

with $a = \left(1 - e^{-\alpha h}\right)\mu$ and $b = e^{-\alpha h}$. We can then run an OLS to obtain the parameters. The regression coefficient, b, can be used to estimate α, and then the constant coefficient, a, can be used for μ. Finally the MSE of the regression model can be used to estimate σ. The t statistics of α and μ can be derived from the t statistics of a and b. With Taylors series expansion, we have:

$$\begin{cases} b \approx 1 - \alpha h \\ a \approx \alpha\mu h \end{cases} \tag{8.28}$$

It then follows that:

$$\begin{cases} \text{var}[b] \approx h^2 \, \text{var}[\alpha] \\ \text{var}[a] \approx (\alpha h)^2 \, \text{var}[\mu] \end{cases} \tag{8.29}$$

and hence the t statistics can be easily calculated. Finally, the asymptotic standard error of the SSE of the regression model is $1/(2\sigma^4)$. Hence, the standard error of the MSE is $1/(2N\sigma^4)$ where N is the sample size.

The problem of the above method is that it fails to estiamte the market price of risk. The chosen interest rate series are regarded as instantaneous rate and hence carries no risk premium. Certainly such substitution, while convenient and often times enough to obtain major intuition of the interest rate dynamics, is theoretically flawed.

To include the market price of risk into the estimation process, we must use the closed form of the Vasicek model. The closed-form solution from equation is derived by Vasicek as follows:

$$P_{t,T} = e^{-r_t F_{T-t} - G_{T-t}} \tag{8.30}$$

where

$$F_{T-t} = \frac{1 - e^{-\alpha(T-t)}}{\alpha}$$
$$G_{T-t} = \left(\mu - \sigma\lambda - \tfrac{\sigma^2}{2\alpha^2}\right)(T - t - F_{T-t}) + \frac{\sigma^2 F_{T-t}^2}{4\alpha}$$

where λ is the market price of risk. Inverting the pricing formula gives the instantaneous rate:

$$r_t = -\frac{\ln P_{t,T} + G_{T-t}}{F_{T-t}} \tag{8.31}$$

Note that the above formula must hold regardless of the choice of T. Also note that at the next period, $t+h$, we can write (8.31) as:

$$r_{t+h} = -\frac{\ln P_{t+h,T+h} + G_{T-t}}{F_{T-t}} \tag{8.32}$$

and the functions F and G remain the same as $T+h-(t+h) = T-t$. This result provides tremendous convenience in data collection. We can choose CMT (constant maturity Treasury) rates for the estimation. A 3-month CMT rate have a constant rolling maturity of 3 months. Hence $T-t=1/4$. We can use daily (i.e. $h=1/365$) CMT rates for estimation.

Substituting (8.31) and (8.32) back into (8.34), we obtain:

$$\frac{-\ln P_{t+h,T+h}}{T-t} = \left(\frac{1-e^{-\alpha h}}{T-t}\right)(\mu F_{T-t}+G_{T-t}) + e^{-\alpha h}\frac{-\ln P_{t,T}}{T-t} + \frac{F_{T-t}}{T-t}e_t \tag{8.33}$$

An OLS can be performed to arrive at the following estimates:

$$\begin{aligned} \text{Slope} &= e^{-\alpha h} \\ \text{Intercept} &= \left(\frac{1-e^{-\alpha h}}{T-t}\right)(\mu F_{T-t}+G_{T-t}) \\ \text{SSE} &= \frac{F_{T-t}^2}{(T-t)^2}\frac{\sigma^2\left(1-e^{-2\alpha h}\right)}{2\alpha} \end{aligned} \tag{8.34}$$

where h is data frequency and $T-t$ is the maturity of the bond. Note that (8.34) has three equations but has four unknowns. The market price of risk contained in function G cannot be identified. As a result, we must choose another bond series and run a simultaneous regression as follows:

$$\begin{cases} \dfrac{-\ln P_{t+h,T+h}}{T-t} = \left(\dfrac{1-e^{-\alpha h}}{T-t}\right)(\mu F_{T-t}+G_{T-t}) + e^{-\alpha h}\dfrac{-\ln P_{t,T}}{T-t} + \dfrac{F_{T-t}}{T-t}e_t \\ \dfrac{-\ln P_{t+h,T'+h}}{T'-t} = \left(\dfrac{1-e^{-\alpha h}}{T'-t}\right)(\mu F_{T'-t}+G_{T'-t}) + e^{-\alpha h}\dfrac{-\ln P_{t,T'}}{T'-t} + \dfrac{F_{T'-t}}{T'-t}e_t \end{cases} \tag{8.35}$$

where T' represents a different maturity. In the simultaneous regression the regressors in the two separate equations must be restricted to have the same coefficient. This allows us to solve for a unique set of α, μ, and λ. However, there are two SSEs from separate regressions and the choice of σ is arbitrary.

From this exercise, we know that it is impossible to estimate market price of risk without crossectional data. A univariate time series data cannot be used to estimate market price of risk.

Finally, via Taylors series expansion, we can derive the standard errors of the parameters:

$$
\begin{aligned}
&\operatorname{var}\left[e^{-\alpha h}\right] \approx \operatorname{var}[1-\alpha h] = h^2 \operatorname{var}[\alpha] \\
&\operatorname{var}\left[\left(\frac{1-e^{-\alpha h}}{T-t}\right)(\mu F_{T-t} + G_{T-t})\right] \\
&\quad \approx \operatorname{var}\left[\left(\frac{1-e^{-\alpha h}}{T-t}\right)\left[\mu F_{T-t} + \left(\mu - \tfrac{\sigma^2}{2\alpha^2}\right)(T-t-F_{T-t}) + \frac{\sigma^2 F_{T-t}^2}{4\alpha}\right]\right] \\
&\quad = \left(\frac{1-e^{-\alpha h}}{T-t}\right)^2 \operatorname{var}\left[\mu(T-t)\right] = \left(1-e^{-\alpha h}\right)^2 \operatorname{var}[\mu]
\end{aligned}
\tag{8.36}
$$

8.3.2 The Cox-Ingersoll-Ross Model

The CIR model is a one-factor mean-reverting square-root model with the following short rate dynamics:

$$
dr = \alpha(\mu - r)dt + \sigma\sqrt{r}dW \tag{8.37}
$$

under the real measure. This is a complex distribution for r as introduced in Chapter 7. The transition density for r is:

$$
f(r(s)|r(t)) = ce^{-c(r(s)+\xi)}\left(\frac{r(s)}{\xi}\right)^{d/2} I_d(2c\sqrt{r(s)\xi}) \tag{8.38}
$$

where

$$
\begin{aligned}
c &= \frac{2\alpha}{\sigma^2\left(1-e^{-\alpha(s-t)}\right)} \\
\xi &= r(t)e^{-\alpha(s-t)} \\
d &= \frac{2\alpha\mu}{\sigma^2} - 1
\end{aligned}
$$

which after change of variable $x = 2cr$:

$$
f(x(s)|x(t)) = \frac{1}{2}e^{-1/2(x(s)+\Lambda)}\left(\frac{x(s)}{\Lambda}\right)^{d/2} I_d(\sqrt{\Lambda x(s)}) \tag{8.39}
$$

is a non-central chi-square distribution with $\Lambda = 2c\xi$ degrees of non-centrality and $2d+2 = \frac{4\alpha\mu}{\sigma^2}$ degrees of freedom. We know that the mean and variance of a non-central chi-squared distribution are, respectively, $\upsilon + \Lambda$ and $2(\upsilon + 2\Lambda)$, which lead to (see Chapter 7 for details):

$$\mathbb{E}[r_s|r_t] = \left(1 - e^{-\alpha(s-t)}\right)\mu + e^{-\alpha(s-t)} r_t \tag{8.40}$$

and

$$\mathbb{V}[r_s|r_t] = \frac{\sigma^2}{2\alpha}\left[\left(1 - e^{-\alpha(s-t)}\right)^2 \mu + r_t\left(e^{-2\alpha(s-t)} - e^{-\alpha(s-t)}\right)\right] \tag{8.41}$$

It is clear that the mean of r under the CIR model is the same as the mean under the Vasicek model. However, unlike the Vasicek model, the variance of r is no longer independent of r. The variance of r under the CIR model is an AR(1) process, which is hetroskedestic if we run the regression model of (8.34). To overcome the problem of hetroskedesticity, we run a two-step regression:

- Run OLS of (8.34):

  $$\begin{aligned} r_t &= a + b r_{t-h} + e_t \text{ where} \\ a &= \left(1 - e^{-\alpha(s-t)}\right)\mu \\ b &= e^{-\alpha(s-t)} \end{aligned}$$

- The variance of the error term must satisfy:
  $\mathbb{E}[e_t^2] = \frac{\sigma^2}{2\alpha}\left[\left(1 - e^{-\alpha(s-t)}\right)^2 \mu + r_t\left(e^{-2\alpha(s-t)} - e^{-\alpha(s-t)}\right)\right]$

- Run $e_t^2 = a' + b' r_{t-1} + u_t$ where

  $$\begin{aligned} a' &= \frac{\sigma^2}{2\alpha}\left(1 - e^{-\alpha(s-t)}\right)^2 \mu \\ b' &= \frac{\sigma^2}{2\alpha}\left(e^{-2\alpha(s-t)} - e^{-\alpha(s-t)}\right) \end{aligned}$$

- Use the first OLS to solve for α and μ and the use the second OLS to solve for σ.

Since both a' and b' can be used to solve for σ, one can use a more reasonable estimate.

8.3.3 The Ho-Lee Model

Note that from the Ho-Lee model introduced in Chapter 12, we define the zero-coupon bond price as $P(t,T,j)$ where t is current time, T is maturity time, and j represents the state of economy. The pricing formula in a binomial model is as follows:

$$\begin{cases} P(u,s,j) = \dfrac{P(t,s,j)}{P(t,u,j)} d(s-u) \\ P(u,s,j+1) = \dfrac{P(t,s,j)}{P(t,u,j)} u(s-u) \end{cases} \tag{8.42}$$

where $t < u < s$ and

$$\begin{aligned} u(\tau) &= \frac{1}{p + (1-p)\delta^\tau} \\ d(\tau) &= \frac{\delta^\tau}{p + (1-p)\delta^\tau} \end{aligned}$$

The frequency of the data is $u-t$ and the maturity is τ. Note that $P(t,s,j)/P(t,u,j)$ is a forward price. If the yield curve is flat, then $P(t,s,j)/P(t,u,j)$ becomes $e^{-r(s-u)}$ under state j, which is $P(t,t+s-u,j)$. Note that this bond has the same time to maturity as $P(u,s,j)$ or $P(u,s,j+1)$. As a result, we can write (8.42) as:

$$\begin{cases} P(u,u+\tau,j) = P(t,t+\tau,j)d(\tau) \\ P(u,u+\tau,j+1) = P(t,t+\tau,j)u(\tau) \end{cases} \tag{8.43}$$

Hence, we can collect a series of τ-maturity bond prices (CMT series would be ideal). Write such bond series as $D_t(\tau)$. Then,

$$D_{t+1}(\tau) = \begin{cases} D_t(\tau)\dfrac{\delta^\tau}{p + (1-p)\delta^\tau} & \text{if rate rises} \\ D_t(\tau)\dfrac{1}{p + (1-p)\delta^\tau} & \text{if rate falls} \end{cases} \tag{8.44}$$

where $D_t(\tau)$ represents the zero-coupon bond price at time t for time to maturity τ in data series. We can choose a liquid Treasury series, e.g. 3-month T bills, for the estimation. Equation (8.44) can be written as:

$$\ln D_{t+1}(\tau) - \ln D_t(\tau) = -\ln[p + (1-p)\delta^\tau] + \tau \ln \delta \mathbb{I}_{t+1} + u_{t+1} \tag{8.45}$$

where the indicator function:

$$\mathbb{I}_t = \begin{cases} 1 & \text{if rate rises} \\ 0 & \text{if rate falls} \end{cases}$$

and $u_t \sim N(0,1)$. In a simple regression as follows

$$y_t = a + bx_t + u_t \tag{8.46}$$

we can set $\delta = e^{b/\tau}$ and $p = \frac{e^{-a}-\delta^{\tau}}{1-\delta^{\tau}}$.

Chapter 9

Numerical Methods

This chapter is heavily based upon Hull. I also add applications from Glasserman (2003), Chen and Scott (1992), and Chen, Chung, and Yang (2002).

9.1 Introduction

Fast numerical solutions are regarded same as closed form solutions. Many difficult numerical methods have been translated in a readable form to the finance audience. Hull and Glasserman have contributed hugely in this area.

9.2 Random Numbers

All simulations start with uniform random numbers. Then an inverse cumulative density function is applied to obtain the desired distribution. Hence it is necessary to know various cumulative density functions.

9.2.1 Normal/log Normal

Simulating a normal distribution is the most basic. Although many software packages have such random number generator, it is really easy to do it by oneself.

Inverse normal probability function has an accurate closed form approximation formula and its code is enclosed at the end of this Chapter. In Excel, simply use `NORMSINV(RAND())` to obtain a normal random variable (as `RAND()` generates a uniform random variable).

Log normal random numbers are simply exponential transforms of normal random numbers.

9.2.2 Poisson

Assume the intensity parameter to be λ. Then the Poisson density function (i.e. probability of j events occurring) is:

$$\Pr(X = j) = \frac{(\lambda \Delta t)^j e^{-\lambda \Delta t}}{j!} \tag{9.1}$$

The most popular probability used in finance is $j = 0$ which represents no default. In that case, the probability is $e^{-\lambda \Delta t}$, which is known as the survival probability over the time period Δt.

Bernulli events

There are two ways to simulate a Poisson process. One is to simulate directly. Note that a Poisson process is a collection of Bernulli events. As a result, we can simulate a sequence of 0/1 events (0 for event not happening and 1 for event happening) over the *dt* period.

This simulation, however, is very time consuming. But it can take care of any flexible dynamics in the state variable. For example, if the intensity parameter is random, then it is only possible to do direct simulation.

Exponential time

An alternative simulation is to simulate exponential time. Note that the probability of no default is identical to the probability that default time τ occurs after the fixed specified time t.

$$\Pr(\tau > t) = e^{-\lambda t} \tag{9.2}$$

Hence, a uniform random number, say u, can be transformed into an exponential random number by $t = -\ln u/\lambda$.

9.2.3 Non-central Chi-square

Non-central Chi-square (nc-χ^2) cumulative distribution function is not closed form, hence an inversion function is not easy to calculate. As a result, simulating nc-χ^2 distribution is not easy. Here is how.

The non-central chi-square cumulative density function, CDF, is (e.g. see Johnson and Kotz):

$$F_{\mathrm{nc}\chi^2}(x) = \sum_{j=0}^{\infty} \frac{e^{-1/2\lambda}(1/2\lambda)^j/j!}{2^{1/2\nu+j}\Gamma(1/2\nu+j)} \int_0^x z^{1/2\nu+j-1}e^{-1/2z}dz \tag{9.3}$$

where ν and λ are degrees of freedom and non centrality respectively.

Note that a chi-square density is:

$$f_{\chi^2}(z;k) = \frac{1}{2^{1/2k}\Gamma(1/2k)} z^{1/2k-1}e^{-1/2z}dz \tag{9.4}$$

where k is the degrees of freedom. Also note that the Poisson density is:

$$f_P(x;h) = \sum_{j=0}^{\infty} \frac{e^{-h}h^x}{x!} \tag{9.5}$$

where h is the intensity parameter. Hence, we can write the CDF of the non-central, chi-square distribution as:

$$F_{\mathrm{nc}\chi^2}(x) = \sum_{j=0}^{\infty} f_P(j;1/2\lambda) \int_0^x f_{\chi^2}(z;\nu+2j)dz \tag{9.6}$$

This implies that the CDF of a non-central, chi-square distribution is a CDF of Poisson weighted chi square distribution. Hence, to simulate the non-central, chi-square random variable, we follow the following steps:

- simulate Poisson random variable, x, with an intensity of $1/2\lambda$
- simulate chi-square random variable, y, with the degrees of freedom as $2x + 4\alpha\mu/\sigma^2$
- To simulate the short rate in the CIR model, $r(t+\Delta t)$, we first set $\lambda = 2cr(t)e^{-\alpha\Delta t}$ where $c = 2\alpha/\sigma^2\left(1-e^{-\alpha\Delta t}\right)$. Then we note that in the CIR model, the short term interest rate does not follow non-central, chi-square distribution but a scaled non-central, chi-square distribution. The scaled amount is $r(t+\Delta t) = y/(2c)$.

While the above algorithm always works, a simpler algorithm can be employed if the degrees of freedom is greater than 1. When $\nu > 1$, the following recursive result holds (see Johnson and Kotz):

$$Z_{nc\chi^2}(x;\nu,\lambda) = (Z_N + \sqrt{\lambda})^2 + Z_{\chi^2}(x;\nu-1) \tag{9.7}$$

where $Z_{nc\chi^2}(x;\nu,\lambda$, Z_N, and $Z_{\chi^2}(x;\nu-1)$ represent the random variables from non-central, chi-square, normal, and chi square respectively. In this case, we can simulate the non-central, chi-square variable as follows:

- simulate standard normal
- simulate chi square with $\nu - 1$
- degrees of freedom
- add them together.

9.2.4 Multi-variate Gaussian

To simulate multiple *correlated* random variables, it is important to understand Cholesky decomposition. Cholesky decomposition makes it possible that we first simulate independent random variables and then correlate them. As a result, simulating many correlated random variables is extremely straightforward.

Given the correlation matrix, **R**, we perform the Cholesky decomposition as follows:

$$\mathbf{C}\mathbf{C}' = \mathbf{R} \tag{9.8}$$

where **C** is the Cholesky matrix. In other words, the Cholesky matrix *decomposes* the correlation matrix. This is Cholesky decomposition. Let **X** (a vector) follow the diffusion:

$$d\mathbf{X} = \mathbf{M}dt + \Sigma\mathbf{C}dW \tag{9.9}$$

where **M** is the drift and Σ is a diagnal matrix of standard deviations. Then,

$$\begin{aligned}\Omega &= (d\mathbf{X})(d\mathbf{X}') \\ &= (\Sigma\mathbf{C}dW)(\Sigma\mathbf{C}dW)' \\ &= \Sigma\mathbf{C}\mathbf{C}'\Sigma' \\ &= \Sigma\mathbf{R}\Sigma'\end{aligned} \tag{9.10}$$

To simulate the joint Gaussian distribution, it is straightforward:

$$\begin{bmatrix} dX_1 \\ dX_2 \\ \vdots \\ dX_n \end{bmatrix} = \begin{bmatrix} \mu_1 \\ \mu_2 \\ \vdots \\ \mu_n \end{bmatrix} dt + \begin{bmatrix} \sigma_1^2 & & & \\ & \sigma_2^2 & & \\ & & \ddots & \\ & & & \sigma_n^2 \end{bmatrix}$$
$$\times \begin{bmatrix} & & & & & x_{1,n} \\ & & & & & x_{2,n} \\ & & & & & \vdots \\ 0 & 0 & 0 & x_{n-2,n-2} & x_{n-2,n-1} & x_{n-2,n} \\ 0 & 0 & 0 & 0 & x_{n-1,n-1} & x_{n-1,n} \\ 0 & 0 & 0 & 0 & 0 & 1 \end{bmatrix} \begin{bmatrix} dW_1 \\ dW_2 \\ \vdots \\ dW_{n-2} \\ dW_{n-1} \\ dW_n \end{bmatrix} \tag{9.11}$$

where

$$\begin{aligned} x_{j,n} &= \rho_{j,n} \text{ for all } j \leqslant n \text{ (note that } \rho_{n,n} = 1) \\ x_{i,j} &= \frac{\rho_{i,j} - \sum_{k=j+1}^{n} x_{i,k} x_{jk}}{x_{j,j}} \text{ for } i < j \\ x_{j,j} &= \sqrt{1 - \sum_{k=1}^{n-j} x_{j,j+k}^2} \text{ for } j < n \end{aligned}$$

Take a 4-variable example and expand it as:

$$\begin{bmatrix} dX_1 \\ dX_2 \\ dX_3 \\ dX_4 \end{bmatrix} = \begin{bmatrix} \mu_1 \\ \mu_2 \\ \mu_3 \\ \mu_4 \end{bmatrix} dt + \begin{bmatrix} \sigma_1^2 & & & \\ & \sigma_2^2 & & \\ & & \sigma_3^2 & \\ & & & \sigma_4^2 \end{bmatrix}$$
$$\times \begin{bmatrix} \sqrt{1 - x_{12}^2 - x_{13}^2 - x_{14}^2} & \frac{\rho_{12} - \rho_{14}\rho_{24} - x_{13}x_{23}}{\sqrt{1 - x_{23}^2 - x_{24}^2}} & \frac{\rho_{13} - \rho_{14}\rho_{34}}{\sqrt{1 - \rho_{34}^2}} & \rho_{14} \\ 0 & \sqrt{1 - x_{23}^2 - x_{24}^2} & \frac{\rho_{23} - \rho_{24}\rho_{34}}{\sqrt{1 - \rho_{34}^2}} & \rho_{24} \\ 0 & 0 & \sqrt{1 - \rho_{34}^2} & \rho_{34} \\ 0 & 0 & 0 & 1 \end{bmatrix} \begin{bmatrix} dW_1 \\ dW_2 \\ dW_3 \\ dW_4 \end{bmatrix} \tag{9.12}$$

The model may not be Gaussian. It could be a different statistical distribution.

9.3 Monte Carlo

The most straightforward example to practice Monte Carlo is the Black-Scholes model, which we shall demonstrate here. The Black-Scholes model employs the following log normal process for the stock price:

$$dS = rSdt + \sigma SdW \tag{9.13}$$

One way to simulate the stock price is to simulate the following stochastic Euler equation:

$$\Delta S = rS\Delta t + \sigma S\sqrt{\Delta t}\varepsilon \tag{9.14}$$

where $\varepsilon \sim N(0,1)$. This equation can be written as:

$$S_{t+\Delta t} - S_t = rS_t\Delta t + \sigma S_t\sqrt{\Delta t}\varepsilon \tag{9.15}$$

or

$$S_{t+\Delta t} = S_t(1 + r\Delta t) + \sigma S_t\sqrt{\Delta t}\varepsilon \tag{9.16}$$

where Δt is the small time step (e.g. 1/252 for one daily simulations). Note that subscripts here are time indexes. After a number of steps as the option reaches maturity (e.g. 3 months, or 0.25 year), we can compute the option price as:

$$\max\{S_T - K, 0\}$$

for call and

$$\max\{K - S_T, 0\}$$

for put. The option value today is:

$$\begin{aligned} C_0 &= e^{-rT}\hat{\mathbb{E}}_t[\max\{S_T - K, 0\} \\ &= e^{-rT}\sum_{j=1}^{N}\frac{1}{N}\max\{S_{T,j} - K, 0\} \end{aligned} \tag{9.17}$$

The following table is an Excel exercise of 8 weekly Monte Carlo simulation result. The basic information of this example is as follows:

Inputs	
initial stock price (S_0)	100
Δt (week)	1/52
risk-free rate (r)	0.04
volatility (σ)	0.3
strike (K)	98

The simulations of 1 path are done in Excel and shown below:

Outputs 1		
week	rand num	stock price
0		100.0000
1	-0.76519	96.8935
2	-0.70608	94.1218
3	1.41599	99.7388
4	-0.39640	98.1707
5	-0.42778	96.4991
6	-0.24764	95.5792
7	-0.58516	93.3259
8	-0.20290	92.6099

On this path, at the end of the 8th week, the stock price is \$92.61 which is out of the money for the call option. Simulate a different path as follows:

Outputs 2		
week	rand num	stock price
0		100.0000
1	1.59777	106.7241
2	0.24837	107.9089
3	1.45871	114.5405
4	1.47683	121.6660
5	-1.20669	115.6518
6	0.26589	117.0200
7	-0.44651	114.9363
8	0.61087	117.9457

and the stock price at the end of the 8th week is \$117.95 which is in the money of the call option, and the payoff is \$17.95. Simulating a number of paths, say 1000, we obtain 1000 values for the call option. Following (9.17), we take an average of the 1000 option values and then discount at the risk-free rate to arrive at the estimated call option premium.

Such simulations have two problems. First, there is a large number of normal random numbers to generate. Along each path, there need to be n steps and then there are N paths. Hence, there are a total of $n \times N$ random numbers to generate. This is very numerically costly.

An alternative, and more efficient, method is to solve for the stock value at the terminal time – S_T. For this we need the tool from Chapter 4: $\ln S_T = \ln S_t + \left(r - \frac{1}{2}\sigma^2\right)(T-t) + \sigma\sqrt{T-t}\varepsilon$. After obtaining $\ln S_T$, then taking an exponential we obtain the stock price at time T. Then proceed with the option payoff calculation as described and compute the option price. This approach only requires N simulations which is more efficient.

9.4 Lattice

9.4.1 Binomial Model

The binomial model is the most basic lattice model. We introduced it in the previous chapter, hence we shall not repeat here. But we shall note that according to Cox, Ross and Rubinstein (1979) the up and down need to be set as:

$$\begin{aligned} u &= e^{\sigma\sqrt{\Delta t}} \\ d &= e^{-\sigma\sqrt{\Delta t}} \end{aligned} \tag{9.18}$$

with the risk-neutral probabilities as (where r is the risk-free rate):

$$\begin{aligned} \hat{p} &= \frac{e^{r\Delta t} - d}{u - d} \\ 1 - \hat{p} &= \frac{u - e^{r\Delta t}}{u - d} \end{aligned} \tag{9.19}$$

so that it can converge to the Black-Scholes model in continuous time as $\Delta t \to 0$.

9.4.2 Finite Difference

The finite difference methods are powerful in solving partial differential equations. Take the Black-Scholes model as an example where C is the derivative asset (e.g. call option) and S is the underlying asset (e.g. stock):

$$\frac{1}{2}\sigma^2 S^2 C_{SS} + rSC_S + C_t = rC \tag{9.20}$$

Explicit Method

The explicit method is a backward induction as follows:

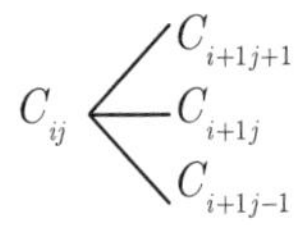

Figure 9.1: Explicit Method

Discretize the partial derivatives as follows:

$$C_S = \frac{C_{i+1j+1} - C_{i+1j-1}}{2\Delta S} \tag{9.21}$$

$$C_{SS} = \frac{C_{i+1j+1} - 2C_{i+1j} + C_{i+1j-1}}{(\Delta S)^2} \tag{9.22}$$

$$C_t = \frac{C_{i+1j} - C_{ij}}{\Delta t} \tag{9.23}$$

Substitute (9.21) $\sim$ (9.23) into (9.20) to arrive at the following difference equation,

$$\frac{1}{2}\sigma^2 S_{ij}^2 \frac{C_{i+1j+1} - 2C_{i+1j} + C_{i+1j-1}}{(\Delta S)^2} + rS_{ij}\frac{C_{i+1j+1} - C_{i+1j-1}}{2\Delta S} + \frac{C_{i+1j} - C_{ij}}{\Delta t} = rC_{ij} \tag{9.24}$$

Solving for C_{ij}, we can arrive at the following solution:

$$\begin{aligned} &\left[\frac{\sigma^2 S_{ij}^2}{2(\Delta S)^2}\Delta t + \frac{rS_{ij}}{2\Delta S}\Delta t\right] C_{i+1j+1} + \left[1 - \frac{\sigma^2 S_{ij}^2}{(\Delta S)^2}\Delta t\right] C_{i+1j} \\ &\quad + \left[\frac{\sigma^2 S_{ij}^2}{2(\Delta S)^2}\Delta t - \frac{rS_{ij}}{2\Delta S}\Delta t\right] C_{i+1j-1} = (1 + r\Delta t)C_{ij} \end{aligned} \tag{9.25}$$

which can be simplified (with the substitutions of $S_{ij} = S_0 j\Delta S$ and cancelations of S_0) as follows:

$$\left[\frac{\sigma^2 j^2}{2}\Delta t + \frac{rj}{2}\Delta t\right] C_{i+1j+1} + \left[1 - \sigma^2 j^2 \Delta t\right] C_{i+1j} + \left[\frac{\sigma^2 j^2}{2}\Delta t - \frac{rj}{2}\Delta t\right] C_{i+1j-1}$$
$$= (1 + r\Delta t)C_{ij} \tag{9.26}$$

This result of the explicit finite difference method has an economic implication. The coefficients:

$$\left[\frac{\sigma^2 j^2}{2}\Delta t - \frac{rj}{2}\Delta t\right] \tag{9.27}$$

$$\left[1 - \sigma^2 j^2 \Delta t\right] \tag{9.28}$$

$$\left[\frac{\sigma^2 j^2}{2}\Delta t + \frac{rj}{2}\Delta t\right] \tag{9.29}$$

are three risk-neutral probabilities. Hull and White (1990) show that the no arbitrage condition in asset pricing is that all three probabilities are strictly between 0 and 1 and this is also the condition in finite difference to guarantee convergence.

Implicit Method

Opposite to the explicit method, the implicit method is a forward induction as follows:

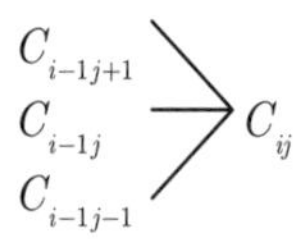

Figure 9.2: Implicit Method

As a result, we then define the partial derivatives as follows:

$$C_S = \frac{C_{i-1j+1} - C_{i-1j-1}}{2\Delta S} \tag{9.30}$$

$$C_{SS} = \frac{C_{i-1j+1} - 2C_{i-1j} + C_{i-1j-1}}{(\Delta S)^2} \tag{9.31}$$

$$C_t = \frac{C_{ij} - C_{i-1j}}{\Delta t} \tag{9.32}$$

Substitute (9.30) $\sim$ (9.32) back to (9.20) to arrive at,

$$\begin{aligned}\frac{1}{2}\sigma^2 S_{ij}^2 \frac{C_{i-1j+1} - 2C_{i-1j} + C_{i-1j-1}}{(\Delta S)^2} + rS_{ij}\frac{C_{i-1j+1} - C_{i-1j-1}}{2\Delta S} + \frac{C_{ij} - C_{i-1j}}{\Delta t} \\ = r\boxed{C_{i-1j}}\end{aligned} \tag{9.33}$$

Solving for C_{ij} (i.e. moving C_{ij} to the right hand side), we obtain:

$$\begin{aligned}\left[\frac{\sigma^2 S_{ij}^2}{2(\Delta S)^2}\Delta t + \frac{rS_{ij}}{2\Delta S}\Delta t\right] C_{i-1j+1} + \left[-1 - r\Delta t - \frac{\sigma^2 S_{ij}^2}{(\Delta S)^2}\Delta t\right] C_{i-1j} \\ + \left[\frac{\sigma^2 S_{ij}^2}{2(\Delta S)^2}\Delta t - \frac{rS_{ij}}{2\Delta S}\Delta t\right] C_{i-1j-1} = -C_{ij}\end{aligned} \tag{9.34}$$

which can be simplified as:

$$\begin{aligned}\left[-\frac{\sigma^2 j^2}{2}\Delta t - \frac{rj}{2}\Delta t\right] C_{i-1j+1} + \left[1 + r\Delta t + \sigma^2 j^2 \Delta t\right] C_{i-1j} \\ + \left[-\frac{\sigma^2 j^2}{2}\Delta t + \frac{rj}{2}\Delta t\right] C_{i-1j-1} = C_{ij}\end{aligned} \tag{9.35}$$

which we can re-write with simpler notation as:

$$a_j C_{i-1j+1} + b_j C_{i-1j} + c_j C_{i-1j-1} = C_{ij} \tag{9.36}$$

where

$$a_j = -\frac{\sigma^2 j^2}{2}\Delta t - \frac{rj}{2}\Delta t \tag{9.37}$$

$$b_j = 1 + r\Delta t + \sigma^2 j^2 \Delta t \tag{9.38}$$

$$c_j = -\frac{\sigma^2 j^2}{2}\Delta t + \frac{rj}{2}\Delta t \tag{9.39}$$

hich are mirror images of the probabilities in the case of explicit finite difference method. However, these coefficients are NOT probabilities and can be negative. Because of this, the implicit method is better. The explicit method, on the other hand, only works when the probabilities are positive.[1]

To solve for the option value today, one must repeatedly use equation (9.36) as we move backwards in time (i.e. along i in equation (9.36)). There are two ways to do it. One is matrix inversion and the other is recursive substitution.

The method of matrix inversion is straightforward. We simply write out all the terms at any given point in time. At time n (i.e. maturity), we have:

$$\begin{bmatrix} b_n & c_n & & & & \\ a_{n-1} & b_{n-1} & c_{n-1} & & & \\ & & & \ddots & & \\ & & & a_1 & b_1 & c_1 \\ & & & & a_0 & b_0 \end{bmatrix} \begin{bmatrix} C_{i-1,n} \\ C_{i-1,n-1} \\ \vdots \\ C_{i-1,1} \\ C_{i-1,0} \end{bmatrix} = \begin{bmatrix} C_{i,n} \\ C_{i,n-1} \\ \vdots \\ C_{i,1} \\ C_{i,0} \end{bmatrix} \tag{9.40}$$

where $a_0 = 0$ and $b_0 = 1$, and $b_n = 1$ and $c_n = 0$ are values at the boundaries. Note that this matrix is a tri-diagonal matrix, meaning that the except for the three central diagonal terms, the rest is all 0.

In matrix notation, we have:

$$\begin{aligned} \mathbf{X}\mathbf{C}_{i-1} &= \mathbf{C}_i \\ \mathbf{C}_{i-1} &= \mathbf{X}^{-1}\mathbf{C}_i \end{aligned}$$

The matrix iversion method is convenient. However, usually inverting a large matrix is extremely time consuming. Luckily, here the particular matrix is a tri-diagonal matrix which has a fast algorithm to obtain its inversion. In the Appendix, we show a piece of code from Numerical Recipe that one can easily implement.

To evaluate an option (say put option), we simply set $C_{n,j} = \max\{K - S_{n,j}, 0\}$ and apply equation (9.40) repeatedly until:

[1]Hull and White show that maintaining positive probabilities is the very criterion for the explicit method to converge.

$$\mathbf{C}_0 = \mathbf{X}^{-1}\mathbf{C}_1$$

Note that $\mathbf{C}_0$ is an $(n+1) \times 1$ vector and hence it is a whole set of solutions for various stock values at current time.

If one does not care for matrix inversion, there is an alternative method that achieves the same result by repetitive substitutions. To proceed, we lay out steps from the lowest state to the highest state. At any given time i and at $j = 1$, we have:

$$\begin{aligned} a_1 C_{i-12} + b_1 C_{i-11} + c_1 C_{i-10} &= C_{i1} \\ C_{i-12} &= \frac{1}{a_1}(\boxed{C_{i1}} - b_1 C_{i-11} - c_1 C_{i-10}) \end{aligned} \tag{9.41}$$

At $j = 2$,

$$\begin{aligned} a_2 C_{i-13} + b_2 C_{i-12} + c_2 C_{i-11} &= C_{i2} \\ C_{i-13} &= \frac{1}{a_2}(\boxed{C_{i2}} - b_2 C_{i-12} - c_2 C_{i-11}) \end{aligned} \tag{9.42}$$

At an arbitrary state $j = i$

$$\begin{aligned} a_{i-1} C_{i-1i} + b_{i-1} C_{i-1i-1} + c_{i-1} C_{i-1i-2} &= C_{ii-1} \\ \boxed{C_{i-1i}} &= \frac{1}{a_{i-1}}(\boxed{C_{ii-1}} - b_{i-1} C_{i-1i-1} - c_{i-1} C_{i-1i-2}) \end{aligned} \tag{9.43}$$

In the case of call we set the highest option price as the stock price $C_{i-1i} = S_i$ and in the case of put we set the highest option price as the strike price $C_{i-1i} = K$. The recursive system repeats itself and solve for (trial and error) C_{ii-1} until $C_{i-1i} = S_i$.

The convergence of the implicit method is shown in Figure 9.3. The example is taken from Hull where $S = K = 50, \sigma = 0.4, r = 0.1$ and $T - t = 5$ months.

9.5 Multi-variate Equity Models

Many options are written on multiple equity assets. For example, an exchange option is the right to exchange one stock for another. A min/max option is the right to choose the minimum or maximum among multiple stocks. These options are then

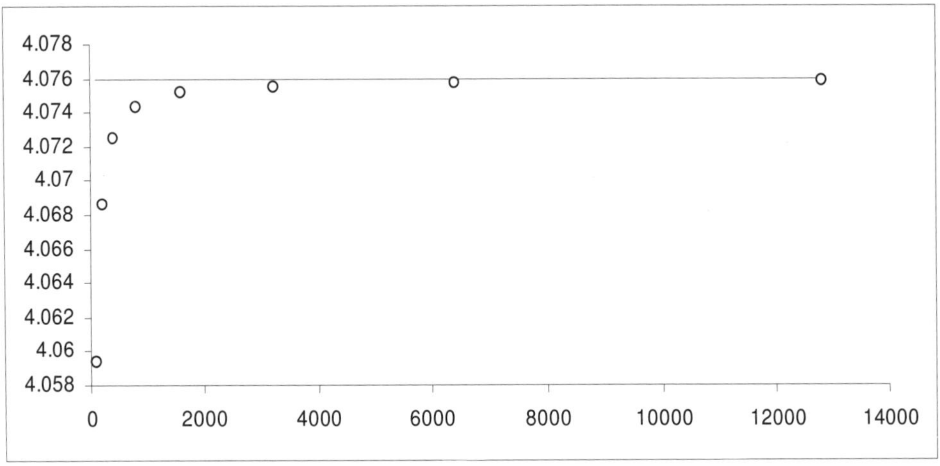

Figure 9.3: Convergence of the Implicit Method

exposed to the correlation risk. That is, the option value is highly dependent on the correlation among the underlying stocks.

Some options can be solved for with closed form solutions but many cannot. And we need numerical solutions. In this chapter, we discuss the popular explicit finite difference method (lattice). There are three typical lattice models available.

In this section, we demonstrate three interesting ways to formulate such a model. The first model by Boyle (1988) and then modified by Kamrad and Ritchken (1991) is a five-branch model. The second model by Boyle, Evnine and Gibbs (1989) is a four-branch model. Finally the model by Chen, Chung, and Yang (2002) is a three-branch model. Through the three models, we can see how one can alter the lattice design to properly characterize the stochastic processes.

9.5.1 A Trivial Case of Independent Assets

To understand the following three models, we first illustrate a trivial case of two independent assets. Then it is straightforward to see how each of the following models handles correlation. The following diagram is a two period two asset case.

In the above diagram, X represents one stock price and Y represents the other. As we can see, in period 1, there are four possibilities:

- both stocks up upper right corner X_u, Y_u with probability p_1

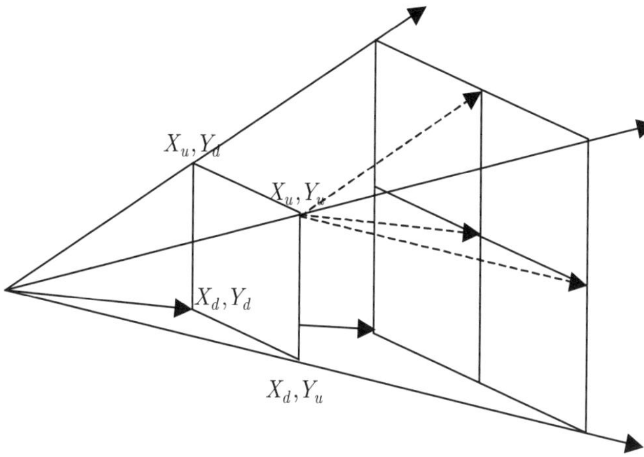

Figure 9.4: Two-dimensional Lattice

- X stock up but Y stock down upper left corner X_u, Y_d with probability p_2
- X stock down but Y stock up lower right corner X_d, Y_u with probability p_3
- both stocks down lower left corner X_d, Y_d with probability p_4

In the uni variate binomial model we saw earlier, the probabilities of up and down are computed as $p_j = \left(e^{r\Delta t} - e^{-\sigma_j\sqrt{\Delta t}}\right) / \left(e^{\sigma_j\sqrt{\Delta t}} - e^{-\sigma_j\sqrt{\Delta t}}\right)$ where $j = X, Y$. Hence it is trivial that:

$$\begin{aligned} p_1 &= p_X p_Y \\ p_2 &= p_X(1 - p_Y) \\ p_3 &= (1 - p_X)p_Y \\ p_4 &= (1 - p_X)(1 - p_Y) \end{aligned} \tag{9.44}$$

It is guaranteed that the sum of all four probabilities is equal to 1.

The challenge here is how we can correlate the two stocks. The following three models deal with the issue differently. Boyle (1988) and Kamrad and Ritchken (1991) add a middle branch for this extra condition. Boyle, Evnine, and Gibbs reweigh the probabilities. Finally, Chen, Chung and Yang use the complete market condition.

9.5.2 Boyle (1988, JFQA) and Kamrad-Ritchken (1991, Mgmt.Sci.)

This is a 5-branch model, as depicted in the following picture.

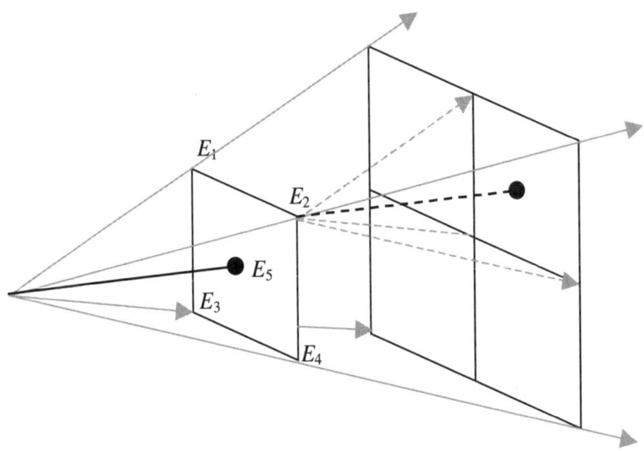

Figure 9.5: Two-dimensional Lattice

Each branch is associated with a probability p_j and $\Sigma p_j = 1$. The corner branches have the same stock prices as before and the middle branch assumes the same stock prices as the current. By matching moments, there are six equations and five unknowns:

$$\begin{aligned}
&(p_1+p_2)X_u + p_5X_0 + (p_3+p_4)X_d = X_0e^{r\Delta t}\\
&(p_1+p_4)Y_u + p_5Y_0 + (p_2+p_3)Y_d = Y_0e^{r\Delta t}\\
&(p_1+p_2)(X_u^2 - X_0^2e^{2r\Delta t}) + p_5(X_0^2 - X_0^2e^{2r\Delta t}) + (p_3+p_4)(X_d^2 - X_0^2e^{2r\Delta t})\\
&= X_0^2e^{2r\Delta t}\left(e^{\sigma_X^2} - 1\right)\\
&(p_1+p_4)(Y_u^2 - Y_0^2e^{2r\Delta t}) + p_5(Y_0^2 - Y_0^2e^{2r\Delta t}) + (p_2+p_3)(Y_d^2 - Y_0^2e^{2r\Delta t})\\
&= Y_0^2e^{2r\Delta t}\left(e^{\sigma_Y^2} - 1\right)\\
&p_1X_uY_u + p_2X_uY_d + p_3X_dY_d + p_4X_dY_u + p_5 = X_0Y_0e^{(2r+\rho\sigma_X\sigma_Y)\Delta t}\\
&\sum_{j=1}^{5} p_j = 1
\end{aligned} \tag{9.45}$$

Hence, the solution is not so straightforward. Boyle shows that the usual binomial setup, that is, $X_u = X_0u_X = X_0e^{\sigma_X\sqrt{\Delta t}}$ and $X_d = X_0d_X = X_0e^{-\sigma_X\sqrt{\Delta t}}$

and similarly for Y does not work probably. He must alter the assumption to $u_X = e^{\lambda\sigma_X\sqrt{\Delta t}}$ and $u_Y = e^{\lambda\sigma_Y\sqrt{\Delta t}}$ where λ is the same. Then he solve for one stock first and then search for the second stock.

Given that the solution is not simple, interested readers should refer to Boyles original work for details. The next two models have simple solutions.

9.5.3 Boyle-Evnine-Gibbs (1989, Rev. of Fin. Studies)

As we can see that if we use four branches, we can only have four equations, in which case, we will not be able to match unknowns and equations. Hence, Boyle, Evnine and Gibbs turn to characteristic functions.

The characteristic function of a bi variate normal distribution is:

$$\begin{aligned}
\Phi(u_X, u_Y) &= \mathbb{E}\left[e^{iu_X \ln X_1/X_0 + iu_Y \ln Y_1/Y_0}\right] \\
&= p_1 e^{i\sqrt{\Delta t}(u_X\sigma_X + u_Y\sigma_Y)} + p_2 e^{i\sqrt{\Delta t}(u_X\sigma_X - u_Y\sigma_Y)} \\
&\quad + p_3 e^{i\sqrt{\Delta t}(-u_X\sigma_X + u_Y\sigma_Y)} + p_4 e^{i\sqrt{\Delta t}(-u_X\sigma_X - u_Y\sigma_Y)} \\
&= p_1\left(1 + i\sqrt{\Delta t}(u_X\sigma_X + u_Y\sigma_Y) - \frac{\Delta t}{2}(u_X\sigma_X + u_Y\sigma_Y)^2\right) \\
&\quad + p_2\left(1 + i\sqrt{\Delta t}(u_X\sigma_X - u_Y\sigma_Y) - \frac{\Delta t}{2}(u_X\sigma_X - u_Y\sigma_Y)^2\right) \\
&\quad + p_3\left(1 + i\sqrt{\Delta t}(-u_X\sigma_X + u_Y\sigma_Y) - \frac{\Delta t}{2}(-u_X\sigma_X + u_Y\sigma_Y)^2\right) \\
&\quad + p_4\left(1 + i\sqrt{\Delta t}(-u_X\sigma_X - u_Y\sigma_Y) - \frac{\Delta t}{2}(-u_X\sigma_X - u_Y\sigma_Y)^2\right)
\end{aligned} \tag{9.46}$$

The exact solution of the characteristic function is:

$$\begin{aligned}
&\Phi(u_X, u_Y) = \\
&1 + i\sqrt{\Delta t}(u_X\sigma_X + u_Y\sigma_Y) - \frac{\Delta t}{2}(u_X^2\sigma_x^2 + 2u_Xu_Y\rho\sigma_X\sigma_Y + u_X^2\sigma_Y^2) + o(\Delta t)
\end{aligned} \tag{9.47}$$

Matching the two equations, we can solve for the probabilities exactly as follows:

$$\begin{aligned} p_1 &= \frac{1}{4}\left(1+\rho+\sqrt{\Delta t}\left(\frac{\mu_X}{\sigma_X}+\frac{\mu_Y}{\sigma_Y}\right)\right) \\ p_2 &= \frac{1}{4}\left(1-\rho+\sqrt{\Delta t}\left(\frac{\mu_X}{\sigma_X}-\frac{\mu_Y}{\sigma_Y}\right)\right) \\ p_3 &= \frac{1}{4}\left(1-\rho+\sqrt{\Delta t}\left(-\frac{\mu_X}{\sigma_X}+\frac{\mu_Y}{\sigma_Y}\right)\right) \\ p_4 &= \frac{1}{4}\left(1+\rho+\sqrt{\Delta t}\left(-\frac{\mu_X}{\sigma_X}-\frac{\mu_Y}{\sigma_Y}\right)\right) \end{aligned} \tag{9.48}$$

Boyle, Evnine and Gibbs note that the above probabilities can all be nonnegative only if the time step Δt is sufficiently small. Hence, this method is not very efficient. Also, they note that as $\Delta t \to 0$, $p_1 = p_4 = {}^1\!/_4(1+\rho)$ and $p_2 = p_3 = {}^1\!/_4(1-\rho)$.

9.5.4 Chen-Chung-Yang (2002, JFQA)

Before formally introducing the model, we shall first present a two-asset example where the two-dimensional "binomial tree" can be visualized. In the two-asset model, the dynamics for the two assets are:

$$\begin{cases} d\ln S_1 = \left(r - \frac{\sigma_1^2}{2}\right) dt + \sigma_1 dW_1 \\ d\ln S_2 = \left(r - \frac{\sigma_2^2}{2}\right) dt + \sigma_2 dW_2 \end{cases} \tag{9.49}$$

where r is the risk-free rate, σ_1 and σ_2 represent the volatility, and W_1 and W_2 are two correlated Wiener processes where $dW_1 dW_2 = \rho dt$.

To maintain a complete market, the tree structure of the above 2-factor model can be graphically depicted as follows:

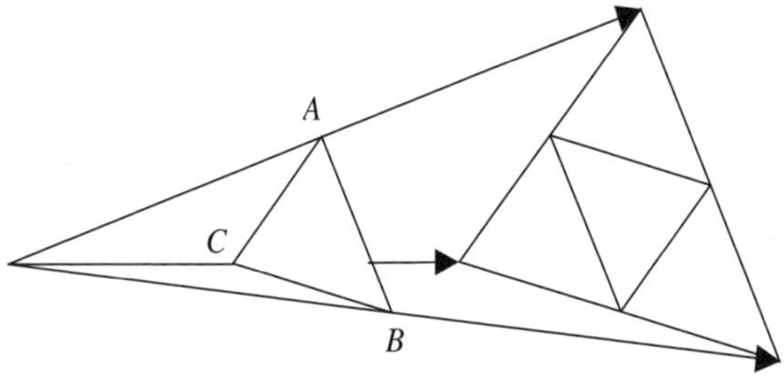

Figure 9.6: Two-dimensional Lattice

Like He (1990), the complete market assumption allows for only 3 states, instead of 4 or 5 states in previous studies. Rather than solving the simultaneous equations, we establish our tree structure by setting up Figure 9.6 for the correlated Brownians, dW_1 and dW_2. Then by adding the drift, we can complete the 2-factor "binomial model" for the stock prices.

With only two correlated Brownian motions, it is possible to use the properties of the ellipse. We shall show that a sufficient condition for the three points, A, B, and C in Figure 9.6 to satisfy equation (9.49) is that they locate evenly on the circumference of an ellipse.

To demonstrate, we first start with two uncorrelated Brownian motions with equal variances. To do that, the three points, A, B, and C, in Figure 9.6 need to be "equally", i.e. 120°, apart from each other, located on the circumference of a circle, as shown in Figure 9.7. It is not important, however, where the three points are. For example, A, B and C are another set of choices that also represent two uncorrelated Brownian motions. Property 1 describes a general property of any ellipse in which a circle is a special case.

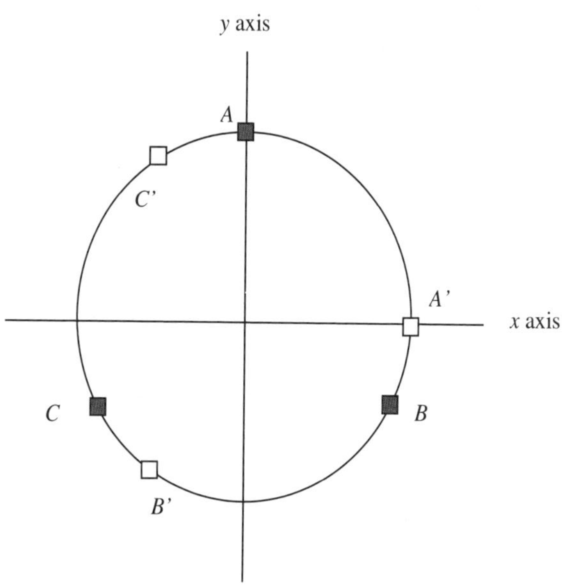

Figure 9.7: Two-dimensional Lattice

Property 1. The x and y coordinates of equally spread n points where $n \geqslant 3$ along the circumference of a "straight-up" ellipse defined by:

$$\frac{x^2}{r_x^2} + \frac{y^2}{r_y^2} = 1 \tag{9.50}$$

which have zero means, variances that are square-proportional to the radiuses, i.e. $\sigma_j^2 = {}^1\!/_2 r_j^2$ where $j = x$ or y, and, zero correlation regardless of where they are located.

Proof. See the appendix A.

Property 1 can be generalized to high dimensional cases as follows. The following Corollary is a straightforward extension of Property 1 and stated without proof.

Corollary 1. The evenly spread 4 points on the surface of an upright football have variances that equal $\frac{1}{3}r_i^2$ where $i = 1, 2, 3$ and correlations equal to 0. This result can be generalized to a k-dimensional hypercube where the variances equal $\frac{1}{k}r_i^2$ and $i = 1, \cdots, k$.

Property 1 is helpful in setting up the 3 ending points in Figure 9.7. Following Property 1, we choose $n = 3$ and $r_x = r_y = \sqrt{2dt}$. The choice of equal radiuses makes the ellipse of the above property a circle. While it is equivalent to work with unequal radiuses, it is easier to visualize how correlation can be incorporated in a circle. Another choice of the three points along the circle is to fulfill the assumption of complete markets. The choice of 4 points is the case of Boyle et. al. (1989). And the choice of 5 points is the case of Boyle (1988).

In order to allow for more intuitive understanding, we rewrite (9.50) under the polar coordinate as follows:

$$\cos^2\theta + \sin^2\theta = 2dt \tag{9.51}$$

As shown in Figure 9.7, Property 1 holds for θ to be $90°, 210°$, and $330°$ (solid points) or $0\text{o}, 120°, 240°$ (hollow points). To incorporate the correlation between the two Brownian motions, we then rotate the axes, as shown in Figure (9.8). Note that in Figure (9.8) * represents points taken from Cholesky transformation and ** represents points taken from an alternative transformation.

In Figure 9.8, the x axis rotates toward y (counter clockwise) and the y axis rotates toward x (clockwise) by ϕ degrees. Any three points (say A*, B*, and C* or A**, B**, and C**) on the new ellipse, as long as they are equally spread according to the new axes, have the identical variances as the unrotated upright ellipse and the required correlation. In other words, the rotation of axes by a certain degree uniquely defines a correlation while leaving individual variances unchanged. In an extreme case where $\phi = 45°$, the two axes collide and the correlation is 100%. In

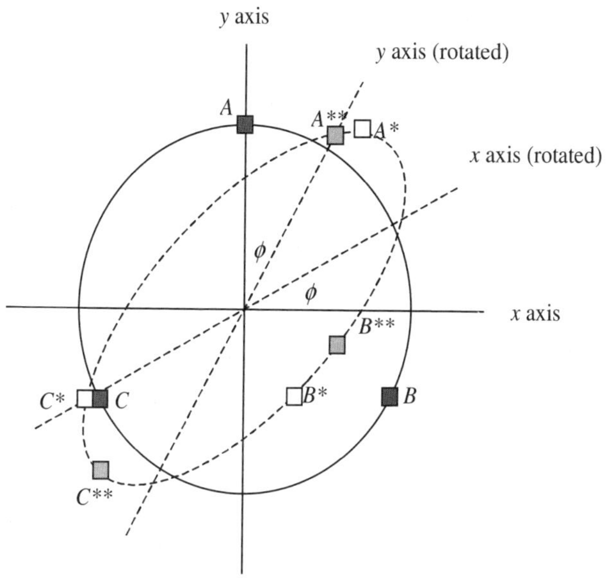

Figure 9.8: Rotated Points

the other extreme, $\phi = -45°$ gives the correlation of 100%.

The result that the correlation between any two assets is achieved by purely the rotation angle allows us to establish a higher dimension model easily. We no longer suffer from the problem that the number of correlations mismatches the number of assets. For k assets, now we always need to define C_2^k angles to uniquely match the same number of correlations.

It is important to note that the variances of the two Brownians, dt, remained unchanged after rotation. In other words, the rotation of the axes only incorporates in the correlation does not change the variances of the Brownians; but there are infinite number of matrices that can transform evenly spread points on the original "upright" ellipse on to the rotated ellipse. One of these solutions is the well-known Cholesky transformation. Following the Cholesky transformation in 2 dimensions, we write the correlated Brownian motions in as follows:

$$\begin{bmatrix} dW_1 \\ dW_2 \end{bmatrix} = \begin{bmatrix} \sqrt{1-\rho^2} & \rho \\ 0 & 1 \end{bmatrix} \begin{bmatrix} dZ_1 \\ dZ_2 \end{bmatrix} \tag{9.52}$$

where $dZ_1 dZ_2 = 0$. It is shown in the Appendix that the equation for the ellipse is:

$$\hat{x}^2 + \hat{y}^2 - 4\hat{x}\hat{y}\cos\phi\sin\phi = 2dt \tag{9.53}$$

where

$$\rho = {}^1\!/\!_2{}^1\!/\!_2 \sin\phi$$

is the correlation between x and y. It can be verified that points taken from A*, B*, and C* in Figure 9.8 satisfy the above equation. In addition to the Cholesky method, there are infinitely number of alternative solutions. For example, we can define the following transformation matrix:

$$\begin{bmatrix} dW_1 \\ dW_2 \end{bmatrix} = \begin{bmatrix} \cos[\sin^{-1}\frac{\rho}{2}] & \frac{\rho}{2} \\ \frac{\rho}{2} & \cos[\sin^{-1}\frac{\rho}{2}] \end{bmatrix} \begin{bmatrix} dZ_1 \\ dZ_2 \end{bmatrix} \tag{9.54}$$

which will serve the same purpose. These points are labeled as (A**, B**, and C** in Figure 9.8). The Cholesky transformation fixes one set of coordinates while the alternative transformation keeps the same polar coordinates of A, B, and C as the coordinates before the rotation. To demonstrate numerically, we first give the coordinates of A, B, and C on an upright ellipse as follows:

	Coordinates		
	Point 1	Point 2	Point 3
x	1	0.8660	-0.8660
y	0	-0.5	-0.5

These are the points that taken from orthogonal Brownians: dZ_1 and dZ_2. To achieve $\rho = 0.5$, we obtain A*, B*, and C* by the Cholesky method as:

$$\begin{bmatrix} x^* \\ y^* \end{bmatrix} = \begin{bmatrix} \sqrt{1-0.25} & 0.5 \\ 0 & 1 \end{bmatrix} \begin{bmatrix} 0 & 0.8660 & -0.8660 \\ 1 & -0.5 & -0.5 \end{bmatrix}$$
$$= \begin{bmatrix} 0.5 & 0.5 & -1 \\ 1 & -0.5 & -0.5 \end{bmatrix}$$

By the alternative method, we get A**, B**, and C** instead.

$$\begin{bmatrix} x^{**} \\ y^{**} \end{bmatrix} = \begin{bmatrix} 0.9682 & 0.25 \\ 0.25 & 0.9682 \end{bmatrix} \begin{bmatrix} 0 & 0.8660 & -0.8660 \\ 1 & -0.5 & -0.5 \end{bmatrix}$$
$$= \begin{bmatrix} 0.25 & 0.7135 & -0.9635 \\ 0.9682 & -0.2676 & -0.7006 \end{bmatrix}$$

One can verify that A*, B*, C*, A**, B**, C** all fall on the same ellipse. Finally, applying this property, the stock prices can be identified by scaling the axes with corresponding stock variances and then adding a drift to the rotated points. The diffusion equation is:

$$\begin{bmatrix} d\ln S_1 \\ d\ln S_2 \end{bmatrix} = \begin{bmatrix} r - \frac{1}{2}\sigma_1^2 \\ r - \frac{1}{2}\sigma_2^2 \end{bmatrix} dt + \begin{bmatrix} \sigma_1 & 0 \\ 0 & \sigma_2 \end{bmatrix} \begin{bmatrix} dW_1 \\ dW_2 \end{bmatrix} \tag{9.55}$$

and hence the positions of the stock prices are

$$\begin{bmatrix} \ln S_{11}(t+dt) & \ln S_{12}(t+dt) & \ln S_{13}(t+dt) \\ \ln S_{21}(t+dt) & \ln S_{22}(t+dt) & \ln S_{23}(t+dt) \end{bmatrix} = \begin{bmatrix} \ln S_{11}(t) & \ln S_{12}(t) & \ln S_{13}(t) \\ \ln S_{21}(t) & \ln S_{22}(t) & \ln S_{23}(t) \end{bmatrix}$$
$$+ \begin{bmatrix} r - \frac{1}{2}\sigma_1^2 \\ r - \frac{1}{2}\sigma_2^2 \end{bmatrix} dt + \begin{bmatrix} \sigma_1 & 0 \\ 0 & \sigma_2 \end{bmatrix} \begin{bmatrix} \sqrt{1-\rho^2} & \rho \\ 0 & 1 \end{bmatrix} \begin{bmatrix} x_A & x_B & x_C \\ y_A & y_B & y_C \end{bmatrix} \tag{9.56}$$

where $< x_j, y_j >$ is the coordinate of the three points: j = A, B, C.

To sum up, in this section, by using a simple 2-factor model, we demonstrate the following:

- The rotated unity ellipse defines correlation. In particular, the angle of rotation is defined by the correlation between two assets. To extend this to k assets, there will be C_2^k.

- rotation angles that are defined exactly by the same number of correlations.

- Our methodology allows for separation between the correlation and the volatility. In other words, rotating the axes gives the capability of handling the correlation and scaling the axes gives the capability of handling the volatility.

- Cholesky transformation is one solution to the coordinates of the ellipse. It provides a solution to multiple dimensions.

To extend the model to three and higher dimensions, readers are asked to refer to the original paper.

9.6 Appendix

9.6.1 VBA Code for inverting a tri-diagonal matrix

```
'find u(j) using numerical recipe tridag
bet = b(1)
u(1) = r(1) / bet
For j = 2 To n + 1
gam(j) = c(j - 1) / bet
bet = b(j) - a(j) * gam(j)
If (bet = 0) Then res = 9999
u(j) = (r(j) - a(j) * u(j - 1)) / bet
Next
For j = n + 1 - 1 To 1 Step -1
u(j) = u(j) - gam(j + 1) * u(j + 1)
Next
```

9.6.2 VBA Code for inverse normal

```
Function xNormSInv(p)
Const a1 = -39.6968302866538, a2 = 220.946098424521, a3 = -275.928510446969
Const a4 = 138.357751867269, a5 = -30.6647980661472, a6 = 2.50662827745924
Const b1 = -54.4760987982241, b2 = 161.585836858041, b3 = -155.698979859887
Const b4 = 66.8013118877197, b5 = -13.2806815528857, c1 = -7.78489400243029E-03
Const c2 = -0.322396458041136, c3 = -2.40075827716184, c4 = -2.54973253934373
Const c5 = 4.37466414146497, c6 = 2.93816398269878, d1 = 7.78469570904146E-03
Const d2 = 0.32246712907004, d3 = 2.445134137143, d4 = 3.75440866190742
Const p_low = 0.02425, p_high = 1 - p_low
Dim q As Double, r As Double
If p < 0 Or p > 1 Then
Err.Raise vbObjectError, , "NormSInv:  Argument out of range."
ElseIf p < p_low Then
q = Sqr(-2 * Log(p))
xNormSInv = (((((c1 * q + c2) * q + c3) * q + c4) * q + c5) * q + c6) / _
((((d1 * q + d2) * q + d3) * q + d4) * q + 1)
ElseIf p <= p_high Then
q = p - 0.5
r = q * q
xNormSInv = (((((a1 * r + a2) * r + a3) * r + a4) * r + a5) * r + a6) * q / _
(((((b1 * r + b2) * r + b3) * r + b4) * r + b5) * r + 1)
Else
q = Sqr(-2 * Log(1 - p))
xNormSInv = -(((((c1 * q + c2) * q + c3) * q + c4) * q + c5) * q + c6) / _
((((d1 * q + d2) * q + d3) * q + d4) * q + 1)
End If
End Function
```

Chapter 10

Interest Rate options

There are numerous interest rate options in the marketplace. Since this is not a fixed income text book, I only introduce three popular interest rate options. Interestingly, each of these options represents a unique pricing methodology.

10.1 Treasury bond futures embedded options

The delivery options in Treasury bond futures are generally known as the quality option and the three timing options. The quality option gives the short the right to deliver any eligible bond (no less than 15 years to maturity or first call) and the various timing options give the short the flexibility of making the delivery decision any time in the delivery month. The end-of-month timing option refers to the deliveries occurring at the last 7 business days in the delivery month when the futures market is closed to trading. For the remaining about 15 business days of the delivery month, the wild card timing option refers to the period from 2:00 p.m. to 7:00 p.m. (Chicago time) every day when the futures market is closed but the bond market is open while the accrued interest timing option refers to the period from 7:20 a.m. to 2:00 p.m. when both futures and its underlying bond markets are open.

Delivery options in Treasury bond futures are difficult to price. A recursive algorithm based upon a lattice model must be used for valuing such options accurately, as Boyle (1989) demonstrates that the futures price is effectively a forward price. To this date, the literature has been approximating the forward price valuation with a futures price valuation due to the high computational cost. Aided with todays computing power, we can now compute accurately the Treasury bond futures price with the proper recursive algorithm. In this paper, we estimate the value of the quality option as well as the value of the end-of-month timing option. Using a two-factor Cox-Ingersoll-Ross model calibrated to the cheapest-to-deliver bond and weekly futures prices data from January 1992 till December 2000, we discover a substantial difference between the risk neutral expectation used in the literature and the forward expectation that requires a recursive algorithm. We find that the correctly estimated futures price with the quality option is 1% lower on average than the futures price estimated in the literature. We also estimate the futures price with both the quality option and the end-of-month timing option to be 23 basis points on average. This indicates that the end-of-month timing option value has been over-estimated in the literature because of a wrongly estimated quality option value.

The delivery option that has the most economic value is the quality option that gives the short of the futures contract the right to choose the cheapest bond to deliver at the delivery date. Other delivery options that are embedded in Treasury bond futures are known as the three timing options. The short can choose any time in the delivery month to make a delivery. The short can make a delivery even when the futures market is closed. At the end of the delivery month, for 7 business days, the futures market is closed but the short can still make a delivery. This is understood as the end-of-month timing option. For the remaining about 15 business days in the delivery month, the short can deliver either between 7:20 a.m. and 2:00 p.m. (Chicago time) when both the futures market and the underlying bond market are open or after 2:00 p.m. when the futures market is closed. The former timing option is called accrued interest timing option and the latter timing option is also known as the daily wild card play. The following picture explains graphically various timing options.

The period of the last 7 business days is the end-of-month period. Throughout the paper we use v for the starting time and T for the ending time of this period. For the rest of the delivery month, there are two sections of each day, the accrued interest period and the wild card period. For a regular futures trading day i between 7:20 a.m. and 2 p.m. Chicago time, both bond and futures markets are open simultaneously. The futures market closes at 2 p.m. but there is no official closing time for the bond market. Since the short has till 8 p.m. to make the delivery decision, the wild card period is defined over 3 p.m. (u_i) to 8 p.m. ($u_i + h$) Eastern time.

For notation and symbols used here, please refer back to Chapter 1. In addition,

$a_i(t) =$ accrued interest of the ith bond
$q_i =$ conversion factor of the ith bond
$Q_i(t) =$ quoted coupon bond price of the ith bond
$\Phi(t) =$ quoted futures price with all delivery options

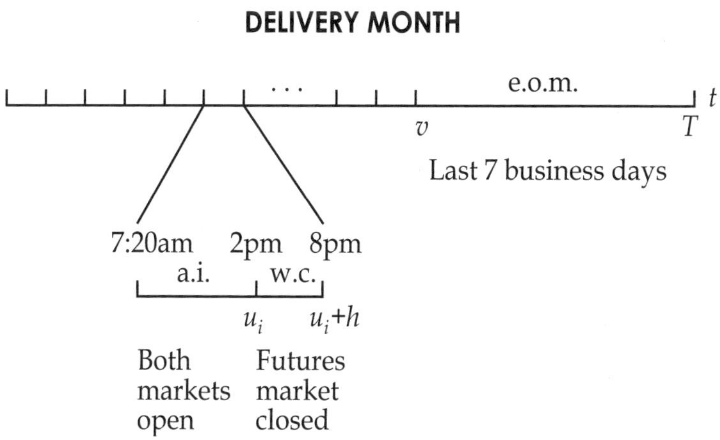

Figure 10.1: Delivery Time Line

$\Phi^*(t)$ = futures price with the quality option and continuous marking to market

$\Phi^{**}(t)$ = futures price with the quality option at the absence of continuous MTM

Note that the transaction price of a coupon bond is the quoted price plus the accrued interest. In other words, for the ith coupon bond in the delivery set for the futures contract, $Q_i(t) = B(t, \underline{T}; c_i) - a_i(t)$ where $B(t, \underline{T}; c_i)$ is the coupon bond notation defined in Chapter 1 that has the coupon and maturity of the i-th bond. Also note that $\Phi(t)$ is the quoted futures price of the cheapest-to-deliver bond. The actual futures price needs to be adjusted by the conversion factor and the accrued interest of the cheapest-to-deliver bond.

10.1.1 The Quality Option with Continuous Marking to Market

In the absence of all timing options, the quality option gives the short the right to deliver the cheapest bond only at maturity, T, and the short receives the following payoff:

$$\begin{aligned}&\max\left\{q_i\Phi(T)+a_i(T)-[Q_i(T)+a_i(T)]\right\}\\&=\max\left\{q_i\Phi(T)-Q_i(T)\right\}\end{aligned} \tag{10.1}$$

Note that the accrued interests of both bond and futures contracts are equal and canceled. Since the delivery value of (10.1) has to be identically 0 for all states, we can solve for the futures price at maturity as:

$$\Phi(T)=\min\left\{\frac{Q_i(T)}{q_i}\right\} \tag{10.2}$$

and todays futures price is merely a risk-neutral expectation of this payoff:

$$\begin{aligned}\Phi^*(t)&=\hat{\mathbb{E}}_t\left[\min\left\{\frac{Q_i(T)}{q_i}\right\}\right]\\&=\frac{\hat{\mathbb{E}}_t[Q_1(T)]}{q_1}-\hat{\mathbb{E}}_t\left[\max\left\{\frac{Q_1(T)}{q_1}-\frac{Q_i(T)}{q_i}\right\}\right]\\&=\frac{\Phi_1(t)}{q_1}-\hat{\mathbb{E}}_t\left[\max\left\{\frac{Q_1(T)}{q_1}-\frac{Q_i(T)}{q_i}\right\}\right]\end{aligned} \tag{10.3}$$

Note $\Phi_1(t)=\hat{\mathbb{E}}_t[Q_1(T)]$ is the futures price of the first bond with no option and $\Phi^*(t)$ is the futures price of the cheapest bond at maturity. This result has been shown previously by Carr (1988) and other authors. This equation says that the futures contract with the quality option is equivalent to a futures contract without the quality option (only bond 1 is eligible for delivery) with an exchange option held by the short. With a specific term structure model, equation (10.3) could have an analytical solution. [1]

10.1.2 The Quality Option with no Marking to Market When the Futures Market Is Closed

Equation (10.3) is correct only if marking to market is applied continuously throughout the life of the futures contract. Unfortunately, in the last 7 business days of the delivery month, the futures market is not open and the futures contract is not marked to market. The futures price used for settlement in this period is the last settlement price at the beginning of the 7-day period. Since the futures price is already determined, the actual payoff at the last delivery day, T, is not necessarily 0. The short can actually gain or lose. To avoid arbitrage, the futures price at the

[1] See Carr (1988).

beginning of the 7-day period should be set so that the expected present value of payoffs at maturity is 0. Under this circumstance, the futures price at the beginning of the 7-day period is a forward price, not a futures price. Formally, label the futures price as $\Phi^{**}(v)$ to represent the futures price at the beginning of the end-of-month period, v, should be so set that:

$$\hat{\mathbb{E}}_v\left[\exp\left(-\int_v^T r(u)du\right)\max\{\Phi^{**}(v)q_i - Q_i(T)\}\right] = 0 \tag{10.4}$$

where $r(u)$ is the instantaneous rate. Using the forward measure, we can then rewrite the above equation as:

$$\tilde{\mathbb{E}}_v^{(T)}[\max\{\Phi^{**}(v)q_i - Q_i(T)\}] = 0 \tag{10.5}$$

which can be expanded as follows:

$$\begin{aligned}
0 &= \tilde{\mathbb{E}}_v^{(T)}[\max\{\Phi^{**}(v)q_i - Q_i(T)\}] \\
0 &= \tilde{\mathbb{E}}_v^{(T)}[\Phi^{**}(v)q_1 - Q_1(T) + \max\{\Phi^{**}(v)(q_i - q_1) - (Q_i(T) - Q_1(T)), 0\}] \\
0 &= \Phi^{**}(v)q_1 - \Psi_1(v) + \tilde{\mathbb{E}}_v^{(T)}[\max\{Q_1(T) - Q_i(T) - \Phi^{**}(v)(q_1 - q_i), 0\}]
\end{aligned} \tag{10.6}$$

and the futures price at time v can be written as:

$$\Phi^{**}(v) = \frac{\Psi_1(v)}{q_1} - \frac{1}{q_1}\tilde{\mathbb{E}}_v^{(T)}\left[\max\left\{Q_1(T) - Q_i(T) - K_i^{**}\right\}\right] \tag{10.7}$$

where $K_i^{**} = (q_1 - q_i)\Phi^{**}(v)$. Note that $\Psi_1(v) = \tilde{E}_v^{(T)}[Q_1(T)]$ is the forward price of the first bond. The interpretation of this result is similar to that of (10.3), except that the risk neutral measure is replaced by the forward measure and the futures price becomes the forward price. However, unlike (10.3), the futures price at time v has no easy solution, because it appears on both sides of the equation. This futures price has to be solved recursively using a numerical method. In a lattice framework suggested by Boyle (1989), we first choose an initial value for the futures price at time v, calculate payoffs at various states at maturity T, and then work backwards along the lattice. We adjust the futures price until the discounted payoff computed from the lattice is 0. Once the futures price at time v is set, we can then travel back along the lattice and use the risk neutral probabilities till the end of the last wild card period, $u_n + h$. Then the similar procedure for the end-of-month period is repeated for the last wild card period to arrive at the futures price at the beginning of the wild card period u_n. Again, the risk neutral expectation is taken at $u_{n-1} + h$ and a recursive search is to compute the futures price at u_{n-1}. The process is repeated until the delivery month is over. Since the futures price becomes a forward price which cannot be obtained without a recursive search. The search for the forward price takes place at every node at all the times (i.e., u_1, u_2, $\cdots$, u_n, v). As a result, to compute the futures price with the quality option is prohibitively expensive.

10.1.3 The Timing Options

Besides the quality option, there are three timing options embedded in the Treasury bond futures price. The most valuable one is the end-of-month (EOM) timing option. Without the EOM timing option, we know that the futures price should be set according to (10.3). With the EOM timing option, deliveries can occur any time in the end-of-month period as long as the current delivery payoff is more than the present value of the expected payoff. This is similar to the early exercise decision of an American option. There is no closed-form solution to compute American option prices. Precisely as Boyle (1989) has observed, the pricing of quality and timing options would need a lattice model.

As in an American option, early delivery (i.e. early exercise) is activated if the delivery payoff is larger than the continuation value (expected value of future payoffs). However, this delivery decision is intertwined with the recursive process in computing the quality option value. That is, every time we start with a trial value for the futures price at the beginning of the EOM period. This futures price will not change throughout the EOM period since the futures market is closed. We then work backwards from the end of the EOM period with an early delivery decision checked at every node until we reach the beginning of the EOM period. If the expected payoff computed via this backward induction at the beginning of the EOM period is not 0, then the trial futures price must be revised. The process continues until the payoff at the beginning of the EOM period is 0. The computation cost of such a recursive algorithm in a two-dimensional lattice is high.

In addition to the EOM timing option that refers the last 7 trading days of the delivery month, there are two other timing options in the rest 15 days of the delivery month the accrued interest timing option and the wild card timing option. The accrued interest timing option refers to the flexibility for the short to deliver the cheapest bond any time in the delivery month when both futures and spot markets are open. This is everyday from 7:20 a.m. to 2:00 p.m. (Chicago time) from the first day of the delivery month to right before the end-of-month period. Since the futures market is open, the futures contract is marked to market and deliveries can take place any time. As a result, the futures price can never be greater than the cheapest-to-deliver bond price. If the futures price were greater than the cheapest bond price, then deliveries would take place instantly. The short will sell the futures, buy the cheapest bond, make the delivery, and earn an arbitrage profit. Formally, for $t < v$,

$$\Phi(t) > \min\left\{\frac{Q_i(t)}{q_i}\right\} \Leftrightarrow \max\left\{\Phi(t)q_i - Q_i(t)\right\} > 0 \tag{10.8}$$

Therefore, the futures price in the period where both markets are open must be less than the cheapest-to-deliver bond price to avoid arbitrage. On the other hand, if the futures price is lower, one can long futures and short spot but the delivery will not occur because the short position of the futures contract will lose money if he makes a delivery. Consequently, the delivery will be postponed and there is no arbitrage profit to be made. If the futures price is always less than the cheapest-to-deliver bond price (adjusted by its conversion factor), the delivery payoff now is negative as opposed to 0 at the end. As a result, the short will never deliver until the last day. Consequently, the accrued interest timing option has no value. We restate this result in the following proposition.

Proposition.

The accrued interest timing option without the wild card and end-of-month options has no value. The existence of the other timing options will lower the current futures price, further reducing the incentive for the short to deliver early. We state this result in the following Corollary.

□

Corollary.

The accrued interest timing option with the end-of-month options has no value.

□

While the accrued interest timing option is worthless, the wild card timing option is not. When the futures market is closed, there is no marking to market in the futures market and the futures contract becomes a forward contract. Boyle (1989) has demonstrated that in a case of forward contracts timing options will have value. We shall extend Boyles analysis to the wild card option in Treasury bond futures. Similar to the end-of-month option, the wild card option refers to the flexibility in delivery in a 5-hour period every day for about 15 days where the futures market is closed but the bond market is open. However, the wild card option is different from the end-of-month option in that the futures market will reopen after each wild card period and the futures contract will be marked to market. If bond prices drop in the wild card period, given that the futures price is fixed, the short can benefit from delivering the cheapest bond. However, the short can equally benefit from the marking to market when the futures market reopens on the next day. As a result, the incentive for the short to deliver in the wild card period is minimal. Delivery can take place in a wild card period only when the payoff from immediate delivery exceeds the expected present value of marking to market on the next day.

The modeling of the wild card period requires a very fine grid. To model the wild card option properly, we need at least two steps in each wild card period to allow for early exercise. As a result, it requires a minimum of four steps per day. Given that practically the wild card option has very little value due to the next-day marking-to-market, we will not evaluate the wild card option in this paper.

10.2 Caps and Floors

Hull explains very well how a cap (or floor/collar) works. A cap is a collection of European rate call options (or called caplets); it is quite easy to price. The difficulty is that the floating rate is not a traded asset but an index so the usual Black-Scholes does not apply (note that many financial firms use the Black-Scholes model anyway ignoring the fact that interest rates are not traded assets.) Luckily, with a simple transformation, we can turn the rate call to a bond put which is possible to use a known term structure model (such as CIR or Vasicek). Similarly, floor is a collection of rate puts that should be turned into a series of bond calls.

A cap is a limitation on the borrower that at each payment date he cannot charge the market rate if the market rate is higher than the contractual cap rate. Hence, we can write the cap payoff as a series of rate call to be received at the end of period j.

$$\max\{R_j - \bar{R}, 0\}a \tag{10.9}$$

for $j = 1, \cdots, n$ where a is the accrued period usually $a = 1/4$ in swaps. Given that R_j is not a traded asset, we must turn it into a bond price. To do that,

$$\begin{aligned}
&\max\{R_j - \bar{R}, 0\}a \\
&= \max\left\{\frac{1}{a}\left(\frac{1}{P(t_j, t_j + a)} - 1\right) - \bar{R}, 0\right\} a \\
&= \max\left\{\frac{1}{P(t_j, t_j + a)} - (1 + a\bar{R}), 0\right\} \\
&= \frac{(1 + a\bar{R})}{P(t_j, t_j + a)} \max\{K - P(t_j, t_j + a), 0\}
\end{aligned} \tag{10.10}$$

where $K = 1/(1 + a\bar{R})$. The present value of this payoff is therefore:

$$\begin{aligned}
V_j &= \hat{\mathbb{E}}_t \left[\exp\left(-\int_t^{t_j+a} r_u du \right) \max\{R_j - \bar{R}, 0\} a \right] \\
&= (1 + a\bar{R}) \hat{\mathbb{E}}_t \left[\exp\left(-\int_t^{t_j+a} r_u du \right) \frac{1}{P(t_j, t_j + a)} \max\{K - P(t_j, t_j + a), 0\} \right] \\
&= \frac{1}{K} \hat{\mathbb{E}}_t \left[\exp\left(-\int_t^{t_j} r_u du \right) \max\{K - P(t_j, t_j + a), 0\} \right]
\end{aligned} \tag{10.11}$$

and the cap value is:

$$V_{\text{cap}} = \sum_{j=1}^{n} V_j \tag{10.12}$$

Equation (10.11) is a standard option on the zero coupon bond and hence has closed-form solutions if certain term structure models are adopted. For example, in Chapter 7, both Vacicek and Cox-Ingersoll-Ross models have closed-form solutions to the zero coupon bond option. Using the Vasicek model, we can write the formula for a caplet as:

$$\begin{aligned}
V_j = P(0, t_j) N &\left(\frac{\ln P(0,t_j) - \ln K - \ln P(0,t_j+a) - \tfrac{1}{2} V_{P,j}}{\sqrt{V_{P,j}}} \right) \\
&- \frac{P(0, t_j + a)}{K} N \left(\frac{\ln P(0,t_j) - \ln K - \ln P(0,t_j+a) + \tfrac{1}{2} V_{P,j}}{\sqrt{V_{P,j}}} \right)
\end{aligned} \tag{10.13}$$

where

$$V_{P,j} = \frac{1 - e^{-2\alpha t_j}}{2\alpha} \sigma^2 \left(\frac{1 - e^{-\alpha a}}{\alpha} \right)^2$$

The floor can be dealt with the same way:

$$\begin{aligned}
&\max\{\bar{R} - R_j, 0\} aN \\
&= \max\left\{ \bar{R} - \frac{1}{a} \left(\frac{1}{P(t, t + a)} - 1 \right), 0 \right\} aN \\
&= \max\left\{ (a\bar{R} + 1) - \frac{1}{P(t, t + a)}, 0 \right\} N \\
&= \frac{(a\bar{R} + 1)N}{P(t, t + a)} \max\{P(t, t + a) - K, 0\}
\end{aligned} \tag{10.14}$$

and the pricing formula is

$$\begin{aligned}V_j &= \hat{\mathbb{E}}_t\left[\exp\left(-\int_t^{t_j+a} r_u du\right)\max\{\bar{R}-R_j,0\}aN\right]\\ &= (a\bar{R}+1)N\hat{\mathbb{E}}_t\left[\exp\left(-\int_t^{t_j+a} r_u du\right)\frac{1}{P(t_j,t_j+a)}\max\{P(t_j,t_j+a)-K,0\}\right]\\ &= \frac{N}{K}\hat{\mathbb{E}}_t\left[\exp\left(-\int_t^{t_j} r_u du\right)\max\{P(t_j,t_j+a)-K,0\}\right]\end{aligned} \tag{10.15}$$

and

$$V_{\text{floor}} = \sum_{j=1}^{n} V_j \tag{10.16}$$

The bond price can be obtained from any term structure model of choice.

Note that usually lenders do not charge an upfront free for the cap and vice versa for the floor. What is a common practice is that the charged interest rate contains a mark-up, or spread. That is, $R_j = \frac{1}{P(t_j,t_j+a)} + s$ where the spread is constant throughout the life of the loan. This being the case, a cap or a floor contract is to solve for the spread so that the NPV is 0 at inception. To accommodate such practice, we compute the value of the spread payments:

$$V_{\text{spread}} = asN\sum_{j=1}^{n} P(0,t_j) \tag{10.17}$$

where N is the notional of the contract. By equating the value of the spread payments with the value of the cap, we solve for the spread as follows:

$$s = \frac{\sum_{j=1}^{n} V_j}{aN\sum_{j=1}^{n} P(0,t_j)} \tag{10.18}$$

where V_j can be either caplet of floorlet.

10.3 Swaptions

10.3.1 Swap

A swap contract is an exchange of a series of fixed cash flows (swap rate times notional times accrual period) with a series of floating cash flows usually indexed to LIBOR (LIBOR times notional times accrual period). At inception, the value of the swap must equal 0. Hence,

$$\hat{\mathbb{E}}_t\left[\sum_{j=1}^{n} \exp\left(-\int_t^{t_j} r_u du\right)(L(t_j)-w_t)\right]=0 \tag{10.19}$$

Using the standard change of measure, we can write:

$$\sum_{j=1}^{n} \hat{\mathbb{E}}_t\left[\exp\left(-\int_t^{t_j} r_u du\right)\right]\tilde{E}_t^{(t_j)}\left[(L(t_j)-w_t)\right]=0 \tag{10.20}$$

which can be re-arranged as:

$$\begin{aligned}\sum_{j=1}^{n} P(t,t_j)\tilde{E}_t^{(t_j)}[L(t_{j-1},t_j)] &= \sum_{j=1}^{n} P(t,t_j)w_t\\ \sum_{j=1}^{n} P(t,t_j)f(t,t_{j-1},t_j) &= \sum_{j=1}^{n} P(t,t_j)w_t\end{aligned} \tag{10.21}$$

which gives rise to the following result:

$$\begin{aligned}w_t &= \frac{\sum_{j=1}^{n} P(t_j)f(t,t_{j-1},t_j)}{\sum_{j=1}^{n} P(t_j)}\\ &= \frac{1-P(t,t_n)}{\sum_{j=1}^{n} P(t_j)}\end{aligned} \tag{10.22}$$

The value of the swap at any point in time after time t is (say u) is:

$$\begin{aligned}V &= (w_u - w_t)\sum_{j=1}^{n} P(u,t_j)\\ &= \left[w_u\sum_{j=1}^{n} P(u,t_j)+P(u,t_n)\right]-\left[w_t\sum_{j=1}^{n} P(u,t_j)+P(u,t_n)\right]\\ &= \left[\sum_{j=1}^{n} P(u,t_j)f(u,t_{j-1},t_j)+P(u,t_n)\right]-\left[w_t\sum_{j=1}^{n} P(u,t_j)+P(u,t_n)\right]\\ &= \text{floating rate bond} \ - \ \text{fixed rate bond}\\ &= 1-B(u,\underline{t};w_t)\end{aligned} \tag{10.23}$$

where $\underline{t} =< t_1,\cdots,t_n >$ and $t < u < t_1$.

10.3.2 Swaption

A swaption is an option to enter (or cancel) a swap. For an option to enter, it is like a call option, as the option is in the money as the swap rate goes up. The payoff is:

$$\begin{aligned}
&\max\{w_t - w_K, 0\} \sum_{j=1}^{n} P(t, t_j) \\
&= \max\left\{ w_t \sum_{j=1}^{n} P(t, t_j) - w_K \sum_{j=1}^{n} P(t, t_j) \right\} \\
&= \max\{1 - B(t, \underline{t}; w_K), 0\}
\end{aligned} \tag{10.24}$$

Hence, it is a put option on a coupon bond. The option of a coupon bond has a true closed-form solution under a factor model described earlier in Chapter 7. If the term structure model is multi-factor, then we lose the closed-form solution and must approximate the exercise boundary.

Chapter 11

FX Math

Symmetry is the key to study FX pricing. Symmetry results in multiple pricing measures. To avoid confusion, I recommend readers always hold one currency constant (as domestic).

11.1 Notation

The drift term of the exchange rate dynamics must follow interest rate parity. In this chapter, we demonstrate the necessary drift adjustment to assure cross currency consistencies.

Throughout out this document, Greek alphabets are constants and English alphabets are functions. To simplify notation and increase readability, I drop the time arguments. For example, $P(t,T)$ the risk-free zero coupon bond price is shown as just P.

There are two risk-neutral probability measures in FX math the domestic measure and the foreign measure. As a result for any variable, there are always a pair of values from both measures. For example,

$$
\begin{aligned}
P &\equiv P(t,T) = \text{domestic price of domestic risk-free zero coupon bond} \\
Q &= \text{domestic price of foreign risk-free zero coupon bond} \\
Y &= \text{domestic price of foreign currency (FX rate under domestic measure)} \\
P^* &= \text{foreign price of foreign risk-free zero coupon bond} \\
Q^* &= \text{foreign price of domestic risk-free zero coupon bond} \\
Y^* &= \text{foreign price of domestic currency (FX rate under foreign measure)}
\end{aligned}
$$

In general, any variable with an asterisk superscript represents a foreign variable. For example, r is the domestic risk-free rate and r^* is the foreign risk-free rate. The subscript of a Brownian represents the state variable. The superscript of a Brownian represents currency measure. For example dW_{P*} is the Brownian for the "*" currency denominated foreign bond under the domestic measure and dW^*_{P*} is the same bond under its own currency measure.

11.2 Basics

We first define the exchange rate and the domestic zero coupon bond price processes under the domestic risk-neutral measure as follows:

$$\begin{aligned} dP/P &= rdt + vd\hat{W}_P \\ dY/Y &= (r - r^*)dt + \sigma d\hat{W}_Y \end{aligned} \tag{11.1}$$

where $v = v(t, T)$ is a T-dependent function and $d\hat{W}_P d\hat{W}_Y = \rho_{YP^*}$. Then, a foreign bond, P^*, measured in the domestic currency is defined as follows:

$$Q = P^*Y \tag{11.2}$$

where

$$dP^*/P^* = \mu^* dt + v^* d\hat{W}_{P^*} \tag{11.3}$$

and $v^* = v^*(t, T)$ is a T-dependent function. Note that (11.3) is the foreign bond price under the domestic measure. [1] By Ito's lemma, we can write:

$$\begin{aligned} dQ/Q &= (\mu^* + r - r^* + \rho_{YP^*}\sigma v^*)dt + v^* dW_{P^*} + \sigma d\hat{W}_Y \\ &= rdt + v^* d\hat{W}_{P^*} + \sigma d\hat{W}_Y \end{aligned} \tag{11.4}$$

The second equality follows because any asset under the risk-neutral measure of the domestic currency must earn the domestic risk-free rate. This implies that:

$$\mu^* = r^* - \rho_{YP^*}\sigma v^* \tag{11.5}$$

[1]Note that the foreign bond under the measure of its own currency should yield the risk-free rate of that currency (which, in our notation, is r^*). We shall come to this later.

where the second term $\rho_{YP^*}\sigma v^*$ is known as the quanto effect. Since this foreign bond should earn the risk free-rate in its own currency, we can define the risk neutral measure in the foreign economy as:

$$dW^*_{P^*} = dW_{P^*} - \rho_{YP^*}\sigma dt \tag{11.6}$$

As we can see, the change of measure between two countries is dependent upon the quanto effect, which is in turn dependent upon the correlation between the exchange rate and the foreign interest rate. In other words, if the correlation is zero or the foreign interest rate in deterministic, then the risk-neutral measure in two countries is the same.

11.3 Forward Measure (domestic)

Given that the interest rates in both countries are random, we must introduce the forward measure in both countries. The forward measure of the domestic currency uses the stochastic discount factor as the numeraire for change of measure (please see Chapter 7 for this result).

$$\begin{aligned} dP/P &= rdt + vd\hat{W}_P \\ &= (r+v^2)dt + vd\tilde{W}_P^{(T)} \end{aligned} \tag{11.7}$$

where $d\tilde{W}_P^{(T)}$ is the forward measure Brownian with time T which corresponds to the maturity of the bond: $P = P(t,T)$. Since $dW_i dW_j = \rho_{ij}dt$, this change of measure will affect the other two processes. In a matrix form, we write a joint process:

$$\begin{aligned} \begin{bmatrix} dP/P \\ dY/Y \\ dP^*/P^* \end{bmatrix} &= \begin{bmatrix} r \\ r - r^* \\ r^* - \rho_{YP^*}\sigma v^* \end{bmatrix} dt + \begin{bmatrix} v & & \\ & \sigma & \\ & & v^* \end{bmatrix} \\ &\times \begin{bmatrix} 1 & 0 & 0 \\ \rho_{YP} & \sqrt{1-\rho_{YP}^2} & 0 \\ \rho_{PP^*} & \frac{\rho_{YP^*} - \rho_{YP}\rho_{PP^*}}{\sqrt{1-\rho_{YP}^2}} & \sqrt{1 - \rho_{PP^*}^2 - \left(\frac{\rho_{YP^*} - \rho_{YP}\rho_{PP^*}}{\sqrt{1-\rho_{YP}^2}}\right)^2} \end{bmatrix} \begin{bmatrix} dZ_P \\ dZ_Y \\ dZ_{P^*} \end{bmatrix} \end{aligned} \tag{11.8}$$

where $dZ_i dZ_j = 0$ and $W_P = Z_P$. The change of measure implies the following (note that these are all domestic measures):

$$\begin{bmatrix} dP/P \\ dY/Y \\ dP^*/P^* \end{bmatrix} = \begin{bmatrix} r+v^2 \\ r-r^*+\rho_{YP}\sigma v \\ r^*-\rho_{YP^*}\sigma v^*+\rho_{PP^*}vv^* \end{bmatrix} dt + \begin{bmatrix} v & & \\ & \sigma & \\ & & v^* \end{bmatrix}$$
$$\times \begin{bmatrix} 1 & 0 & 0 \\ \rho_{YP} & \sqrt{1-\rho_{YP}^2} & 0 \\ \rho_{PP^*} & \frac{\rho_{YP^*}-\rho_{YP}\rho_{PP^*}}{\sqrt{1-\rho_{YP}^2}} & \sqrt{1-\rho_{PP^*}^2-\left(\frac{\rho_{YP^*}-\rho_{YP}\rho_{PP^*}}{\sqrt{1-\rho_{YP}^2}}\right)^2} \end{bmatrix} \begin{bmatrix} d\tilde{Z}_P \\ dZ_Y \\ dZ_{P^*} \end{bmatrix} \tag{11.9}$$

The use of (11.9) is its drift terms that are used in deriving various derivative pricing formulas.

11.4 Viewpoint of a Foreign Nation (Change of Measure)

Let $Y(t)^* = 1/Y(t)$. Then, $\partial Y^*/\partial Y = -1/Y^2$ and $\partial^2 Y^*/\partial Y^2 = 2/Y^3$ and hence,

$$\begin{aligned} dY^* &= \tfrac{-1}{Y^2}((r-r^*)Ydt+\sigma Y dW_Y)+\tfrac{1}{2}\tfrac{2}{Y^3}\sigma^2Y^2dt \\ &= -Y^*((r-r^*-\sigma^2)dt+\sigma d\hat{W}_Y) \end{aligned} \tag{11.10}$$

Given the symmetry (that is (11.10) must be symmetrical (identical but from the perspective of a foreign country) to (11.1)), we know that the change of measure must be:

$$dW_Y^* = -dW_Y + \sigma dt \tag{11.11}$$

Since W's are correlated, the change of measure of the exchange rate will affect other two random variables as follows:

$$\begin{aligned} dW_P^* &= dW_P - \rho_{YP}\sigma dt \\ dW_{P^*}^* &= dW_{P^*} - \rho_{YP^*}\sigma dt \end{aligned} \tag{11.12}$$

This defines the foreign measure. The derivation of (11.12) is straightforward and left for an exercise.

11.5 Interest Rate Parity Theorem

The interest rate parity theorem (IRP) is a no-arbitrage relationship between two trading strategies. It states that the outcome of either investing domestically and foreign should be the same. Formally, we can employ the following definition.

Definition

$$\hat{\mathbb{E}}^*\left[e^{-\int_0^t r(u)^* du}\right] Y_0 = \hat{\mathbb{E}}\left[e^{-\int_0^t r(u)du} Y(t)\right] \tag{11.13}$$

Proof

The LHS of the definition yields $P^*Y(0)$. Using the forward measure result in Chapter 7, we can write the RHS of the definition as $P\tilde{\mathbb{E}}[Y(t)]$. Hence,

$$\tilde{\mathbb{E}}[Y(t)] = \frac{P^*}{P} Y(0) \tag{11.14}$$

which is IRP.

□

Algebraically, we can show that:

$$\hat{\mathbb{E}}\left[e^{-\int_0^t r(u)du} Y(t)\right] = P\tilde{\mathbb{E}}^{(t)}[Y(t)] \tag{11.15}$$

with the following change of measure:

$$d\tilde{W}_P^{(t)} = d\hat{W}_P - vdt \tag{11.16}$$

and (t corresponds to the maturity of the bond $P = P(0,t)$):

$$\begin{aligned} dY/Y &= (r - r^* + \rho_{YP}\sigma v)dt + \sigma\left[\rho_{YP} d\tilde{Z}_1 + \sqrt{1-\rho_{YP}^2} dZ_Y\right] \\ &= (r - r^* + \rho_{YP}\sigma v)dt + \sigma d\tilde{W}_Y^{(t)} \end{aligned} \tag{11.17}$$

To simplify notation and derivation, we let $dM = (r - r^*)dt + \sigma d\tilde{W}_Y$. Note that both r and r^* are random. Hence,

$$
\begin{aligned}
Y(t) = Y(0)\exp\bigg\{\int_0^t (r(u) - r(u)^* - \rho_{YP}\sigma v)\,du - {}^1\!/\!_2\,\mathrm{var}\left[\int_0^t dM\right] \\
+\sigma\rho_{YP}\int_0^t d\tilde{Z}_P + \sigma\sqrt{1-\rho_{YP}^2}\int_0^t dZ_Y\bigg\}
\end{aligned}
\tag{11.18}
$$

Hence,

$$
\begin{aligned}
\tilde{\mathbb{E}}^{(t)}[Y(t)] &= Y(0)\exp\left\{\tilde{\mathbb{E}}\left[\int_0^t (r(u) - r(u)^* - \rho_{YP}\sigma v)\,du\right]\right\} \\
&= Y(0)\exp\left\{\int_0^t \tilde{\mathbb{E}}^{(t)}[-r(u)^* - \rho_{YP}\sigma v]du\right\}\exp\left\{\tilde{\mathbb{E}}^{(t)}\left[\int_0^t r(u)du\right]\right\} \\
&= Y(0)\exp\left\{\int_0^t \tilde{\mathbb{E}}^{*(t)}[-r_u^*]du\right\}\exp\left\{\tilde{\mathbb{E}}^{(t)}\left[\int_0^t r_u du\right]\right\} \\
&= Y(0)\exp\left\{-\int_0^t f(0,u)^* du\right\}\exp\left\{\int_0^t f(0,u)du\right\} \\
&= Y(0)\frac{P^*}{P}
\end{aligned}
\tag{11.19}
$$

Note that from the second line to the third line in the above equation is the necessary change of measure from one currency to another: $dW_P^* = dW_P - \rho_{12}\sigma dt$. From the third to the fourth line is the forward measure result (i.e. the forward measure expectation of the short rate (under its own countrys measure).

11.6 Including Basis

Although the interest rate parity theorem must hold in theory, in reality, due to market frictions, it often does not. The difference is called basis. Empirically, market participants have found that this basis amount varies over time, becoming a substantial risk management issue.

In order to accommodate this market friction, often we modify the exchange rate process to include a quantity for basis α. Basis can be viewed as a financing cost that investors have to pay in order to own the other country's currency. Hence, due to this extra cost, IRP does not hold. As investors continue to hold the foreign currency, the investors need to continue to pay such cost.

$$\begin{aligned} dP/P &= rdt + vdW_P \\ dY/Y &= (r - r^* - \alpha)dt + \sigma dW_Y \end{aligned} \tag{11.20}$$

The domestic price of a foreign bond can be adjusted as follows:

$$\begin{aligned} dQ/Q &= (\mu^* + r - r^* - \alpha + \rho_{23}\sigma v^*)dt + v^*(t,T)dW_{P^*} + \sigma dW_Y \\ &= (r - \alpha)dt + v^*(t,T)dW_{P^*} + \sigma dW_Y \end{aligned} \tag{11.21}$$

Hence, interestingly, the follow result does not change (same as equation (11.5):

$$\mu^* = r^* - \rho_{23}\sigma v^* \tag{11.22}$$

and also the following change of measure remains the same:

$$dW^*_{P^*} = dW_{P^*} - \rho_{23}\sigma dt \tag{11.23}$$

It follows really easily as a result that:

$$\begin{aligned} dY^* &= -Y^*((r - r^* - \alpha - \sigma^2)dt + \sigma dW_Y) \\ &= (r^* - r - \alpha)Y^*dt + \sigma Y^* dW^*_Y \end{aligned} \tag{11.24}$$

Interest Rate Parity Theorem (with basis)

Following the exactly same derivation as in IRP, we naturally arrive at:

$$\tilde{\mathbb{E}}[Y(t)] = Y(0)\frac{P^* e^{-\alpha t}}{P} \tag{11.25}$$

If we view the yield of the foreign bond (in foreign currency) as: $P^* = e^{-y^* t}$, the basis adjustment is a addition to the yield: $P^* e^{-\alpha t} = e^{-(y^* + \alpha)t}$. In other words, the basis is a compensation of owning foreign currency.

The FX forward markets are liquid and we observe $\tilde{\mathbb{E}}[Y_t]$ for all t. This forward FX rate should have incorporated the basis.

11.7 Adding a third country

To complete the analysis, we must add the third country. With three countries, the economy is complete. Adding a third country does not change the analysis but only with an augmented space. And under the domestic measure, foreign bonds are priced as:

$$\begin{aligned} Q &= P^*Y \\ \Theta &= P^\dagger \Upsilon \end{aligned} \tag{11.26}$$

The newly added currency, "†", repeats the exact same result as the "*" currency derived before:

$$\begin{aligned} d\Theta/\Theta &= (\mu^\dagger + r - r^\dagger + \rho_{\Upsilon P^\dagger}\sigma_\Upsilon v^\dagger)dt + v^\dagger dW_{P^\dagger} + \sigma_\Upsilon dW_\Upsilon \\ &= rdt + v^\dagger dW_{P^\dagger} + \sigma_\Upsilon dW_\Upsilon \end{aligned} \tag{11.27}$$

where $v^\dagger = v^\dagger(t, T)$ and

$$\mu^\dagger = r^\dagger - \rho_{\Upsilon P^\dagger}\sigma_\Upsilon v^\dagger \tag{11.28}$$

and

$$dW^\dagger_{P^\dagger} = dW_{P^\dagger} - \rho_{\Upsilon P^\dagger}\sigma_\Upsilon dt \tag{11.29}$$

Note that we have two exchange rates now and therefore the volatilities for the exchange rates now carry a subscript, σ_Υ and the σ_Y for the two currencies respectively.

11.8 Cross Currency FX Rate

The exchange rate of currency "†" per currency "*" is:

$$X = \Upsilon / Y \tag{11.30}$$

As a result, by Ito's lemma, we obtain the following:

$$\begin{aligned}
dX/X &= [X_Y dY + X_\Upsilon d\Upsilon + {}^1\!/\!_2 X_{YY} dY^2 + X_{Y\Upsilon} dY d\Upsilon]/X \\
&= -(r - r^* + \alpha_Y)dt - \sigma_Y dW_Y + (r - r^\dagger + \alpha_\Upsilon)dt + \sigma_\Upsilon dW_\Upsilon + \sigma_Y^2 dt \\
&\quad - \rho_{Y\Upsilon}\sigma_Y\sigma_\Upsilon dt \\
&= (r^* - r^\dagger + \alpha_\Upsilon - \alpha_Y + \sigma_Y^2 - \rho_{Y\Upsilon}\sigma_Y\sigma_\Upsilon)dt - \sigma_Y dW_Y + \sigma_\Upsilon dW_\Upsilon \\
&= (r^* - r^\dagger + \alpha_\Upsilon - \alpha_Y + \sigma_Y^2 - \rho_{Y\Upsilon}\sigma_Y\sigma_\Upsilon)dt + \sigma_X dW_X \\
&= (r^* - r^\dagger + \alpha_\Upsilon - \alpha_Y)dt + \sigma_X dW_X^*
\end{aligned} \tag{11.31}$$

where $\sigma_X = \sqrt{\sigma_Y^2 + \sigma_\Upsilon^2 - 2\rho_{Y\Upsilon}\sigma_Y\sigma_\Upsilon}$ and the change of measure to guarantee no arbitrage is:

$$dW_X = \frac{-\sigma_Y dW_Y + \sigma_\Upsilon dW_\Upsilon}{\sqrt{\sigma_Y^2 + \sigma_\Upsilon^2 - 2\rho_{Y\Upsilon}\sigma_Y\sigma_\Upsilon}} \tag{11.32}$$

And also, by comparing the last two lines, we can see that

$$\sigma_X dW_X^* = \sigma_X dW_X + (\sigma_Y^2 - \rho_{Y\Upsilon}\sigma_Y\sigma_\Upsilon)dt \tag{11.33}$$

Note that from before, due to symmetry, the following result must hold:

$$\begin{aligned}
dW_X^\dagger &= -dW_X^* + \sigma_X dt \\
&= -dW_X^* + \sqrt{\sigma_Y^2 + \sigma_\Upsilon^2 - 2\rho_{Y\Upsilon}\sigma_Y\sigma_\Upsilon}dt
\end{aligned} \tag{11.34}$$

which results in:

$$\begin{aligned}
dW_X^\dagger &= -dW_X^* + \sigma_X dt \\
&= -dW_X - (\sigma_Y^2 - \rho_{Y\Upsilon}\sigma_Y\sigma_\Upsilon)dt + \sqrt{\sigma_Y^2 + \sigma_\Upsilon^2 - 2\rho_{Y\Upsilon}\sigma_Y\sigma_\Upsilon}dt
\end{aligned} \tag{11.35}$$

Also due to symmetry, we know that $\sigma_X dW_X = -\sigma_Y dW_Y + \sigma_\Upsilon dW_\Upsilon$ and $dW_Y^* = -dW_Y + \sigma_Y dt$ must hold (see equation (11.11). Hence,

$$\begin{aligned}
\sigma_X dW_X &= -\sigma_Y dW_Y + \sigma_\Upsilon dW_\Upsilon \\
\sigma_X dW_X^* - (\sigma_Y^2 - \rho_{Y\Upsilon}\sigma_Y\sigma_\Upsilon)dt &= \sigma_Y(dW_Y^* - \sigma_Y dt) + \sigma_\Upsilon dW_\Upsilon \\
\sigma_X dW_X^* &= \sigma_Y dW_Y^* + \sigma_\Upsilon dW_\Upsilon + \rho_{Y\Upsilon}\sigma_Y\sigma_\Upsilon dt
\end{aligned} \tag{11.36}$$

It is reasonable to assume that $dW^*_\Upsilon = dW_\Upsilon + \rho_{Y\Upsilon}\sigma_Y dt$ (this is the result earlier), and then we have:

$$\sigma_X dW^*_X = \sigma_Y dW^*_Y + \sigma_\Upsilon dW^*_\Upsilon \tag{11.37}$$

Note that the above result implies that if the bases of the two exchange rates are the same, then the cross currency exchange rate should have no basis of its own.

Interest Rate Parity (currency † per currency *)

We know IRP for Y (currency * per domestic). The same IRP result can be derived for (currency † per domestic). The reverse must be true (domestic currency per currency * and per currency †).

Now we need to derive IRP for X (currency † per currency *). We guess the result to be:

$$\tilde{\mathbb{E}}^*[X_t] = X_0 \frac{P^\dagger}{P^*} e^{(\alpha_\Upsilon - \alpha_Y)t} \tag{11.38}$$

Lets start with the following joint process. From the table in the Appendix,

$$\begin{aligned}
\begin{bmatrix} dY/Y \\ d\Upsilon/\Upsilon \\ dP^*/P^* \end{bmatrix} &= \begin{bmatrix} r - r^* - \alpha_Y + \sigma_Y^2 \\ r - r^\dagger - \alpha_\Upsilon - \rho_{Y\Upsilon}\sigma_Y\sigma_\Upsilon \\ r^* \end{bmatrix} dt + \begin{bmatrix} \sigma_Y dW^*_Y \\ \sigma_\Upsilon dW^*_\Upsilon \\ \sigma_{P*} dW^*_{P*} \end{bmatrix} \\
&= \begin{bmatrix} r - r^* - \alpha_Y + \sigma_Y^2 \\ r - r^\dagger - \alpha_\Upsilon - \rho_{Y\Upsilon}\sigma_Y\sigma_\Upsilon \\ r^* \end{bmatrix} dt + \Omega\Sigma \begin{bmatrix} dZ^*_Y \\ dZ^*_\Upsilon \\ dZ^*_{P*} \end{bmatrix} \\
&= \begin{bmatrix} r - r^* - \alpha_Y + \sigma_Y^2 + \rho_{YP*}\sigma_Y\sigma_{P*}v^* \\ r - r^\dagger - \alpha_\Upsilon - \rho_{Y\Upsilon}\sigma_Y\sigma_\Upsilon + \rho_{\Upsilon P*}\sigma_\Upsilon\sigma_{P*}v^* \\ r^* + v^{*2} \end{bmatrix} dt + \Omega\Sigma \begin{bmatrix} dZ^*_Y \\ dZ^*_\Upsilon \\ d\tilde{Z}^*_{P*} \end{bmatrix}
\end{aligned} \tag{11.39}$$

where the variances are:

$$\Omega = \begin{bmatrix} \sigma_Y & & \\ & \sigma_\Upsilon & \\ & & v^* \end{bmatrix}$$

and the Cholesky matrix is:

$$\Sigma = \begin{bmatrix} x_{1,1} & x_{1,2} & x_{1,3} \\ 0 & x_{2,2} & x_{2,3} \\ 0 & 0 & 1 \end{bmatrix} = \begin{bmatrix} \sqrt{1-\rho_{YP*}^2 - \left(\frac{\rho_{\Upsilon P*}-\rho_{Y\Upsilon}\rho_{YP*}}{\sqrt{1-\rho_{Y\Upsilon}^2}}\right)^2} & \frac{\rho_{\Upsilon P*}-\rho_{12}\rho_{YP*}}{\sqrt{1-\rho_{Y\Upsilon}^2}} & \rho_{YP*} \\ 0 & \sqrt{1-\rho_{\Upsilon P*}^2} & \rho_{\Upsilon P*} \\ 0 & 0 & 1 \end{bmatrix} \tag{11.40}$$

From before, we know that

$$\begin{aligned} dX/X &= (r^* - r^\dagger + \alpha_\Upsilon - \alpha_Y)dt + \sigma_X dW_X^* \\ &= (r^* - r^\dagger + \alpha_\Upsilon - \alpha_Y)dt + \sigma_Y dW_Y^* + \sigma_\Upsilon dW_\Upsilon^* \\ &= (r^* - r^\dagger + \alpha_\Upsilon - \alpha_Y)dt + \sigma_Y(x_{1,1}dZ_Y^* + x_{1,2}dZ_\Upsilon^2 + x_{1,3}dZ_{P*}^*) \\ &\quad + \sigma_\Upsilon\left(\sqrt{1-\rho_{\Upsilon P*}^2}\,dZ_\Upsilon^* + \rho_{\Upsilon P*}dZ_{P*}^*\right) \\ &= (r^* - r^\dagger + \alpha_\Upsilon - \alpha_Y + \rho_{YP*}\sigma_Y\sigma_{P*}v^* + \rho_{\Upsilon P*}\sigma_\Upsilon\sigma_{P*}v^*)dt \\ &\quad + \sigma_Y(x_{1,1}dZ_Y^* + x_{1,2}dZ_\Upsilon^2 + x_{1,3}d\tilde{Z}_{P*}^*) \\ &\quad + \sigma_\Upsilon\left(\sqrt{1-\rho_{\Upsilon P*}^2}\,dZ_\Upsilon^* + \rho_{\Upsilon P*}d\tilde{Z}_{P*}^*\right) \end{aligned} \tag{11.41}$$

Then repeating the same process as the previous case for dY/Y and using the second table in the Appendix, we can show that

$$\begin{aligned} &E^\dagger\left[e^{-\int_0^t r_u^\dagger du}\right]X_0 = E^*\left[e^{-\int_0^t r_u^* du}X_t\right] \\ &P^\dagger X_0 = P^*\tilde{\mathbb{E}}^*[X_t] \end{aligned} \tag{11.42}$$

and the Proposition is proved.

□

Remark

The results presented here are useful for simulating FX rates and calibration in multiple countries. It is often more convenient to simulate FX and interest rates under one (domestic) measure and then calibrate to the yield curves from various countries (foreign measures). Also, the results are useful in deriving various derivative pricing models.

11.9 Appendix

11.9.1 FX Option Formula – Garman and Kohlhagen, 1983

The FX option pricing model is easy to derive when the two interest rates are constant. From the second equation of (11.1)

$$dY/Y = (r - r^*)dt + \sigma dW \tag{11.43}$$

we know that

$$Y(T) = Y(t)\exp\left\{(r - r^* - 1/2\sigma^2)(T-t) + \sigma\int_t^T dW_u\right\} \tag{11.44}$$

which implies

$$\begin{aligned}\hat{\mathbb{E}}_t[Y(T)] &= Y(t)e^{(r-r^*)(T-t)} \\ \Pr[Y(T) > K] &= N(d_2)\end{aligned} \tag{11.45}$$

where

$$d_2 = \frac{\ln Y(t)/K + (r - r^* - 1/2\sigma^2)(T-t)}{\sigma\sqrt{T-t}}$$

The derivation can be found identically to the stock option case in Chapter 5 Section 5.4.

$$\begin{aligned}C(t) &= e^{-r(T-t)}\hat{\mathbb{E}}_t[\max\{Y(T) - K, 0\}] \\ &= e^{-r(T-t)}\hat{\mathbb{E}}_t[Y(T)\mathbb{I}_{\{Y>K\}}] - e^{-r(T-t)}K\hat{\mathbb{E}}_t[\mathbb{I}_{\{Y>K\}}] \\ &= e^{-r(T-t)}\hat{\mathbb{E}}_t[Y(T)]\hat{\mathbb{E}}_t[\mathbb{I}_{\{Y>K\}}] - e^{-r(T-t)}K\Pr[Y > K] \\ &= e^{-r^*(T-t)}Y(t)N(d_1) - e^{-r(T-t)}KN(d_2)\end{aligned} \tag{11.46}$$

where

$$d_1 = \frac{\ln Y(t)/K + (r - r^* + 1/2\sigma^2)(T-t)}{\sigma\sqrt{T-t}}$$

Note that this is the same as constant dividend yield option model and futures option model where the foreign rate is replaced by the dividend yield and domestic rate respectively.

11.9.2 Summary Table for Measure Change

To make the exposition easy, we use the following tables. The first table shows the changes of measure, as follows:

Measure Change		
	Currency *	Currency †
Domestic	$dW_Y^* = -dW_Y + \sigma_Y dt$ $dW_P^* = dW_P - \rho_{YP}\sigma_Y dt$ $dW_{P*}^* = dW_{P*} - \rho_{YP*}\sigma_Y dt$ $dW_{P\dagger}^* = dW_{P\dagger} - \rho_{YP\dagger}\sigma_Y dt$ $dW_\Upsilon^* = dW_\Upsilon + \rho_{Y\Upsilon}\sigma_Y dt$	$dW_\Upsilon^\dagger = -dW_\Upsilon + \sigma_\Upsilon dt$ $dW_P^\dagger = dW_P - \rho_{\Upsilon P}\sigma_\Upsilon dt$ $dW_{P*}^\dagger = dW_{P*} - \rho_{\Upsilon P*}\sigma_\Upsilon dt$ $dW_{P\dagger}^\dagger = dW_{P\dagger} - \rho_{\Upsilon P\dagger}\sigma_\Upsilon dt$ $dW_Y^\dagger = -dW_Y + \rho_{Y\Upsilon}\sigma_\Upsilon dt$
Currency *		$dW_X^\dagger = -dW_X^* + \sigma_X dt$ $dW_P^\dagger = dW_P^* + \rho_{XP}\sigma_X dt$ $dW_{P*}^\dagger = dW_{P*}^* - \rho_{XP*}\sigma_X dt$ $dW_{P\dagger}^\dagger = dW_{P\dagger}^* - \rho_{XP\dagger}\sigma_X dt$

Note that $\rho_{XP} = dXdP = (-\sigma_Y dW_Y + \sigma_\Upsilon dW_\Upsilon)(vdW_P) = -\rho_{YP}\sigma_Y v + \rho_{\Upsilon P}\sigma_\Upsilon v$ and similarly for ρ_{XP*} and $\rho_{XP\dagger}$.

The second table describes the stochastic processes under each currency, as shown below:

Dynamics		
	Domestic measure	Currency *
dP/P	$rdt + vdW_P$	$(r - \rho_{YP}\sigma_Y v)dt + vdW_P^*$
dY/Y	$(r - r^* - \alpha_Y)dt + \sigma_Y dW_Y$	$(r - r^* - \alpha_Y + \sigma_Y^2)dt + \sigma_Y dW_Y^*$
dP^*/P^*	$(r^* - \rho_{YP*}\sigma_Y v^*)dt + v^* dW_{P*}$	$r^* dt + v^* dW_{P*}^*$
$d\Upsilon/\Upsilon$	$(r - r^\dagger - \alpha_\Upsilon)dt + \sigma_\Upsilon dW_\Upsilon$	$(r - r^\dagger - \alpha_\Upsilon - \rho_{Y\Upsilon}\sigma_Y\sigma_\Upsilon)dt + \sigma_\Upsilon dW_\Upsilon^*$
$dP^\dagger/P^\dagger$	$(r^\dagger - \rho_{YP}\sigma_Y v^\dagger)dt + v^\dagger dW_{P\dagger}$	$(r^\dagger - \rho_{\Upsilon P}\sigma_\Upsilon v^\dagger)dt + v^\dagger dW_{P\dagger}^*$
dY^*/Y^*	$(r^* - r + \alpha_Y + \sigma_Y^2)dt + \sigma_Y dW_Y$	$(r^* - r + \alpha_Y)dt + \sigma_Y dW_Y^*$
$d\Upsilon^\dagger/\Upsilon^\dagger$	$(r^\dagger - r + \alpha_\Upsilon - \sigma_\Upsilon^2)dt + \sigma_\Upsilon dW_\Upsilon^\dagger$	$(r^\dagger - r + \alpha_\Upsilon - \rho_{Y\Upsilon}\sigma_Y\sigma_\Upsilon)dt + \sigma_\Upsilon dW_\Upsilon^\dagger$

Dynamics (continued)		
	Currency †	Comment
dP/P	$(r - \rho_{\Upsilon P}\sigma_\Upsilon v)dt + vdW_P^\dagger$	domestic bond price
dY/Y	$(r - r^* - \alpha_Y - \rho_{Y\Upsilon}\sigma_Y\sigma_\Upsilon)dt + \sigma_Y dW_Y^\dagger$	FX rate to *
dP^*/P^*	$(r^* - \rho_{XP*}\sigma_X v^*)dt + v^* W_{P*}^\dagger$	foreign bond price *
$d\Upsilon/\Upsilon$	$(r - r^\dagger - \alpha_\Upsilon + \sigma_\Upsilon^2)dt + \sigma_\Upsilon dW_\Upsilon^\dagger$	FX rate to †
$dP^\dagger/P^\dagger$	$r^\dagger dt + v^\dagger dW_{P\dagger}^\dagger$	foreign bond price †
dY^*/Y^*	$(r^* - r + \alpha_Y - \rho_{Y\Upsilon}\sigma_Y\sigma_\Upsilon)dt + \sigma_Y dW_Y^*$	FX rate to domestic
$d\Upsilon^\dagger/\Upsilon^\dagger$	$(r^\dagger - r + \alpha_\Upsilon)dt + \sigma_\Upsilon dW_\Upsilon^\dagger$	FX rate to domestic

Chapter 12

No-arbitrage Interest Rate Models

no-arbitrage interest rate models generally have no closed-form solutions and must use lattice models for valuation. In this chapter, we introduce two examples the Ho-Lee model and the Black-Derman-Toy model to represent the forward rate lattice and short rate lattice respectively. The Heath-Jarrow-Morton model is only briefly discussed as its lattice model is quite complex.

12.1 Introduction

No-arbitrage models take the current interest rates as given (either spot rates or forward rates). In this chapter, we introduce two representative models the Ho-Lee model in 1986 and the Black-Derman-Toy model in 1989. The former is a forward rate model and the latter is a spot rate model. No-arbitrage models usually do not have closed-form solutions and must use lattice algorithms. Without showing details of the Heath-Jarrow-Morton model in 1992, I highlight the key issues of the model.

12.2 Ho-Lee Model

The Ho-Lee model is a forward rate model and hence belongs to the HJM family (although the Ho-Lee model was published 6 years prior to the HJM model). However, the original version of the Ho-Lee model is a forward price model. It was Phil Dybvig who then extended the Ho-Lee model to continuous time, forward rate model in 1989.

Define the zero-coupon bond price as $P(i, n, j)$ where i is current time, n is maturity time, and j represents state of economy. The Ho-Lee model is a simple formula as follows:

$$\begin{cases} P(i,i+k,j) = \dfrac{P(i-1,i+k,j)}{P(i-1,i,j)}d(k) \\ P(i,i+k,j+1) = \dfrac{P(i-1,i+k,j)}{P(i-1,i,j)}u(k) \end{cases} \tag{12.1}$$

where

$$u(k) = \frac{1}{p+(1-p)\delta^k}$$
$$d(k) = \frac{\delta^k}{p+(1-p)\delta^k}$$

and p is the risk-neutral probability and δ is the volatility parameter (that is, δ itself is not volatility but it is directly related to volatility.) As we can see, when $\delta = 1$, then $u(k) = d(k)$ for all k. Then there is no volatility. To maintain $u(k) > d(k) > 0$, it must be that $0 < \delta < 1$. As δ becomes small the volatility becomes large.

12.2.1 A Discrete Time Example

Take a four-year yield curve as an example:

Yield Curve		
time to maturity	discount factor	yield to maturity
1	0.9524	0.049979
2	0.8900	0.059998
3	0.8278	0.065021
4	0.7686	0.068009

Given the current term structure of discount factors, we then can compute the forward prices,

$$0.934481 = \frac{0.8900}{0.9524}$$
$$0.869173 = \frac{0.8278}{0.9524}$$
$$0.807014 = \frac{0.7686}{0.9524}$$

Note that forward rates are returns of forward prices. For example:

$$f_{0,1,2} = \frac{1}{0.934481} - 1 = \frac{(1+5.9998\%)^2}{1+4.9979\%} - 1 = 7.0112\%$$

$$f_{0,1,3} = \sqrt{\frac{1}{0.869173}} - 1 = \sqrt{\frac{(1+6.5021\%)^3}{1+4.9979\%}} - 1 = 7.2623\%$$

$$f_{0,1,4} = \sqrt[3]{\frac{1}{0.807014}} - 1 = \sqrt[3]{\frac{(1+6.8009\%)^3}{1+4.9979\%}} - 1 = 7.4088\%$$

Different from the equity binomial that has only one pair of u and d. In our example, we set $p = 0.6$ and $\delta = 0.9$ and we have:

Ho-Lee Pert Funcs		
k	$d(k)$	$u(k)$
0	1.000000	1.000000
1	0.937500	1.041667
2	0.876623	1.082251
3	0.817631	1.121579
4	0.760749	1.159501

In the HL model, the next periods up term structure and down term structure are computed by applying the proper u and d on the forward price. For convenience, we introduce the following labeling system. $P(i,n,j)$ represents the discount factor value at current time i, for maturity time n, and in state j. For convenience, we also label the lowest state 0 and 1, 2, as we go up. So for today, we have four discount factors $P(0,1,0), P(0,2,0), P(0,3,0), P(0,4,0)$ which are 0.9524, 0.8900, 0.8278, and 0.7686 respectively. For next year, we apply the forward prices computed above and multiply them by corresponding us and ds. For example,

$$0.9734 = \underbrace{P(1,2,1)}_{\text{up}} = \frac{P(0,2,0)}{P(0,1,0)}u(1) = \frac{0.8900}{0.9524} \times 1.041667$$

$$0.8761 = \underbrace{P(1,2,0)}_{\text{down}} = \frac{P(0,2,0)}{P(0,1,0)}d(1) = \frac{0.8900}{0.9524} \times 0.9375$$

$$0.9407 = \underbrace{P(1,3,1)}_{\text{up}} = \frac{P(0,3,0)}{P(0,1,0)}u(2) = \frac{0.8278}{0.9524} \times 1.082251$$

$$0.7619 = \underbrace{P(1,3,0)}_{\text{down}} = \frac{P(0,3,0)}{P(0,1,0)}d(2) = \frac{0.8278}{0.9524} \times 0.876623$$

$$0.9051 = \underbrace{P(1,4,1)}_{\text{up}} = \frac{P(0,4,0)}{P(0,1,0)}u(3) = \frac{0.8278}{0.9524} \times 1.121579$$

$$0.6598 = \underbrace{P(1,4,0)}_{\text{down}} = \frac{P(0,4,0)}{P(0,1,0)}d(3) = \frac{0.7686}{0.9524} \times 0.817631$$

Put in the table,

Year = 1						
current time	maturity time	state 0	state 1	state 2	state 3	state 4
1	1	1	1			
	2	0.8761	0.9734			
	3	0.7619	0.9407			
	4	0.6598	0.9051			

Now we have two term structures of the next year (time 1). The task continues to time 2. For each term structure in time 1, we shall compute two term structures in time 2 (up and down) by applying the same principle. The up and down term structures of the left are:

$$0.9060 = P(2,3,1) = \frac{P(1,3,0)}{P(1,2,0)}u(1) = \frac{0.7619}{0.8761} \times 1.041667$$

$$0.8154 = P(2,3,0) = \frac{P(1,3,0)}{P(1,2,0)}d(1) = \frac{0.7619}{0.8761} \times 0.9375$$

$$0.8151 = P(2,4,1) = \frac{P(1,4,0)}{P(1,2,0)}u(2) = \frac{0.6598}{0.8761} \times 1.082251$$

$$0.6603 = P(2,4,0) = \frac{P(1,4,0)}{P(1,2,0)}d(2) = \frac{0.6598}{0.8761} \times 0.876623$$

and of the right are:

$$1.0066 = P(2,3,2) = \frac{P(1,3,1)}{P(1,2,1)}u(1) = \frac{0.9407}{0.9734} \times 1.041667$$
$$0.9060 = P(2,3,1) = \frac{P(1,3,1)}{P(1,2,1)}d(1) = \frac{0.9407}{0.9734} \times 0.9375$$
$$1.0063 = P(2,4,2) = \frac{P(1,4,1)}{P(1,2,1)}u(2) = \frac{0.9051}{0.9734} \times 1.082251$$
$$0.8151 = P(2,4,1) = \frac{P(1,4,1)}{P(1,2,1)}d(2) = \frac{0.9051}{0.9734} \times 0.876623$$

and put in table,

Year = 2						
current time	maturity time	state 0	state 1	state 2	state 3	state 4
2	2	1	1	1		
	3	0.8154	0.9060	1.0066		
	4	0.6603	0.8151	1.0063		

It can be seen that from the left (applying $u(k)$) we arrive at $P(2,3,1) = 0.9060$ which is the same from the right (applying $d(k)$). This also applies to $P(2,4,1) = 0.8151$. This is known as the re-combination assumption. Note that this assumption must be maintained or the binomial model will explode.

Continuing this process going forward, we can derive the complete the 4-year table for the Ho-Lee model:

Complete Ho-Lee Model						
current time	maturity time	state 0	state 1	state 2	state 3	state 4
0	0	1				
	1	0.9524				
	2	0.89				
	3	0.8278				
	4	0.7686				
1	1	1	1			
	2	0.8761	0.9734			
	3	0.7619	0.9407			
	4	0.6598	0.9051			
2	2	1	1	1		
	3	0.8154	0.906	1.0066		
	4	0.6603	0.8151	1.0063		
3	3	1	1	1	1	
	4	0.7592	0.8435	0.9372	1.0414	
4	4	1	1	1	1	1

This concludes the discrete example.

12.2.2 The Continuous Time Limit

Phil Dybvig in 1989 showed that the Ho-Lee model, in its continuous time limit, is a time-dependent parameter model.

$$f(i,n,n+1,j) = -\ln \frac{P(i,n+1,j)}{P(i,n,j)} \tag{12.2}$$

where i is the current time, n and $n+1$ represent the forward period, and j is the state. From the down state we have:

$$\begin{aligned} f(i,n,n+1,j) &= -\ln\frac{P(i,n+1,j)}{P(i,n,j)} \\ &= -\ln\frac{u(n+1-i)P(i-1,n+1,j-1)/P(i-1,i,j-1,)}{u(n-i)P(i-1,n,j-1,)/P(i-1,i,j-1,)} \\ &= -\ln\frac{u(n+1-i)}{u(n-i)} + f(i-1,n,n+1,j-1) \\ &= f(i-1,n,n+1,j-1) + \ln\frac{p+(1-p)\delta^{n+1-i}}{p+(1-p)\delta^{n-i}} \end{aligned} \tag{12.3}$$

From the up state we have:

$$\begin{aligned} f(i,n,n+1,j) &= -\ln\frac{d(n+1-i)}{d(n-i)} + f(i-1,n,n+1,j-1) \\ &= f(i-1,n,n+1,j-1) + \ln\frac{p+(1-p)\delta^{n+1-i}}{p+(1-p)\delta^{n-i}} - \ln\delta \end{aligned} \tag{12.4}$$

Hence, at any time i and any state j, we can write the forward rate for any forward period between n and $n+1$ as:

$$f(i,n,n+1,j) = f(i-1,n,n+1,j-1) + \ln\frac{p+(1-p)\delta^{n+1-i}}{p+(1-p)\delta^{n-i}} - (1-p)\ln\delta + \varepsilon_{i,j} \tag{12.5}$$

where

$$\varepsilon_{i,j} = \begin{cases} (1-p)\ln\delta & \text{if up state at time } i \\ -p\ln\delta & \text{if down state at time } i \end{cases}$$

which has an expected value of 0 and volatility of $\ln\delta\sqrt{p(1-p)}$. Also, moving the forward rate at time $i-1$ to the left and letting the time between i and $i-1$ be very small, we shall obtain a forward rate differential equation as:

$$df(t,T) = a(t,T)dt + bdW \tag{12.6}$$

where b is a constant (function of parameters p and δ) and $a(t)$ is a deterministic function of time and maturity (function of i and n). Note that b being a constant implies (1) the forward rate is normally distributed and (2) the corresponding spot rate process also has constant diffusion coefficient. Note that the forward rate differential equation applies for all maturity time T available at current time t. If such a vector of forward rates are derived on a single spot rate, the spot rate process must have a time-dependent parameter. Furthermore, since a does not depend on the interest rate, the drift of the short rate process should not depend on the interest rate either. With this information, we can write the short rate process in general as:

$$dr = \theta(t)dt + \sigma d\hat{W} \tag{12.7}$$

The spot rate is Gaussian process without mean reversion. From Arnold (1974), we can write the solution to be:

$$r(s) = r(t) + \int_t^s \theta(u)du + \int_t^s \sigma d\hat{W}(u) \tag{12.8}$$

Assuming the similar risk premium to the Vasicek model, we can find the price of the discount bond using the same derivation:

$$\begin{aligned} P(t,T) &= \hat{\mathbb{E}}_t \left[\exp\left(- \int_t^T r(u)du \right) \right] \\ &= \exp\left(-r(t)(T-t) - \int_t^T \int_t^s \theta(u)duds + \tfrac{\sigma^2(T-t)^3}{6} \right) \\ &= D(t,T) \exp\left(-r(t)(T-t) + \tfrac{\sigma^2(T-t)^3}{6} \right) \end{aligned} \tag{12.9}$$

Note that function $D(t,T) = \exp\left\{ -\int_t^T \int_t^s \theta(u)du \right\}$ depends on the time dependent function $\theta(t)$. In order to fit perfectly the yield curve, the time dependent parameter $\theta(t)$ needs to generate D for every point on the yield curve. In other words, $\theta(t)$ is cooked in such a way that the discount function above is flexible enough to fit every point of the yield curve. The $\theta(t)$ functional form is unimportant. It is interesting to examine the option formula. Since the spot rate is normally distributed, the option formula remains the same as the Jamshidian model. The only change is the volatility in the Jamshidian model:

$$\text{var}[\ln P(T,s)] = \text{var}[r(T)(s-T)] = (s-T)^2\sigma^2(T-t) \tag{12.10}$$

The option price computed from the discrete Ho-Lee model should converge to the Jamshidian price with the volatility value given above.

12.3 Black-Derman-Toy Model

The material in this section is taken from Analysis of Fixed Income (where it has the most updated version).

12.3.1 Tree (Log Normal)

Standard binomial model in equity looks as follows:

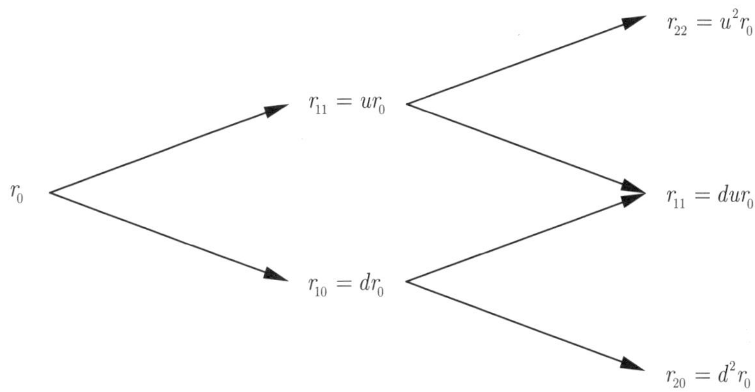

Figure 12.1: Two-period BDT Model

where

$$u = e^{\sigma\sqrt{\Delta t}}$$
$$d = e^{-\sigma\sqrt{\Delta t}}$$

12.3.2 Modified Tree (since $ud = 1$)

In order to fit to the current yield curve, the standard binomial tree needs to be modified. The key change is to allow for the flexibility in each period in setting a mean rate. In the standard binomial tree, the up and down are perturbations around r_0. In the modified binomial model that can fit to the current yield curve, r_0 is replaced by a series of other values: $\bar{r}_1$, $\bar{r}_2, \cdots, \bar{r}_n$ in various periods, as the following diagram shows:

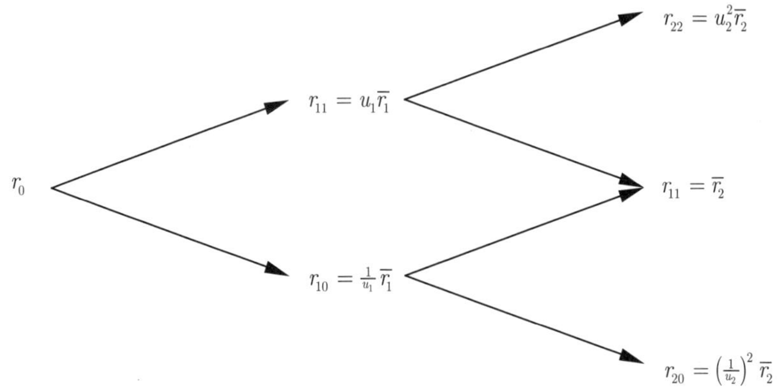

Figure 12.2: Delivery Time Line

Hence, for every period, there are always two unknowns $\bar{r}_t$ and u.

The following example is taken from the original paper published in 1989 by Black, Derman, and Toy in Financial Analyst Journal. The hyperthetical market inputs are yield and volatility (yield vol) curves as follows:

Yield Curve		
maturity	yield	volatility
1	10%	20%
2	11	19
3	12	18
4	12.5	17
5	13	16

Note that yield vols are annualized. The yield curve implies the following discount factors:

$$
\begin{aligned}
D(0,1) &= \frac{1}{1.1} = 0.9091 \\
D(0,2) &= \frac{1}{1.11^2} = 0.8116 \\
D(0,3) &= \frac{1}{1.12^3} = 0.7118 \\
D(0,4) &= \frac{1}{1.125^4} = 0.6243 \\
D(0,5) &= \frac{1}{1.13^5} = 0.5428
\end{aligned}
$$

The following diagram (12.3) depicts the lattice structure for the model.

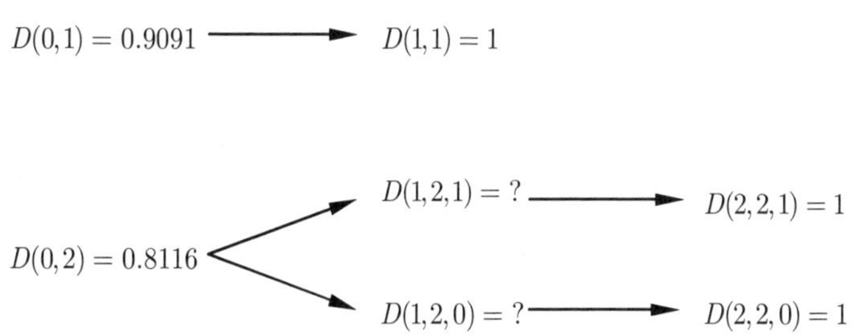

Figure 12.3: Delivery Time Line

Assume probabilities to be and . Then it must be the case that

$$D(0,2) = \frac{1}{2}[D(1,2,1) + D(1,2,0)]D(0,1) \tag{12.11}$$

that is

$$0.8116 = \frac{1}{2}\left(\frac{1}{1+r_{11}} + \frac{1}{1+r_{10}}\right) \times 0.9091$$

Two unknowns cannot be solved with one equation. The other equation we need comes from the volatility. In the standard binomial model, we get:

$$\sigma = \frac{\ln \frac{u}{d}}{2} = \frac{\ln \frac{r_{11}}{r_{10}}}{2} = 0.19 \tag{12.12}$$

Hence we solve for r_{11} and r_{10} as 14.32% and 9.79% respectively. Also, $\bar{r} =$ 11.84% and $u = 1.209$ (or $d = 1/u = 0.827$).

The lattice is now extended to include one more period, as demonstrated as follows:

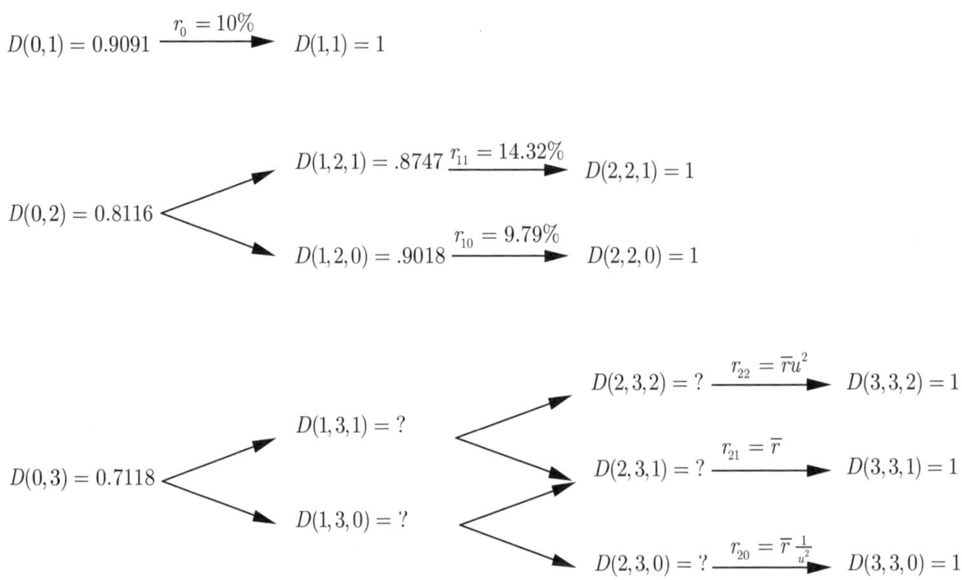

Figure 12.4: Delivery Time Line

To solve for the third period, we need the following two equations:

$$D(0,3) = \frac{1}{2}[D(1,3,1) + D(1,3,0)]D(0,1) = 0.7118 \tag{12.13}$$

and

$$\sigma = \frac{1}{2}\ln\left[\frac{\sqrt{\frac{1}{D(1,3,1)}} - 1}{\sqrt{\frac{1}{D(1,3,0)}} - 1}\right] = 0.18 \tag{12.14}$$

where

$$D(1,3,1) = \tfrac{1}{2}\{D(2,3,2) + D(2,3,1)\}D(1,2,1)$$
$$D(1,3,0) = \tfrac{1}{2}\{D(2,3,1) + D(2,3,0)\}D(1,2,0)$$

and in turn,

$$D(2,3,2) = \frac{1}{1+\bar{r}u^2}$$
$$D(2,3,1) = \frac{1}{1+\bar{r}u}$$
$$D(2,3,0) = \frac{1}{1+\bar{r}/u^2}$$

Again, two equations and two unknowns solve for $\bar{r}$ and u to be 13.77% and 1.1877 respectively. As a result, the short rates in period 2 are $r_{22} = 19.42\%$, $r_{21} = 13.77\%$, and $r_{20} = 9.76\%$. Continuing this process, we can arrive at all the short rates as follows:

$$r_{33} = 21.79\%; r_{32} = 16.06\%; r_{31} = 11.83\%; r_{30} = 8.72\%$$
$$r_{44} = 25.52\%; r_{43} = 19.48\%; r_{42} = 14.86\%; r_{41} = 11.34\%; r_{40} = 8.65\%$$

12.3.3 Option Pricing

Once we solve out the lattice, pricing derivatives such as options is straightforward.

12.4 Heath-Jarrow-Morton Model

The Heath-Jarrow-Morton (HJM) model is not a model but a class of models. The major contribution of the HJM model is its discovery of, now known as, "drift adjustment". Basically, Heath, Jarrow and Morton discovered that the drift term of the forward rate process cannot be arbitrarily determined but a function of the volatility curve. This is drastically different from the short rate process (such as CIR and Vasicek) where the short rate drift can be freely determined. Financial industry finds this to be extremely convenient and it needs a model to be able to calibrate to both yield and volatility curves. The HJM model provides such convenience, that it can fit both curves easily simultaneously.

The HJM model is a forward rate model so the Ho-Lee model is its close cousin. In fact, the Ho-Lee model is a special case of the HJM model by setting the volatility as a constant. They both take the forward curve as given and adopt Gaussian dynamics (although in theory the HJM model can adopt other forms of dynamics but the Gaussian dynamics remain as the most popular choice.)

There is a variation of the HJM model, developed by Miltersen, Sandmann, and Sondermann (1997) and Brace, Gatarek, and Musiela (1997), known as the LIBOR market model (LMM) that builds dynamics on discrete forward rates as opposed to continuous time forward rates. In doing so, they can avoid the Gaussian dynamics and introduce the log normal dynamics that do not permit negative interest rates. Furthermore, by adopting the log normal dynamics, many interest rate options can be priced by the Black-Scholes formula. Unfortunately, the drift adjustment for the discrete forward rates are much more complex than the drift adjustment of the HJM model. European banks tend to use the LMM yet the American banks tend to use the HJM model.

The HJM model is based upon the forward yield curve, as opposed to the spot yield curve. We write the general spot rate process as follows:

$$dr = a(r,t)dt + b(r,t)d\hat{W} \tag{12.15}$$

By Ito (subscripts represent partial derivatives),

$$\begin{aligned} dP &= P_r dr + \tfrac{1}{2} P_{rr}(dr)^2 + P_t \\ &= rPdt + P_r b(r,t)d\hat{W} \\ &= rPdt + v(t,T)Pd\hat{W} \end{aligned} \tag{12.16}$$

or equivalently

$$\begin{aligned} d\ln P(t,T) &= \frac{1}{P(t,T)}dP(t,T) - \frac{1}{2P(t,T)^2}(dP(t,T))^2 \\ &= \frac{1}{P(t,T)}\left(P_r(t,T)dr + \tfrac{1}{2}P_{rr}(t,T)(dr)^2 + P_t(t,T)dt\right) \\ &\quad - \frac{1}{2P(t,T)^2}P_r^2(t,T)(dr)^2 \\ &= rdt + \frac{P_r(t,T)}{P(t,T)}b(r,t)d\hat{W} - \frac{1}{2}\left(\frac{P_r(t,T)}{P(t,T)}b(r,t)\right)^2 dt \\ &= \left(r - \tfrac{v(t,T)^2}{2}\right)dt + v(t,T)d\hat{W} \end{aligned} \tag{12.17}$$

Note that

$$v(t,t) = 0 \tag{12.18}$$

Note that the forward rate is defined as:

$$f(t,T) = -\frac{\partial \ln P(t,T)}{\partial T} \tag{12.19}$$

As a result,

$$\begin{aligned} df(t,T) &= -\frac{\partial}{\partial T} d\ln P(t,T) \\ &= -\frac{\partial}{\partial T}\left[\left(r - \tfrac{v(t,T)^2}{2}\right)dt + v(t,T)d\hat{W}\right] \\ &= v(t,T)\frac{\partial v(t,T)}{\partial T}dt - \frac{\partial v(t,T)}{\partial T}d\hat{W} \end{aligned} \tag{12.20}$$

This proves that the drift term of the forward rate dynamics do not depend on anything but the volatility function. The drift term in the above equation is known as the "drift adjustment".

Usually, people would like to write (12.20) as follows:

$$df(t,T) = \left[w(t,T)\int_t^T w(t,u)du\right]dt - w(t,T)d\hat{W} \tag{12.21}$$

where

$$w(t,T) = \frac{\partial v(t,T)}{\partial T}$$

12.4.1 Example 1 – Ho-Lee Model

The Ho-Lee model has the following continuous time short rate process (under the risk-neutral measure) as follows (same as (12.7)):

$$dr = \theta(t)dt + \sigma d\hat{W} \tag{12.22}$$

Note that from (12.9) we can write

$$-\frac{\partial}{\partial T}\ln P(t,T) = \int_t^T \theta(u)du + r(t) - \frac{1}{2}\sigma^2(T-t)^2 \tag{12.23}$$

As a result,

$$\begin{aligned}
df(t,T) &= d\left[-\frac{\partial}{\partial T}\ln P(t,T)\right] \\
&= -\frac{\partial}{\partial T} d\ln P(t,T) \\
&= -\theta(t)dt + dr(t) + \sigma^2(T-t)dt \\
&= \sigma d\hat{W} + \sigma^2(T-t)dt
\end{aligned} \tag{12.24}$$

This satisfies the HJM drift restriction that $\sigma \int_t^T \sigma du = \sigma^2(T-t)$.

Another interesting result for the Ho-Lee model using the HJM framework is that the short rate process can be expressed in a form of forward curve without the drift term $\theta(t)$. Combining (12.19) and (12.24) with (12.8), we obtain:

$$r(T) = f(t,T) + \frac{1}{2}\sigma^2(T-t)^2 + \int_t^T \sigma d\hat{W}(u) \tag{12.25}$$

This result clearly demonstrates that the Ho-Lee model will generate exploding short rates. That is $\hat{\mathbb{E}}[r(T)] \to \infty$ as $T \to \infty$.

12.4.2 Example 2 – Vasicek Model

Note that the derivation given above does not assume any particular interest rate model. So, it would be interesting to see an example where the drift term of the short rate process (particularly the market price of risk parameter) disappears from the equation. To be simple, we adopt the Vasicek model as our example for demonstration.

The Vasicek model has the following short rate process under the risk-neutral measure:

$$dr = \alpha\left(\mu - \tfrac{\sigma\lambda}{\alpha} - r\right)dt + \sigma d\hat{W} \tag{12.26}$$

where λ is the market price of risk which is constant under the Vasicek model with log utility. The Vasicek model has the following closed-form solution to the discount factor (or zero coupon bond):

$$P(t,T) = e^{-rF(t,T)-G(t,T)} \tag{12.27}$$

where

$$
\begin{aligned}
F(t,T) &= \left(1 - e^{-\alpha(T-t)}\right)/\alpha \\
G(t,T) &= \left(\mu - \tfrac{\sigma^2}{2\alpha^2} - \tfrac{\sigma\lambda}{\alpha}\right)(T - t - F(t,T)) + \tfrac{\sigma^2 F(t,T)^2}{4\alpha}
\end{aligned}
$$

Note that $-\ln P(t,T) = rF(t,T) + G(t,T)$ and $\frac{\partial F(t,T)}{\partial t} = \alpha F(t,T) - 1$. As a result:

$$
\begin{aligned}
-d\ln P(t,T) &= F(t,T)dr + r\frac{\partial F(t,T)}{\partial t}dt + \frac{\partial G(t,T)}{\partial t}dt \\
&= F(t,T)\left[\alpha\left(\mu - \tfrac{\sigma\lambda}{\alpha} - r\right)dt + \sigma d\hat{W}\right] + r(\alpha F(t,T) - 1)dt \\
&\quad + \left(\mu - \tfrac{\sigma^2}{2\sigma^2} - \tfrac{\sigma\lambda}{\alpha}\right)(-\alpha F(t,T))dt + \tfrac{\sigma^2}{2\alpha}F(t,T)(\alpha F(t,T) - 1)dt \\
&= F(t,T)\sigma d\hat{W} - rdt + \frac{1}{2}(F(t,T)\sigma)^2 dt
\end{aligned}
\tag{12.28}
$$

Take derivative with respect to the maturity time T yields the forward rate process as follows:

$$
\begin{aligned}
-\frac{\partial}{\partial T}d\ln P(t,T) &= df(t,T) \\
&= \frac{\partial F(t,T)}{\partial T}\sigma d\hat{W} + F(t,T)\frac{\partial F(t,T)}{\partial T}\sigma^2 dt \\
&= e^{-\alpha(T-t)}\sigma d\hat{W} + e^{-\alpha(T-t)}\left(\tfrac{1-e^{-\alpha(T-t)}}{\alpha}\right)\sigma^2 dt
\end{aligned}
\tag{12.29}
$$

which verifies the HJM drift restriction result. We shall note that the reversion level, μ, disappears from the equation! It is interesting to note that the market price of risk, λ, also disappears from the equation.

We also note that the volatility term is time-decaying, which does not cause exploding short rates. To see that, we first use the result in Chapter 4:

$$
\begin{aligned}
r(T) &= e^{-\alpha(T-t)}\left(r(t) + \int_t^T e^{\alpha(s-t)}\alpha\mu ds + \int_t^T e^{\alpha(s-t)}\sigma d\hat{W}(s)\right) \\
&= r(t)e^{-\alpha(T-t)} + \mu\left(1 - e^{-\alpha(T-t)}\right) + \sigma e^{-\alpha(T-t)}\int_t^T e^{\alpha(s-t)}d\hat{W}(s) \\
&= r(t)\frac{\partial F(t,T)}{\partial T} + \alpha\mu F(t,T) + \sigma e^{-\alpha(T-t)}\int_t^T e^{\alpha(s-t)}d\hat{W}(s)
\end{aligned}
\tag{12.30}
$$

From equation (12.27), we can write:

$$\begin{aligned}
-\ln P(t,T) &= rF(t,T) + G(t,T) \\
f(t,T) &= \frac{\partial}{\partial T}(-\ln P(t,T)) \\
&= r\frac{\partial F(t,T)}{\partial T} + \frac{\partial G(t,T)}{\partial T} \\
&= r\frac{\partial F(t,T)}{\partial T} + \left(\mu - \frac{\sigma^2}{2\alpha^2}\right)\left(1 - \frac{\partial F(t,T)}{\partial T}\right) + \frac{\sigma^2}{2\alpha}F(t,T)\frac{\partial F(t,T)}{\partial T} \\
&= r\frac{\partial F(t,T)}{\partial T} + \mu\alpha F(t,T) - \frac{\sigma^2}{2\alpha^2}F(t,T) + \frac{\sigma^2}{2\alpha}F(t,T)\frac{\partial F(t,T)}{\partial T}
\end{aligned} \tag{12.31}$$

Combining equations (12.30) and (12.31), we can derive the following result:

$$\begin{aligned}
r(T) &= f(t,T) + \frac{\sigma^2}{2\alpha}F(t,T) - \frac{\sigma^2}{2\alpha}F(t,T)\frac{\partial F(t,T)}{\partial T} + \sigma e^{-\alpha(T-t)}\int_t^T e^{\alpha(s-t)}d\hat{W}(s) \\
&= f(t,T) + \frac{\sigma^2}{2\alpha}F(t,T)\left(1 - \frac{\partial F(t,T)}{\partial T}\right) + \sigma e^{-\alpha(T-t)}\int_t^T e^{\alpha(s-t)}d\hat{W}(s) \\
&= f(t,T) + \frac{\sigma^2}{2}F^2(t,T) + \sigma e^{-\alpha(T-t)}\int_t^T e^{\alpha(s-t)}d\hat{W}(s)
\end{aligned} \tag{12.32}$$

Note that the middle term is the same as the extra term under the forward measure for the short rate (see (7-24) in Chapter 7). Unlike the Ho-Lee model, the Vasicek model will not generate exploding short rates. As $T \to \infty$, $\hat{\mathbb{E}}[r(T)] = f(t,T) + \frac{\sigma^2}{2\alpha^2}$.

To build a lattice or a Monte-Carlo model for the Vasicek model (say for American option pricing), it is convenient to compute the variance of the last term of equation (12.32) as:

$$\begin{aligned}
\text{var}\left[\sigma e^{-\alpha(T-t)}\int_t^T e^{\alpha(s-t)}d\hat{W}(s)\right] &= \sigma^2 e^{-2\alpha(T-t)}\int_t^T e^{2\alpha(s-t)}ds \\
&= \sigma^2\frac{1 - e^{-2\alpha(T-t)}}{2\alpha}
\end{aligned} \tag{12.33}$$

and hence (12.32) can be written as:

$$r(T) = f(t,T) + \frac{\sigma^2}{2} F^2(t,T) + \sigma \sqrt{\frac{1-e^{-2\alpha(T-t)}}{2\alpha}} z \tag{12.34}$$

where z is a standard Normal variable. In simulations, for any given future time $T > t$, we can simulate standard Normal and then linearly transform it to obtain the instantaneous short rate. In a lattice, we can build a lattice for z (standard Brownian) first and then add on to it the linear calculation. We demonstrate the use of the equation in Appendix B.

12.4.3 Example 3 Cox-Ingersoll-Ross Model

From Chapter 7, we know that the CIR model has the following short rate process under the risk-neutral measure:

$$dr = (\alpha + \lambda) \left(\tfrac{\sigma\mu}{\alpha+\lambda} - r \right) dt + \sigma \sqrt{r} d\hat{W} \tag{12.35}$$

with the following closed-form solution to the zero coupon bond:

$$P(t,T) = A(t,T) e^{-rB(t,T)} \tag{12.36}$$

where

$$\begin{aligned} A(t,T) &= \left[\frac{2\gamma e^{(\alpha+\lambda+\gamma)(T-t)/2}}{(\alpha+\lambda+\gamma)\left(e^{\gamma(T-t)}-1\right)+2\gamma} \right]^{2\alpha\mu/\sigma^2} \\ B(t,T) &= \frac{2\left(e^{\gamma(T-t)}-1\right)}{(\alpha+\lambda+\gamma)\left(e^{\gamma(T-t)}-1\right)+2\gamma} \\ \gamma &= \sqrt{(\alpha+\lambda)^2 + 2\sigma^2} \end{aligned}$$

Note that $A(t,T)$ and $B(t,T)$ satisfy the following ordinary differential equations:

$$\begin{aligned} &\alpha B(t,T) + \frac{1}{2}\sigma^2 B(t,T) - \frac{\partial B(t,T)}{\partial t} - 1 = 0 \\ &\frac{\partial A(t,T)}{\partial t} \frac{1}{A(t,T)} - \alpha\mu B(t,T) = 0 \end{aligned} \tag{12.37}$$

Since $-\ln P(t,T) = rB(t,T) - \ln A(t,T)$, we have:

$$\begin{aligned}
-d\ln P(t,T) &= B(t,T)dr + r\frac{\partial B(t,T)}{\partial t}dt - \frac{\partial A(t,T)}{\partial t}\frac{1}{A(t,T)}dt \\
&= B(t,T)[(\alpha\mu - (\alpha+\lambda)r)dt + \sigma\sqrt{r}d\hat{W}] \\
&\quad + r\left((\alpha+\lambda)B(t,T) + \frac{1}{2}\sigma^2 B^2(t,T) - 1\right)dt - \alpha\mu B(t,T)dt \\
&= -rdt + \sigma\sqrt{r}B(t,T)d\hat{W} + \frac{1}{2}\sigma^2 r B^2(t,T)dt
\end{aligned} \tag{12.38}$$

Taking the derivative with respect to the maturity time T, we obtain:

$$\begin{aligned}
-\frac{\partial}{\partial T}d\ln P(t,T) &= df(t,T) \\
&= \sigma\sqrt{r}\frac{\partial B(t,T)}{\partial T}d\hat{W} + \sigma^2 r B(t,T)\frac{\partial B(t,T)}{\partial T}dt
\end{aligned} \tag{12.39}$$

Note that similar to the Vasicek model, the reverting level, μ, disappears from the equation. However, the market price of risk still remains.

12.5 Appendix

12.5.1 Stock Option Pricing When the Interest Rate Follows the Ho-Lee Model

In this appendix, we show the equity option pricing model when the interest rate follows the Ho-Lee model. This is a continuation of an Appendix in Chapter 7. In an Appendix of Chapter 7, we show the equity option pricing model when the interest rate follows the Vasicek model. The Vasicek model is an equilibrium model but the Ho-Lee model is a no-arbitrage model. We demonstrate the link between the equilibrium version of the Ho-Lee model in this Chapter. Now, we shall use this version to derive the equity option model.

From Chapter 7,

$$V(t,T) = \int_t^T \left\{ \left[\delta \frac{P_r(u,T)}{P(u,T)}\right]^2 - 2\rho\sigma\delta\frac{P_r(u,T)}{P(u,T)} + \sigma^2 \right\} du \tag{12.40}$$

From equation (12.9):

$$\frac{P_r(u,T)}{P(u,T)} = -(T-t) \tag{12.41}$$

Hence,

$$\begin{aligned} V(t,T) &= \int_t^T \left\{\delta^2(T-u)^2 + 2\rho\sigma\delta(T-u) + \sigma^2\right\} du \\ &= \frac{\delta^2}{3}(T-t)^3 + \rho\sigma\delta(T-t)^2 + \sigma^2(T-t) \end{aligned} \tag{12.42}$$

To complete the formula, we write the option model as follows with the substitution of the variance:

$$C(t) = S(t)N(d_1) - P(t,T)KN(d_2) \tag{12.43}$$

where

$$\begin{aligned} d_1 &= \frac{\ln S(t) - \ln P(t,T) + V(t,T)/2}{\sqrt{V(t,T)}} \\ d_2 &= d_1 - V(t,T) \end{aligned}$$

12.5.2 The Binomial Lattice for the Vasicek Model

Following (12.32), we can design a binomial lattice for the Vasicek model as follows:

```
Sub expl()
alf = 0.25
mu =0.04
sig = 0.2
lam = -0.15
r0 = 0.002
mu = mu - sig * lam / alf
ttm1 = 0.25
ttm2 = 1
strk = 0.85
n = 200
n1 = (ttm1 / ttm2) * n
ReDim p1(2 * n)
ReDim p0(2 * n)
ReDim c1(2 * n)
```

```
ReDim c0(2 * n)
ReDim pu(2 * n)
ReDim pp(2 * n)
ReDim pd(2 * n)
ReDim rj(2 * n)
dt = ttm2 / (n)
eps = 1 'eps > 1
dr = eps * sig * Sqr(dt)
flg = 0
rmax = mu + sig / alf / Sqr(dt)
rmin = mu - sig / alf / Sqr(dt)
For j = 0 To 2 * n
rj(j) = r0 + (-n + j) * dr
p1(j) = 1
pu(j) = 0.5 * (1 + alf * (mu - rj(j)) * Sqr(dt) / sig)
If rj(j) = rmax Then pu(j) = 1
If rj(j) = rmin Then pu(j) = 0
pp(j) = 0
pd(j) = 1 - pu(j)
Next
For i = n - 1 To 0 Step -1
For j = n - i To n + i
p0(j) = (1 / (1 + rj(j) * dt)) * (pu(j) * p1(j + 1) + pp(j) *
p1(j) + pd(j) * p1(j - 1))
If i = n1 Then c1(j) = rrcmax(p0(j) - strk, 0)
If i < n1 Then c0(j) = (1 / (1 + rj(j) * dt)) * (pu(j) * c1(j
+ 1) + pp(j) * c1(j) + pd(j) * c1(j - 1))
Next
For j = n - i To n + i
p1(j) = p0(j)
If i < n1 Then c1(j) = c0(j)
Next
Next
res1 = p0(n)
res2 = c0(n)
End Sub
```

Using the parameter values given in Chapter 7, that is:

Inputs	
α	0.2456
μ	0.0648
σ	0.0289
λ	-0.2718
$r(t)$	0.06

and maturity to be one year, we obtain the closed-form result as 0.937956. The following table demonstrates the convergence:

n	result
75	0.938053
125	0.938014
250	0.937985
500	0.93797
750	0.937965
1000	0.937963
1500	0.937961
2000	0.937959
5000	0.937957
10000	0.937956

Chapter 13

Complex Options

Many of these options need Fourier Inversion to compute prices efficiently. With FI, we can easily turn a multi-variate problem in the time domain to a uni-variate problem in the frequency domain. In this chapter, we introduce various complex option valuation formulas.

13.1 Introduction

In this chapter, we introduce complex options. My notion of the complex options is different from the usual notion of the exotic options where payoffs are complex. My notion of complex options in this chapter is complex calculations, which include (1) Fourier inversion (random volatility option pricing, multi-factor interest rate options, and option bounds), (2) multivariate normal probability function (compound option and American option with known dividends), and (3) the first passage time (barrier option).

13.2 Compound Option Model (Geske)

The Geske model is based upon Geske's publications in 1977, 1979, and 1984. Geske's compound option methodology has a number of applications. In addition to pricing compound option pricing, Geske's model can be used for pricing American options with fixed dividends and credit risky debts.

13.2.1 Two-period Compound Option Formula (JFE, 1979a)

Here we demonstrate the two-period compound option formula derived by Geske in 1979. Given that bi-variate normal probabilities are very easy to compute, the formula is regarded as a closed-form solution.

Let V be the underlying asset value that follows the Black-Scholes model that follows the risk neutral process below:[1]

$$\frac{dV}{V} = rdt + \sigma d\hat{W} \tag{13.1}$$

where r is the risk-free rate and σ is volatility. From Chapter 4, we know that:

$$V(T) = V(t)\exp\left((r - {}^{1}\!/\!{}_{2}\sigma^2)(T-t) + \sigma r - {}^{1}\!/\!{}_{2}\sigma^2 d\hat{W}(u)\right) \tag{13.2}$$

which implies $\ln V(T)$ is normally distributed with mean $\ln V(t) + (r - \frac{1}{2}\sigma^2)(T-t)$ and variance $\sigma^2(T-t)$. As a result,

$$\Pr[\ln V(T) > \ln K] = N\left(\frac{\ln V(t) + (r - {}^{1}\!/\!{}_{2}\sigma^2)(T-t) - \ln K}{\sigma\sqrt{T-t}}\right) \tag{13.3}$$

Define V_0, V_1 and V_2 as short-hand notation for $V(t)$, $V(T_1)$ and $V(T_2)$ respectively to represent the asset prices at future times t, T_1 and T_2 respectively (call price C is labeled similarly too). The two strikes are K_1 and K_2 respectively. Define:

$$h_{ij}^{\pm} = \frac{\ln V_i - \ln \bar{V}_j + (r \pm \sigma^2/2)(T_j - T_i)}{\sigma\sqrt{T_j - T_i}} \tag{13.4}$$

where $T_0 = t$, $\bar{V}_2 = K_2$, and $\bar{V}$ is to be defined later. A two-period compound option is an option that can be exercised into another option. The price of the compound option (C) can be derived as follows:

[1]We use V (as opposed to S) to represent the underlying asset value which is not necessarily stock in previous chapters where the Black-Scholes model is discussed.

$$
\begin{aligned}
C_0 &= \hat{\mathbb{E}}_t[e^{-r(T_1-t)}\max\{C_1 - K_1, 0\}] \\
&= e^{-r(T_1-t)}\int_{K_1}^{\infty}[C_1 - K_1]\phi(C_1)dC_1 \\
&= e^{-r(T_1-t)}\int_{\bar{V}_1}^{\infty}[f(V_1) - K_1]\phi(V_1)dV_1 \\
&= e^{-r(T_1-t)}\int_{\bar{V}_1}^{\infty}f(V_1)\phi(V_1)dV_1 - e^{-r(T_1-t)}K_1\int_{\bar{V}_1}^{\infty}\phi(V_1)dV_1 \\
&= A - e^{-r(T_1-t)}K_1N(h_{01}^-)
\end{aligned}
\tag{13.5}
$$

where $\phi(\cdot)$ is the standard normal density function. The call option value at time T_1:

$$
\begin{aligned}
C_1 &= f(V_1) \\
&= V_1N(h_{12}^+) - e^{-r(T_2-T_1)}K_2N(h_{12}^-)
\end{aligned}
\tag{13.6}
$$

is the Black-Scholes model and $K_1 = f(\bar{V}_1)$. In other words, $\bar{V}_1$ is the solution to equation (13.6) when $C_1 = K_1$. Now, we solve for A:

$$
\begin{aligned}
A &= e^{-r(T_1-t)}\int_{\bar{V}_1}^{\infty}C_1\phi(V_1)dV_1 \\
&= e^{-r(T_1-t)}\int_{\bar{V}_1}^{\infty}e^{-r(T_2-T_1)}\hat{\mathbb{E}}_{T_1}[\max\{V_2 - K_2, 0\}]\phi(V_1)dV_1 \\
&= e^{-r(T_2-t)}\int_{\bar{V}_1}^{\infty}\int_{K_2}^{\infty}(V_2 - K_2)\phi(V_2|V_1)dV_2\phi(V_1)dV_1 \\
&= e^{-r(T_2-t)}\int_{\bar{V}_1}^{\infty}\int_{K_2}^{\infty}(V_2 - K_2)\phi(V_1, V_2)dV_2dV_1 \\
&= e^{-r(T_2-t)}\int_{\bar{V}_1}^{\infty}\int_{K_2}^{\infty}V_2\phi(V_1, V_2)dV_2dV_1 - e^{-r(T_2-t)}K_2\int_{\bar{V}_1}^{\infty}\int_{K_2}^{\infty}\phi(V_1, V_2)dV_2dV_1 \\
&= B - e^{-r(T_2-t)}K_2M(h_{01}^-, h_{02}^-; \rho)
\end{aligned}
\tag{13.7}
$$

where $M(x_1, x_2; \rho)$ is the bi-variate normal probability function with limits x_1 and x_2 and correlation $\rho = \sqrt{(T_1 - t)/(T_2 - t)}$ (which is a correlation between two Browanian values at different times). $K_2 = \bar{V}_2$ in the second period. Lastly, we solve for B. Let $X_1 = \ln V_1$ and $X_2 = \ln V_2$. Then:

$$
\begin{aligned}
B &= e^{-r(T_2-t)} \int_{\bar{V}_1}^{\infty} \int_{K_2}^{\infty} V_2 \phi(V_1, V_2) dV_2 dV_1 \\
&= e^{-r(T_2-t)} \int_{\ln \bar{V}_1}^{\infty} \int_{\ln K_2}^{\infty} e^{X_2} \phi(X_1, X_2) dX_2 dX_1 \\
&= e^{-r(T_2-t)} \int_{\ln \bar{V}_1}^{\infty} \int_{\ln K_2}^{\infty} e^{X_2} \frac{1}{2\pi\sigma_1\sigma_2\rho} \\
&\quad \exp\left\{ -\tfrac{1}{2(1-\rho^2)} \left(\tfrac{X_1-\mu_1}{\sigma_1}\right)^2 - 2\rho \left(\tfrac{X_1-\mu_1}{\sigma_1}\right)\left(\tfrac{X_2-\mu_2}{\sigma_2}\right) + \left(\tfrac{X_2-\mu_2}{\sigma_2}\right)^2 \right\} dX_2 dX_1 \\
&= e^{-r(T_2-t)} \int_{\ln \bar{V}_1}^{\infty} \int_{\ln K_2}^{\infty} \frac{1}{2\pi\sigma_1\sigma_2\sqrt{1-\rho^2}} \exp\left\{ -\tfrac{1}{2(1-\rho^2)} \underbrace{\left(\tfrac{X_1-\mu_1}{\sigma_1}\right)^2}_{B_1} \right. \\
&\quad \left. \underbrace{-2\rho \left(\tfrac{X_1-\mu_1}{\sigma_1}\right)\left(\tfrac{X_2-\mu_2}{\sigma_2}\right)}_{B_3} + \underbrace{\left(\tfrac{X_2-\mu_2}{\sigma_2}\right)^2}_{B_2} - 2(1-\rho^2)X_2 \right\} dX_2 dX_1
\end{aligned}
\tag{13.8}
$$

Let:

$$\mu_j^+ = \mu_j + \sigma_j^2 \tag{13.9}$$

Look at each individual term in square brackets:

$$
\begin{aligned}
B_1 &= \left(\frac{X_1-\mu_1}{\sigma_1}\right)^2 \\
&= \left(\frac{X_1-(\mu_1+\sigma_1^2)}{\sigma_1} + \sigma_1\right)^2 \\
&= \left(\frac{X_1-\mu_1^+}{\sigma_1}\right)^2 + 2(X_1-\mu_1^+) + \sigma_1^2 \\
&= \left(\frac{X_1-\mu_1^+}{\sigma_1}\right)^2 + 2(X_1-\mu_1^+) + \rho^2\sigma_2^2
\end{aligned}
\tag{13.10}
$$

$$
\begin{aligned}
B_2 &= \left(\frac{X_2-\mu_2}{\sigma_2}\right)^2 \\
&= \left(\frac{X_2-(\mu_2+\sigma_2^2)}{\sigma_2}+\sigma_2\right)^2 \\
&= \left(\frac{X_2-\mu_2^+}{\sigma_2}\right)^2 + 2(X_2-\mu_2^+)+\sigma_2^2
\end{aligned} \tag{13.11}
$$

and

$$
\begin{aligned}
B_3 &= -2\rho\left(\frac{X_1-\mu_1}{\sigma_1}\right)\left(\frac{X_2-\mu_2}{\sigma_2}\right) \\
&= -2\rho\left(\frac{X_1-\mu_1^+}{\sigma_1}+\sigma_1\right)\left(\frac{X_2-\mu_2^+}{\sigma_2}+\sigma_2\right) \\
&= -2\rho\left(\frac{X_1-\mu_1^+}{\sigma_1}\right)\left(\frac{X_2-\mu_2^+}{\sigma_2}\right) - 2\rho\sigma_2\left(\frac{X_1-\mu_1^+}{\sigma_1}\right) - 2\rho\sigma_1\left(\frac{X_2-\mu_2^+}{\sigma_2}\right) - 2\rho\sigma_1\sigma_2 \\
&= -2\rho\left(\frac{X_1-\mu_1^+}{\sigma_1}\right)\left(\frac{X_2-\mu_2^+}{\sigma_2}\right) - 2(X_1-\mu_1^+) - 2\rho^2(X_2-\mu_2^+) - 2\rho^2\sigma_2^2
\end{aligned} \tag{13.12}
$$

Putting these terms back into the squared brackets, we get:

$$
\begin{aligned}
B &= e^{-r(T_2-t)}\int_{\ln \bar{V}_1}^{\infty}\int_{\ln K_2}^{\infty}\frac{1}{2\pi\sigma_1\sigma_2\sqrt{1-\rho^2}}e^{\left(\mu_2^+-\sigma_2^2/2\right)} \\
&\quad \exp\left\{-\tfrac{1}{2(1-\rho^2)}\left[\left(\tfrac{X_1-\mu_1^+}{\sigma_1}\right)^2 - 2\rho\left(\tfrac{X_1-\mu_1^+}{\sigma_1}\right)\left(\tfrac{X_2-\mu_2^+}{\sigma_2}\right)+\left(\tfrac{X_2-\mu_2^+}{\sigma_2}\right)^2\right]\right\}dX_2dX_1 \\
&= V_0 M(h_{01}^+, h_{02}^+;\rho)
\end{aligned} \tag{13.13}
$$

Putting all results together, we can arrive at the final pricing equation for equation (13.5), which is a two-period compound option (call on call), as:

$$
C_0 = V_0 M(h_{01}^+, h_{02}^+;\rho) - e^{-r(T_2-t)}K_2 M(h_{01}^-, h_{02}^-;\rho) - e^{-r(T_1-t)}K_1 N(h_{01}^-) \tag{13.14}
$$

As the readers can easily tell, the extension of the two-period compound option model to n periods is straightforward (although tedious). The interested readers can refer to the original work of Geske to see how to write the solution in a form of multi-variate normal probabilities. We note that while one can write down such a solution, the computation of multi-variate normal probabilities requires numerical methods, such as Monte Carlo. Hence, for multi-period compound option pricing, we recommend the lattice method.

The compound option model has a number of important applications, among which the most notably are American option pricing with known (dollar) dividend(s) and (default) risky debt valuation. We shall discuss these two in this Chapter.

13.2.2 American Option Pricing (JFE, 1979b)

American call options will always be exercised at the ex-dividend date, if it is profitable to do so. Hence, it is not necessary to look at other dates. To apply the compound option model to price American options, we assume T_1 is the ex-dividend date. At T_1, the stock will go ex-dividend, so we define S_1 to be the price after the dividend (ex-div price) and V_1 to be the price before the dividend (so call cum-div price), and of course, $V_1 = S_1 + d$. At time T_1, the option will be exercised if the exercise value is greater than the option value after ex-dividend, that is the American option will be exercised if $V_1 - K > C_1$ where C_1 is the European value of the option at time T_1 (which is a function of the ex-dividend stock price S_1). In this case, the payoff is $V_1 - K$. Otherwise, i.e. $V_1 - K < C_1$, the call continues to be held. Hence, todays option value is:

$$
\begin{aligned}
C_0 &= e^{-r(T_1-t)} \max\{V_1 - K, C_1\} \\
&= e^{-r(T_1-t)} \left[\int_{\bar{V}_1}^{\infty} (V_1 - K)\phi(V_1) dV_1 + \int_0^{\bar{V}_1} C_1 \phi(V_1) dV_1 \right]
\end{aligned}
\tag{13.15}
$$

where $\phi(\cdot)$ is the standard normal density function and $\bar{V}_1$ is the critical value over which triggers early exercise and $\bar{V}_1 - K = C_1$. Note that:

$$
C_1 = e^{-r(T_2-T_1)} \int_K^{\infty} (S_2 - K)\phi(S_2|S_1) dS_2 \tag{13.16}
$$

Plugging back,

$$
\begin{aligned}
C_0 = e^{-r(T_1-t)} &\int_{\bar{V}_1}^{\infty} (V_1 - K)\phi(V_1) dV_1 + e^{-r(T_2-t)} \\
&\left[\int_0^{\bar{V}_1} \int_K^{\infty} S_2 \phi(S_2|S_1)\phi(V_1) dS_2 dV_1 - K \int_0^{\bar{V}_1} \int_K^{\infty} \phi(S_2|S_1)\phi(V_1) dS_2 dV_1 \right]
\end{aligned}
\tag{13.17}
$$

A simple change of variable, $S_1 = V_1 - d$, simplifies the above equation into (Jacobian is 1):

$$C_0 = e^{-r(T_1-t)} \int_{\bar{S}_1}^{\infty} (S_1 + d - K)\phi(S_1)dS_1 + e^{-r(T_2-t)}$$
$$\left[\int_0^{\bar{S}_1} \int_K^{\infty} S_2\phi(S_2|S_1)\phi(S_1)dS_2dS_1 - K\int_0^{\bar{S}_1} \int_K^{\infty} \phi(S_2|S_1)\phi(S_1)dS_2dS_1\right] \tag{13.18}$$

where $\bar{S}_1 = \bar{V}_1 - d$. Following the derivation above, we get:

$$\begin{aligned} C_0 = S_0N(h_{1+}) + e^{-r(T_1-t)}(d-K)N(h_{1-}) + S_0M\left(-h_{1+}, h_{2+}; -\tfrac{\sqrt{T_1-t}}{\sqrt{T_2-t}}\right) \\ -e^{-r(T_2-t)}KM\left(-h_{1-}, h_{2-}; -\tfrac{\sqrt{T_1-t}}{\sqrt{T_2-t}}\right) \end{aligned} \tag{13.19}$$

13.2.3 Geske's Risky Debt Model (JFQA, 1977 and 1984)

The Geske model can be applied to evaluate credit risk. Similar to (13.6), define:

$$\begin{aligned} h_{j\pm} &= \frac{\ln V_0 - \ln \bar{V}_j + (r \pm \sigma^2/2)(T_j - t)}{\sigma\sqrt{T_j - t}} \\ y_{j\pm} &= \frac{\ln V_{j-1} - \ln K_j + (r \pm \sigma^2/2)(T_j - T_{j-1})}{\sigma\sqrt{T_j - T_{j-1}}} \end{aligned} \tag{13.20}$$

where $T_0 = t$. Assume the firm issues two zero coupon bonds, expiring at T_1 and T_2, with face values K_1 and K_2 respectively. Default at T_1 is defined by Geske (1984) as the firm value less than the face value of the first debt plus the market value of the second debt, that is $V_1 < K_1 + D_{1,2}$ where $D_{1,2}$ is the market value of K_2 at time T_1.

Note that Geske assumes each time the firm pays off its maturing debt with new equity. As a result, the default condition becomes:

$$\begin{aligned} &V_1 < K_1 + D_{1,2} \\ &V_1 - D_{1,2} < K_1 \\ &E_1^{(a)} < K_1 \end{aligned} \tag{13.21}$$

which is, the equity value after paying off debt K_1 must be greater than K_1. In other words, the equity value after paying off debt includes the new equity that equals

K_1. Hence, this value must be equal to at least K_1. If not, then it indicates that the firm cannot issue new equity, which implies obvious default. The following table demonstrates the balance sheet before and after paying off K_1.

At time T_1			
Before K_1 is paid off		After K_1 is paid off (with equity)	
V_1	K_1 $D_{1,2}(V_1)$ $E_1^{(b)}$	V_1	$D_{1,2}(V_1)$ $E_1^{(a)} = E_1^{(b)} + K_1$

The default barrier can be calculated $\bar{V}_1 = K_1 + D_{1,2}(\bar{V}_1)$ as $D_{1,2}$ is a function of V_1. Note that $\bar{V}_2 = K_2$ as this is a two period model and the firm liquidates at T_2.

When $V_1 > \bar{V}_1 (= K_1 + D_{1,2})$, the market value of the second debt is:

$$\begin{aligned} D_{1,2} &= e^{-r(T_2-T_1)}\hat{\mathbb{E}}_{T_1}[\min\{V_2, \bar{V}_2\}] \\ &= V_1 N(-y_{2+}) - e^{-r(T_2-T_1)}\bar{V}_2 N(y_{2-}) \end{aligned} \tag{13.22}$$

and $D_{1,1} = K_1$ paid by new equity:

$$\begin{aligned} E_1 &= V_1 - D_{1,2} - D_{1,1} \\ &= V_1 N(y_{2+}) - e^{-r(T_2-T_1)}\bar{V}_2 N(y_{2-}) - K_1 \end{aligned} \tag{13.23}$$

When $K_1 < V_1 < \bar{V}_1 (= K_1 + D_{1,2})$, the firm defaults and be liquidated but the first debt (which is more senior) gets full payment. $D_{1,1} = K_1$ paid by new equity, $D_{1,2} = V_1 - K_1$ which is the residual value, and $E_1 = 0$.

When $V_1 < K_1$:

$$\begin{aligned} D_{1,1} &= V_1 \\ D_{1,2} &= 0 \\ E_1 &= 0 \end{aligned} \tag{13.24}$$

To solve for today's values, we have:

$$\begin{aligned} D_{0,1} &= e^{-r(T_1-t)}\hat{\mathbb{E}}_t[\min\{V_1, K_1\}] \\ &= V_0 N(-y_{1+}) + e^{-r(T_1-t)} K_1 N(y_{1-}) \end{aligned} \tag{13.25}$$

and

$$
\begin{aligned}
E_0 &= e^{-r(T_1-t)}\hat{\mathbb{E}}_t[E_1] \\
&= V_0 M(h_{1+}, h_{2+}; \rho) - e^{-r(T_2-t)} K_2 M(h_{1-}, h_{2-}; \rho) - e^{-r(T_1-t)} K_1 N(h_{1-})
\end{aligned} \tag{13.26}
$$

and

$$
\begin{aligned}
D_{0,2} &= V_0 - D_{0,1} - E_0 \\
&= V_0(N(y_{1+}) - M(h_{1+}, h_{2+}; \rho)) - e^{-r(T_1-t)} K_1(N(y_{1-}) - N(h_{1-})) \\
&\quad + e^{-r(T_2-t)} K_2 M(h_{1-}, h_{2-}; \rho)
\end{aligned} \tag{13.27}
$$

Note that $D_{0,2}$ result can be verified by direct integration:

$$
D_{0,2} = e^{-r(T_1-t)}\hat{\mathbb{E}}_t \left[e^{-r(T_2-T_1)}\hat{\mathbb{E}}_{T_1}[\min\{V_2, K_2\}] \times \mathbb{I}_{V_1 > \bar{V}_1} + (V_1 - K_1) \times \mathbb{I}_{K_1 < V_1 < \bar{V}_1} \right] \tag{13.28}
$$

The probability of default for the first period is $1 - N(h_{1-})$. The probability of the second period default is:

$$
\int_0^{\bar{V}_2} \phi(V_2 | V_1 > \bar{V}_1) dV_2 \tag{13.29}
$$

which is the conditional probability of default for the second period, and

$$
\int_{\bar{V}_1}^{\infty} \left[\int_0^{\bar{V}_2} \phi(V_2 | V_1 > \bar{V}_1) dV_2 \right] \phi(V_1) dV_1 = N(h_{1-}) - M(h_{1-}, h_{2-}; \rho)
$$

which is the unconditional probability of the second period default. As a result, the total default probability is:

$$
1 - M(h_{1-}, h_{2-}; \rho) \tag{13.30}
$$

Finally, we present the volatility equation for the equity. By observing the similarity between the Black-Scholes and Geske derivation, it is not hard to guess the volatility formula.

$$\begin{aligned}\sigma_E &= \frac{V}{E}\cdot\frac{\partial E}{\partial V}\cdot\sigma_V \\ &= \frac{V}{E}\cdot M(h_{1+},h_{2+};\rho)\cdot\sigma_V\end{aligned} \tag{13.31}$$

To extend the Geske model to a higher dimensions, please see the Appendix.

13.2.4 An Alternative Bankruptcy Assumption

The no arbitrage default for a firm is defined above. However, this is not how defaults happen in reality. In reality, defaults happen only if companies fail to pay interests. As a result, a firm can continue to survive even if $C_1 < K_1$ (or $V_1 < \bar{V}_1$) as long as $V_1 > K_1$.

To solve the problem of this kind, we first define the face value of the T_j-maturity debt as F_j and total coupon paid at time T_j as c_j. Then re-define:

$$\begin{aligned}h_{j\pm} &= \frac{\ln V_0 - \ln[F_j + \Sigma_{i=j}^{n} F_i c_i] + (r \pm \sigma^2/2)(T_j - t)}{\sigma\sqrt{T_j - t}} \\ y_{j\pm} &= \frac{\ln V_{j-1} - \ln[F_j(1+c_j)] + (r \pm \sigma^2/2)(T_j - T_{j-1})}{\sigma\sqrt{T_j - T_{j-1}}}\end{aligned} \tag{13.32}$$

In Geske (1984), bankruptcy (at T_1) is defined as the firm value being less than the total debt market value. But a more realistic assumption for default is that the firm value less than the total cash flow payments. We further assume coupon bonds rather than zero bonds. Let F_j represent the face value of T_j-maturity debt and c_j the coupon rate of the T_j-maturity debt.

When $V_1 > F_1(1+c_1) + F_2c_2$ (no default):

$$D_{1,1} = F_1(1+c_1) \tag{13.33}$$

$$D_{1,2} = F_2c_2 + V_1N(-y_{2+}) + e^{-r(T_2-T_1)}(F_2(1+c_2))N(y_{2-}) \tag{13.34}$$

$$\begin{aligned}E_1 &= V_1 - D_{1,1} - D_{1,2} \\ &= V_1N(y_{2+}) - e^{-r(T_2-T_1)}(F_2(1+c_2))N(y_{2-}) - F_1(1+c_1) - F_2c_2\end{aligned} \tag{13.35}$$

which can possibly be less than 0!

When $F_1(1+c_1) < V_1 < F_1(1+c_1) + F_2c_2$, the firm defaults and liquidates but the senior debt (T_1-maturity debt) stills receives full payment. The junior debt (T_2-maturity debt) receives the residual. Hence,

$$\begin{aligned} E_1 &= 0 \\ D_{1,1} &= F_1(1+c_1) \\ D_{1,2} &= V_1 - D_{1,1} \end{aligned} \tag{13.36}$$

When $V_1 < F_1(1+c_1)$,

$$\begin{aligned} E_1 &= 0 \\ D_{1,1} &= V_1 \\ D_{1,2} &= 0 \end{aligned} \tag{13.37}$$

Under this circumstance, today's equity value has the following result:

$$\begin{aligned} E_0 = V_0 M(h_{1+}, h_{2+}; \rho) - e^{-r(T_2-t)}(F_2(1+c_2))M(h_{1-}, h_{2-}; \rho) \\ -e^{-r(T_1-t)}(F_1(1+c_1) + F_2c_2)N(h_{1-}) \end{aligned} \tag{13.38}$$

The T_1-maturity debt has the following result:

$$\begin{aligned} D_{0,1} &= e^{-r(T_i-t)}\hat{\mathbb{E}}_t[\min\{F_1(1+c_1), V_1\}] \\ &= e^{-r(T_i-t)}\hat{\mathbb{E}}_t[V_1] - e^{-r(T_i-t)} \max\{V_1 - F_1(1+c_1), 0\}] \\ &= V_0 - V_0 N(y_{1+}) + e^{-r(T_1-t)} F_1(1+c_1) N(y_{1-}) \end{aligned} \tag{13.39}$$

The T_2-maturity debt has the following result:

$$
\begin{aligned}
D_{0,2} &= e^{-r(T_1-t)}\left(\int_{\bar{V}_1}^{\infty}(V_1-E_1-D_{1,1})\phi(V_1)dV_1+\int_{F_1(1+c_1)}^{\bar{V}_1}(V_1-D_{1,1})\phi(V_1)dV_1\right. \\
&\quad\left.+\int_0^{F_1(1+c_1)}V_1\phi(V_1)dV_1\right) \\
&= e^{-r(T_1-t)}\left(\hat{\mathbb{E}}_t[V_1]-\int_{\bar{V}_1}^{\infty}(E_1+D_{1,1})\phi(V_1)dV_1-\int_{F_1(1+c_1)}^{\bar{V}_1}D_{1,1}\phi(V_1)dV_1\right) \\
&= e^{-r(T_1-t)}\left(\hat{\mathbb{E}}_t[V_1]-\int_{\bar{V}_1}^{\infty}E_1\phi(V_1)dV_1-\int_{F_1(1+c_1)}^{\bar{V}_1}D_{1,1}\phi(V_1)dV_1\right) \\
&= V_0-E_0-D_{0,1} \\
&= V_0[N(y_{1+})-M(h_{1+},h_{2+};\rho)]-e^{-r(T_1-t)}F_1(1+c_1)[N(y_{1-})-N(h_{1-})] \\
&\quad+e^{-r(T_2-t)}F_2(1+c_2)M(h_{1-},h_{2-};\rho)+e^{-r(T_1-t)}F_2c_2N(h_{1-})
\end{aligned} \tag{13.40}
$$

Note that if both debts are zero coupon, then $h=y$ and $D_{0,2}$ can be simplified as:

$$
D_{0,2}=V_0[N(h_{1+})-M(h_{1+},h_{2+};\rho)]+e^{-r(T_2-t)}F_2M(h_{1-},h_{2-};\rho) \tag{13.41}
$$

To extend this modified model to a higher dimensions, please see the Appendix.

Alternative Recovery Assumptions

When $V_1 > F_1(1+c_1)+F_2c_2$ (no default):

$$
\begin{aligned}
E_1 &= V_1-D_{1,2} \\
D_{1,1} &= F_1(1+c_1) \\
D_{1,2} &= e^{-r(T_2-T_1)}\hat{\mathbb{E}}_{T_1}[F_2(1+c_2)1_{V_2>F_2(1+c_2)},R_2F_21_{V_2<F_2(1+c_2)}] \\
&= e^{-r(T_2-T_1)}\left(\int_{F_2(1+c_2)}^{\infty}F_2(1+c_2)\phi(V_2)dV_2+\int_0^{F_2(1+c_2)}R_2F_2\phi(V_2)dV_2\right) \\
&= e^{-r(T_2-T_1)}(F_2(1+c_2)N(y_{2-})+R_2F_2N(-y_{2-}))
\end{aligned} \tag{13.42}
$$

where R_2 is recovery rate at T_2. Note that the equity value at T_1 can be negative (low enough V_1 and high enough R_2). To justify this, we need to assume that equity holders have unlimited liabilities (because they have to come out of pocket to pay R_2F_2 even if the asset value is less than that) and hence do not hold an option.

When $F_1(1+c_1) < V_1 < F_1(1+c_1) + F_2c_2$:

$$\begin{aligned} E_1 &= V_1 - D_{1,1} - D_{1,2} \\ D_{1,1} &= F_1(1+c_1) \\ D_{1,2} &= R_2F_2 \end{aligned} \tag{13.43}$$

When $V_1 < F_1(1+c_1)$:

$$\begin{aligned} E_1 &= V_1 - D_{1,1} - D_{1,2} \\ D_{1,1} &= R_1F_1 \\ D_{1,2} &= R_2F_2 \end{aligned} \tag{13.44}$$

As a result, at current time, the T_1-maturity debt has the following result:

$$\begin{aligned} D_{0,1} &= e^{-r(T_1-t)}\hat{\mathbb{E}}_t[F_1(1+c_1)1_{V_1>F_1(1+c_1)} + R_1F_11_{V_1<F_1(1+c_1)}\}] \\ &= e^{-r(T_1-t)}(F_1(1+c_1)N(y_{1-}) + R_1F_1N(-y_{1-})) \end{aligned} \tag{13.45}$$

The T_2-maturity debt has the following result:

$$\begin{aligned} D_{0,2} &= e^{-r(T_1-t)}\left(e^{-r(T_2-T_1)}\int_{F_1(1+c_1)+F_2c_2}^{\infty}\left[F_2(1+c_2)\int_{F_2(1+c_2)}^{\infty}\phi(V_2|V_1)dV_2+\right.\right. \\ &\qquad \left.\left. R_2F_2\int_0^{F_2(1+c_2)}\phi(V_2|V_1)dV_2\right]\phi(V_1)dV_1 + R_2F_2\int_0^{F_1(1+c_1)+F_2c_2}\phi(V_1)dV_1\right) \\ &= e^{-r(T_2-t)}\{F_2(1+c_2)M(h_{1-},h_{2-};\rho) + R_2F_2[N(h_{1-}) - M(h_{1-},h_{2-};\rho)]\} \\ &\quad + e^{-r(T_1-t)}R_2F_2N(h_{1-}) \end{aligned} \tag{13.46}$$

The equity is the residual value:

$$E_0 = V_0 - D_{0,1} - D_{0,2} \tag{13.47}$$

13.3 Random Volatility Model

The random volatility option pricing model is known as the Heston model (1993). But prior to Heston, there had been three important papers addressing this important issue by Hull and White, Wiggins, and Scott, all of which were published in 1987.

13.3.1 The 1987 Model

Define the joint process:

$$\begin{cases} dS = \mu S dt + \sqrt{V} S dW_1 \\ dV = \alpha(\beta - V)dt + \gamma\sqrt{V} dW_2 \end{cases} \tag{13.48}$$

where $dW_1 dW_2 = \rho dt$. The option price change, following Ito's lemma, is:

$$\begin{aligned} dC &= C_S dS + {}^1\!/\!_2 C_{SS}(dS)^2 + C_V dV + {}^1\!/\!_2 C_{VV}(dV)^2 + C_{SV}(dS)(dV) + C_t dt \\ &= \left(\mu S C_S + {}^1\!/\!_2 V S^2 C_{SS} + \alpha(\beta - V)C_V + {}^1\!/\!_2 \gamma^2 V C_{VV} + \rho\gamma V S C_{SV} + C_t\right) dt \\ &\quad + \sqrt{V} S C_S dW_1 + \gamma\sqrt{V} C_V dW_2 \end{aligned} \tag{13.49}$$

Form a portfolio of two options and the stock:

$$\begin{aligned} V &= C^{(1)} + h_1 S + h_2 C^{(2)} \\ dV &= dC^{(1)} + h_1 dS + h_2 dC^{(2)} \end{aligned} \tag{13.50}$$

The uncertainty part must equal 0 to avoid risk. In other words,

$$\begin{aligned} &\sqrt{V} S C_S^{(1)} dW_1 + \gamma\sqrt{V} C_V^{(1)} dW_2 + h_1\sqrt{V} S dW_1 \\ &+ h_2\sqrt{V} S C_S^{(2)} dW_1 + h_2\gamma\sqrt{V} C_V^{(2)} dW_2 = 0 \end{aligned} \tag{13.51}$$

or equivalently,

$$\begin{aligned} \gamma\sqrt{V} C_V^{(1)} dW_{21} + h_2\gamma\sqrt{V} C_V^{(2)} dW_2 = 0 \\ \sqrt{V} S C_S^{(1)} dW_1 + h_1\sqrt{V} S dW + h_2\sqrt{V} S C_S^{(2)} dW_1 = 0 \end{aligned} \tag{13.52}$$

which can be simplified as:

$$h_2 = -\frac{C_V^{(1)}}{C_V^{(2)}} \tag{13.53}$$

$$h_1 = -C_S^{(1)} + \frac{C_V^{(1)}}{C_V^{(2)}} C_S^{(2)} \tag{13.54}$$

Substitude this back to have:

$$\begin{aligned}
dV &= rV dt \\
&\Big(\mu S C_S^{(1)} + {}^1\!/\!_2 V S^2 C_{SS}^{(1)} + \alpha(\beta - V) C_V^{(1)} + {}^1\!/\!_2 \gamma^2 V C_{VV}^{(1)} + \rho\gamma V C_{SV}^{(1)} + C_t^{(1)} \\
&+ \left[-C_S^{(1)} + \frac{C_V^{(1)}}{C_V^{(2)}} C_S^{(2)} \right] \mu S \\
&- \frac{C_V^{(1)}}{C_V^{(2)}} \left[\mu S C_S^{(1)} + {}^1\!/\!_2 V S^2 C_{SS}^{(1)} + \alpha(\beta - V) C_V^{(1)} + {}^1\!/\!_2 \gamma^2 V C_{VV}^{(1)} + \rho\gamma V C_{SV}^{(1)} + C_t^{(1)} \right] \Big) \\
&= r \left(C^{(1)} + \left[-C_S^{(1)} + \frac{C_V^{(1)}}{C_V^{(2)}} C_S^{(2)} \right] S - \frac{C_V^{(1)}}{C_V^{(2)}} C^{(2)} \right)
\end{aligned} \tag{13.55}$$

Simplify the above to get:

$$\begin{aligned}
&\Big(r S C_S^{(1)} + {}^1\!/\!_2 V S^2 C_{SS}^{(1)} + \alpha(\beta - V) C_V^{(1)} + {}^1\!/\!_2 \gamma^2 V C_{VV}^{(1)} + \rho\gamma V C_{SV}^{(1)} + C_t^{(1)} \\
&\quad - \frac{C_V^{(1)}}{C_V^{(2)}} \left[r S C_S^{(2)} + {}^1\!/\!_2 V S^2 C_{SS}^{(2)} + \alpha(\beta - V) C_V^{(2)} + {}^1\!/\!_2 \gamma^2 V C_{VV}^{(2)} + \rho\gamma V C_{SV}^{(2)} + C_t^{(2)} \right] \Big) \\
&= r C^{(1)} - \frac{C_V^{(1)}}{C_V^{(2)}} r C^{(2)}
\end{aligned} \tag{13.56}$$

It is clear that the solution to the Black-Scholes PDE is the solution to the Heston PDE. It is interesting to know that the solution to the following PDE is also the solution to the Heston PDE:

$$r S C_S + {}^1\!/\!_2 V S^2 C_{SS} + \alpha(\beta - V) C_V + {}^1\!/\!_2 \gamma^2 V C_{VV} + \rho\gamma V C_{SV} + C_t + b C_V = r C \tag{13.57}$$

Hence, we need an intertermporal asset pricing model:

$$\mathbb{E}\left[\frac{dC}{C}\right]=\left(r+\frac{C_S}{C}S(\mu-r)+\frac{C_V}{C}\lambda V\right)dt \tag{13.58}$$

which immediately implies that

$$rC=rSC_S+{}^1\!/\!_2VS^2C_{SS}+[\alpha(\beta-V)-\lambda V]C_V+{}^1\!/\!_2\gamma^2VC_{VV}+\rho\gamma VSC_{SV}+C_t \tag{13.59}$$

The problem is of course how to solve for the solution to this PDE. As it turns out this PDE does not have a closed-form solution. Hull and White, Higgins, and Scott all tried in 1986 to solve this problem but could only arrive at numerical solutions. The closed-form solution was not discovered until Heston (1993) and Scott (1997) who used the Fourier Inversion technique.

13.3.2 The Heston Model

We first perform the following usual transformation:

$$x=\ln S \tag{13.60}$$

Then

$$\begin{aligned}&C_S=C_xx_S=\tfrac{1}{S}C_x\\&C_{SS}=C_xx_{SS}+C_{xx}x_S^2=C_x\tfrac{-1}{S^2}+C_{xx}\tfrac{1}{S^2}=\tfrac{1}{S^2}(C_{xx}-C_x)\end{aligned} \tag{13.61}$$

As a result, the original PDE can be simplified as follows:

$$rC=(r-{}^1\!/\!_2V)C_x+{}^1\!/\!_2VC_{xx}+[\alpha(\beta-V)-\lambda V]C_V+{}^1\!/\!_2\gamma^2VC_{VV}+\rho\gamma VC_{xV}+C_t \tag{13.62}$$

Guess the solution of the option pricing model to be:

$$\begin{aligned}C&=e^{-r(T-t)}\hat{\mathbb{E}}_t[\max\{S_T-K,0\}]\\&=e^{-r(T-t)}\{E[S]\Pi_1+K\Pi_2\}\\&=e^{-r(T-t)}\left\{Se^{r(T-t)}\Pi_1+K\Pi_2\right\}\\&=e^{-r(T-t)}U\end{aligned} \tag{13.63}$$

Plug U back to the PDE to arrive at the following Kolmogorov backward equation:

$$0 = (r - 1/2V)U_x + 1/2VU_{xx} + [\alpha(\beta - V) - \lambda V]U_V + 1/2\gamma^2 VU_{VV} + \rho\gamma VU_{xV} + U_t \quad (13.64)$$

Note that this PDE is almost identical to the PDE for the call price except for the left hand side term rC. Taking derivatives:

$$\begin{aligned}
U_x &= e^{x+r(T-t)}\Pi_1 + e^{x+r(T-t)}\Pi_{1x} - K\Pi_{2x} \\
U_{xx} &= e^{x+r(T-t)}\Pi_1 + 2e^{x+r(T-t)}\Pi_{1x} + e^{x+r(T-t)}\Pi_{1xx} - K\Pi_{2xx} \\
U_V &= e^{x+r(T-t)}\Pi_{1V} - K\Pi_{2V} \\
U_{VV} &= e^{x+r(T-t)}\Pi_{1vv} - K\Pi_{2VV} \\
U_{xV} &= e^{x+r(T-t)}\Pi_{1V} + e^{x+r(T-t)}\Pi_{1xV} - K\Pi_{2xV} \\
U_t &= e^{x+r(T-t)}(-r)\Pi_1 + e^{x+r(T-t)}\Pi_{1t} - K\Pi_{2t}
\end{aligned} \quad (13.65)$$

The PDE for U becomes two PDEs for Π_1 and Π_2 as follows:

$$\begin{aligned}
&\frac{V}{2}\Pi_{1xx} + \frac{\gamma^2 V}{2}\Pi_{1VV} + \rho\gamma V\Pi_{1xV} + [\alpha\beta - (\alpha + \lambda - \rho\gamma)V]\Pi_{1V} + \left[r + \frac{V}{2}\right]\Pi_{1x} + \Pi_{1t} = 0 \\
&\frac{V}{2}\Pi_{2xx} + \frac{\gamma^2 V}{2}\Pi_{2VV} + \rho\gamma V\Pi_{2xV} + [\alpha\beta - (\alpha + \lambda)V]\Pi_{2V} + \left[r - \frac{V}{2}\right]\Pi_{2x} + \Pi_{2t} = 0
\end{aligned} \quad (13.66)$$

The characteristic functions satisfy the same PDEs. Guess the form:

$$f_j = e^{C_j(T-t) + D_j(T-t)V + iux} \quad (13.67)$$

The boundary condition as $t \to T$ is $f_j = e^{iux}$; or equivalently $C_j = 0$ and $D_j = 0$.

Computing the corresponding partials:

$$\begin{aligned}
f_{jx} &= f_j iu \\
f_{jxx} &= f_j(iu)^2 = -f_j u^2 \\
f_{jV} &= f_j D_j \\
f_{jVV} &= f_j D_j^2 \\
f_{jxV} &= f_j iuD_j \\
f_{jt} &= f_j(C_{jt} + D_{jt}V)
\end{aligned} \quad (13.68)$$

Note that from equation (4.50) in Chapter 4 that the characteristic function satisfies the same PDE as the probability function. Hence, substitute equation (13.66) back to equation (13.64), the PDE becomes:

$$\frac{V}{2}(-u^2)+\frac{\gamma^2 V}{2}D_1^2+\rho\gamma V iuD_1+[\alpha\beta-(\alpha+\lambda-\rho\gamma)V]D_1+\left[r+\frac{V}{2}\right]iu+(C_{1t}+D_{1t}V)=0$$
$$\frac{V}{2}(-u^2)+\frac{\gamma^2 V}{2}D_2^2+\rho\gamma V iuD_2+[\alpha\beta-(\alpha+\lambda)V]D_2+\left[r-\frac{V}{2}\right]iu+(C_{2t}+D_{2t}V)=0 \tag{13.69}$$

This is a system of two ODEs:

$$\begin{aligned}&-\frac{1}{2}u^2+\frac{\gamma^2}{2}D_1^2-(\alpha+\lambda-\rho\gamma(1+iu))D_1+\frac{1}{2}iu+D_{1t}=0\\&riu+\alpha\beta D_1+C_{1t}=0\end{aligned} \tag{13.70}$$

and

$$\begin{aligned}&-\frac{1}{2}u^2-\frac{1}{2}iu+\frac{\gamma^2}{2}D_2^2-(\alpha+\lambda-\rho\gamma iu)D_2+D_{2t}=0\\&riu+\alpha\beta D_2+C_{2t}=0\end{aligned} \tag{13.71}$$

We can write the solutions to D_1 and D_2 as:

$$D_1=\frac{d_1}{\gamma^2}\left[\frac{\left(\frac{d_1+\rho\gamma(1+iu)-(\alpha+\lambda)}{d_1-\rho\gamma(1+iu)+(\alpha+\lambda)}\right)e^{-d_1(T-t)}-1}{\left(\frac{d_1+\rho\gamma(1+iu)-(\alpha+\lambda)}{d_1-\rho\gamma(1+iu)+(\alpha+\lambda)}\right)e^{-d_1(T-t)}+1}-\frac{\rho\gamma(1+iu)-(\alpha+\lambda)}{d_1}\right] \tag{13.72}$$

where

$$d_1=\sqrt{[\rho\gamma(1+iu)-(\alpha+\lambda)]^2-\gamma^2\left(iu-u^2\right)}$$

and

$$D_2=\frac{d_2}{\gamma^2}\left[\frac{\left(\frac{d_2+\rho\gamma iu-(\alpha+\lambda)}{d_2-\rho\gamma iu+(\alpha+\lambda)}\right)e^{-d_2(T-t)}-1}{\left(\frac{d_2+\rho\gamma iu-(\alpha+\lambda)}{d_2-\rho\gamma iu+(\alpha+\lambda)}\right)e^{-d_2(T-t)}+1}-\frac{\rho\gamma iu-(\alpha+\lambda)}{d_2}\right] \tag{13.73}$$

where

$$d_2 = \sqrt{[\rho\gamma iu - (\alpha+\lambda)]^2 + \gamma^2 \left(iu + u^2\right)}$$

Finally, $C_j = riu(T-t) + \alpha\beta \int_t^T D_j ds$ can be solved as:

$$\begin{aligned} C_1 &= riu(T-t) \\ &\quad -\frac{\alpha\beta}{\gamma^2}\left[-(\alpha+\lambda-\rho\gamma(1+iu)) + \sqrt{(\alpha+\lambda-\rho\gamma(1+iu))^2+\gamma^2(u^2-iu)}\right](T-t) \\ -\frac{2\alpha\beta}{\gamma^2} &\left\{ \begin{array}{l} \ln\left[1+\frac{-(\alpha+\lambda-\rho\gamma(1+iu))+\sqrt{(\alpha+\lambda-\rho\gamma(1+iu))^2+\gamma^2(u^2-iu)}}{(\alpha+\lambda-\rho\gamma(1+iu))+\sqrt{(\alpha+\lambda-\rho\gamma(1+iu))^2+\gamma^2(u^2-iu)}} e^{-\sqrt{(\alpha+\lambda-\rho\gamma(1+iu))^2+\gamma^2(u^2-iu)}(T-t)}\right] \\ -\ln\left[1+\frac{-(\alpha+\lambda-\rho\gamma(1+iu))+\sqrt{(\alpha+\lambda-\rho\gamma(1+iu))^2+\gamma^2(u^2-iu)}}{(\alpha+\lambda-\rho\gamma(1+iu))+\sqrt{(\alpha+\lambda-\rho\gamma(1+iu))^2+\gamma^2(u^2-iu)}}\right] \end{array} \right\} \end{aligned} \tag{13.74}$$

and

$$\begin{aligned} C_2 &= riu(T-t) \\ &-\frac{\alpha\beta}{\gamma^2}\left[-(\alpha+\lambda) + \sqrt{(\alpha+\lambda)^2+\gamma^2(u^2+iu)}\right](T-t) \\ &-\frac{1}{\gamma^2}\left\{\ln\left[1+\frac{-(\alpha+\lambda)+\sqrt{(\alpha+\lambda)^2+\gamma^2(u^2+iu)}}{(\alpha+\lambda)+\sqrt{(\alpha+\lambda)^2+\gamma^2(u^2+iu)}} e^{-\sqrt{(\alpha+\lambda)^2+\gamma^2(u^2+iu)}(T-t)}\right]\right. \\ &\qquad\qquad \left. -\ln\left[1+\frac{-(\alpha+\lambda)+\sqrt{(\alpha+\lambda)^2+\gamma^2(u^2+iu)}}{(\alpha+\lambda)+\sqrt{(\alpha+\lambda)^2+\gamma^2(u^2+iu)}}\right]\right\} \end{aligned} \tag{13.75}$$

It is noted that when $iu = 1$, $D_2 = 0$, and $C_2 = A$ and the characteristic function becomes the futures price (risk neutral expectation of S), as expected.

Finally, the probabilities are Fourier inversions of the characteristic functions:

$$\Pi_j = \frac{1}{2} + \frac{1}{\pi}\int_0^\infty \mathrm{Re}\left[\frac{e^{-iu\ln K} f_j(u)}{iu}\right] du \tag{13.76}$$

13.3.3 The Scott model

The Scott model is more complex. It has random interest rates and jumps:

$$\begin{cases} dS = (\mu - \lambda_J \mu_J)Sdt + \sqrt{V}SdW_1 + dJ \\ dV = \alpha(\beta - V)dt + \gamma\sqrt{V}dW_2 \\ dr = a(b - r)dt + g\sqrt{r}dW_3 \end{cases} \tag{13.77}$$

where dJ represents a Poisson jump process with jump frequency λ_J and the mean of the jump size μ_J. The other processes are similar to the Heston model and the Cox-Ingersoll-Ross model for the term structure. The interest rate process is uncorrelated with either variance or stock price process, i.e. $dW_1 dW_3 = 0$ and $dW_2 dW_3 = 0$. Jumps are uncorrelated with any diffusion. The PDE of this model can be derived similarly, which is left as an exercise.

13.3.4 The Upper Bound for American Option Prices

American options require numerical methods, namely lattice models, to provide accurate price estimates. The computations can become expensive when more than one state variable is involved. Analytical upper bounds can therefore provide a useful guideline for how high the American values can reach . In this paper, we derive analytical (closed-form) upper bounds for American option prices under stochastic interest rates, stochastic volatility, and jumps where American option prices are difficult to compute with accuracy. In a stochastic volatility model (Heston (1993) and Scott (1997)) that has two random factors, we demonstrate that the upper bound only needs a very small fraction of the time of the American option to compute.

We find that An American option is bounded from above by the risk neutral expectation of its maturity payoff if this expectation is greater than the intrinsic value at all times. To show that we shall prove this theorem using a discrete approximation similar to the lattice approach. Let T be the maturity of the option contract. At T, the payoff is given by $X(T)$. At $T - \Delta t$, the payoff is either the discounted terminal value:

$$\hat{\mathbb{E}}_{T-\Delta t}[\delta(T - \Delta t, T)X(T)] \tag{13.78}$$

or the intrinsic value:

$$X(T - \Delta t) \tag{13.79}$$

whichever is larger where $\delta(t, s)$ represents the discount factor from time t to time s. At $T - 2\Delta t$, the larger quantity of the above result is discounted and compared with the early exercise value, i.e. the payoff is either the continuation American value:

$$\hat{\mathbb{E}}_{T-2\Delta t}[\delta(T-2\Delta t, T-\Delta t)\max\{\hat{\mathbb{E}}_{T-\Delta t}[\delta(T-\Delta t, T)X(T)], X(T-\Delta t)\}] \quad (13.80)$$

or the intrinsic value:

$$X(T-2\Delta t) \quad (13.81)$$

whichever is larger. This process is repeated until the current time is reached and the American option value is obtained. Now, consider a function at time $T-\Delta t$, $\hat{\mathbb{E}}_{T-\Delta t}[X(T)]$. It is certainly true that:

$$\hat{\mathbb{E}}_{T-\Delta t}[X(T)] > \hat{\mathbb{E}}_{T-\Delta t}[\delta(T-\Delta t, T)X(T)] \quad (13.82)$$

provided that the values of the discount factor are less than one for all sample paths. If $\hat{\mathbb{E}}_{T-\Delta t}[X(T)] > X(T-\Delta t)$ is also true, then at $T-2\Delta t$, it is true that this expected payoff is greater than the continuation value, i.e.:

$$\begin{aligned} &\hat{\mathbb{E}}_{T-2\Delta t}[X(T)] = E_{T-2\Delta t}[\hat{\mathbb{E}}_{T-\Delta t}[X(T)]] \\ &> \hat{\mathbb{E}}_{T-2\Delta t}[\delta(T-2\Delta t, T-\Delta t)\max\{E_{T-\Delta t}[\delta(T-\Delta t, T)X(T)], X(T-\Delta t)\}] \end{aligned} \quad (13.83)$$

As a result, as long as we can show that $\hat{\mathbb{E}}_t[X(T)] > X(t)$ for all t, then $\hat{\mathbb{E}}_t[X(T)]$ is an upper bound of the American value. The result will hold in continuous time as Δt reaches a limit.

American Calls on Dividend Paying Stocks

Following from the above argument, we know that if the expected value of a certain contingent claims maturity payoff is greater than the early exercise value, this expected value will serve as an upper bound of the American value. For American calls, the expected value of the payoff at an arbitrary time t is shown to be greater than the early exercise value as follows:

$$\begin{aligned} \hat{\mathbb{E}}_t[\max\{S(T)-K, 0\}] &\geqslant \max\{\hat{\mathbb{E}}_t[S(T)]-K, 0\} \\ &> \max\{S(t)-K, 0\} \end{aligned} \quad (13.84)$$

where S is the stock price and K is the strike price. The first line follows from Jensens inequality and the second line holds as long as the expected net growth of the stock price is positive.

American Calls on Futures

The upper bound for futures calls is more straightforward since the futures price is a martingale, that is the futures price at current time t for maturity s is the conditional expectation of the futures price at some future time T for the same maturity s. Put it formally, $\Phi(t,s) = \hat{\mathbb{E}}_t[\Phi(T,s)]$. As a result,

$$\begin{aligned} \hat{\mathbb{E}}_t[\max\{\Phi(T,s) - K, 0\}] &\geqslant \max\{\hat{\mathbb{E}}_t[\Phi(T,s)] - K, 0\} \\ &= \max\{\Phi(t,s) - K, 0\} \end{aligned} \tag{13.85}$$

American Puts on Dividend Paying Stocks

It is interesting to note that American puts are not so easy to analyze as the American calls. The expected terminal payoff is not necessarily greater than the intrinsic value:

$$\begin{aligned} \hat{\mathbb{E}}_t[\max\{K - S(T), 0\}] &\geqslant \max\{K - \hat{\mathbb{E}}[S(T)], 0\} \\ &< \max\{K - S(t), 0\} \end{aligned} \tag{13.86}$$

and therefore the expected payoff of a put is not an upper bound for American puts. Unlike the upper bound result for American calls, the direction of the second line is reversed when the expected net growth of the stock is positive. This is why Chaudhury and Wei (1994) assert that they cannot use the pure futures option formula to bound American puts.

Note that the incentive for the American puts to be exercised early is the higher interest income (than the dividend income). If we eliminate this incentive, the American puts will not be exercised early. As a result, we can obtain an upper bound by setting both interest rate and dividend yield to 0. However, we discover a tighter upper bound for American puts by lowering the stock price. We suggest the stock price to be lowered by the net growth (risk-free rate minus dividend yield). We define an upper bound as follows:

$$\begin{aligned} \hat{\mathbb{E}}_t\left[\max\left\{K - \delta(t,T)S(T), 0\right\}\right] &> \max\left\{K - \hat{\mathbb{E}}_t\left[\delta(t,T)S(T)\right], 0\right\} \\ &= \max\{K - S(t), 0\} \end{aligned} \tag{13.87}$$

Note that the discount factor here includes the risk-free rate and any dividend rate. It is seen that the expected value is greater than the intrinsic value. For this equation to indeed be an upper bound, we still need to show that this equation is greater than a time homogeneous expectation for any t:

$$\begin{aligned}\hat{\mathbb{E}}_{t-\Delta t}\left[\max\left\{K-\delta(t-\Delta t,T)S(T),0\right\}\right] &> \hat{\mathbb{E}}_{t-\Delta t}\left[\max\left\{K-\delta(t,T)S(T),0\right\}\right] \\ &= \hat{\mathbb{E}}_{t-\Delta t}\left[\hat{\mathbb{E}}_{t}\left[\max\left\{K-\delta(t,T)S(T),0\right\}\right]\right]\end{aligned} \tag{13.88}$$

Here, the process must be strictly negative. Since this inequality holds, (13.83) is indeed an upper bound for American puts. The above equation has an interesting interpretation. Note that the cause the triggers early exercise of a put option is when the dividend income is less than the interest income. If we can design a put option that is protected against any loss in the difference between interest and dividend incomes, then this option will never be exercised early. This put option price should be an upper bound of standard American put options.

American Puts on Futures

Although American puts are not bounded by the expectation of the terminal payoff, American futures puts are. Theorem 1 applies to American futures puts:

$$\begin{aligned}\hat{\mathbb{E}}_t[\max\{K-\Phi(T,s),0\}] &\geqslant \max\{K-\hat{\mathbb{E}}[\Phi(T,s)],0\} \\ &= \max\{K-\Phi(t,s),0\}\end{aligned} \tag{13.89}$$

This is because, of course, the futures price is a martingale.

13.4 Multi-factor Interest Rate Options (Chen-Scott)

The multi-factor CIR option pricing model leverages a lot upon the single factor CIR model. Readers need to refer back the single factor CIR option model for details of the non-central chi-square probability function and forward measure. The Chen-Scott version of the multi factor CIR model is as follows.

$$r=\sum\nolimits_{j=1}^{J}y_j \tag{13.90}$$

where

$$dy_j = \alpha_j(\mu_j - y_j)dt + \sigma_j\sqrt{y_j}dW_j \tag{13.91}$$

and $dW_i dW_j = 0$. The zero coupon bond formula is therefore:

$$\begin{aligned}P(t,T) &= \hat{\mathbb{E}}_t\left[\exp\left(-\int_t^T r(u)du\right)\right] \\ &= \hat{\mathbb{E}}_t\left[\exp\left(-\int_t^T \sum\nolimits_{j=1}^J y_j(u)du\right)\right] \\ &= \hat{\mathbb{E}}_t\left[\prod\nolimits_{j=1}^J \left[\exp\left(-\int_t^T y_j(u)du\right)\right]\right] \\ &= \prod\nolimits_{j=1}^J \hat{\mathbb{E}}_t\left[\exp\left(-\int_t^T y_j(u)du\right)\right]\end{aligned} \tag{13.92}$$

Since each expected value is a CIR solution, the bond price is therefore:

$$P(t,T) = \prod\nolimits_{j=1}^J A_j(t,T)e^{-y_j(t)B_j(t,T)} \tag{13.93}$$

where

$$\begin{aligned}A_j(t,T) &= \left[\frac{2\gamma_j e^{(\alpha_j+\lambda_j+\gamma_j)(T-t)/2}}{(\alpha_j+\lambda_j+\gamma_j)\left(e^{\gamma_j(T-t)}-1\right)+2\gamma_j}\right]^{2\alpha_j\mu_j/\sigma_j^2} \\ B_j(t,T) &= \frac{2\left(e^{\gamma_j(T-t)}-1\right)}{(\alpha_j+\lambda_j+\gamma_j)\left(e^{\gamma_j(T-t)}-1\right)+2\gamma_j} \\ \gamma_j &= \sqrt{(\alpha_j+\lambda_j)^2+2\sigma_j^2}\end{aligned} \tag{13.94}$$

A European option on such a zero coupon bond is not easily solvable. The maturity payoff is:

$$\begin{aligned}C(T) &= \max\{P(T,s)-K,0\} \\ &= \max\left\{\prod\nolimits_{j=1}^J A_j(T,s)e^{-y_j(T)B_j(T,s)} - K, 0\right\}\end{aligned} \tag{13.95}$$

The exercise region is where the bond price is greater than the strike price. In log space, this is:

$$\begin{aligned}\ln K &< \ln P(T,s) \\ &= \sum\nolimits_{j=1}^J \ln A_j(T,s) - y_j(T)B_j(T,s)\end{aligned} \tag{13.96}$$

In other words, in the state variable space:

$$y_j(T)B_j(T,s) < \sum_{j=1}^{J} \ln A_j(T,s) - \ln K \tag{13.97}$$

To derive the option price today, we need the forward measure

$$\begin{aligned} C(T) &= \hat{\mathbb{E}}_t\left[\exp\left(-\int_t^T r_u du\right)\max\{P(T,s)-K,0\}\right] \\ &= P(t,T)\tilde{\mathbb{E}}_t\left[\max\left\{\prod_{j=1}^{J} A_j(T,s)e^{-y_j(T)B_j(T,s)} - K, 0\right\}\right] \\ &= P(t,T)\left(\overset{*}{\Pr}\left[\sum_{j=1}^{J} y_j(T)B_j(T,s) < \sum_{j=1}^{J} \ln A_j(T,s) - \ln K\right]\right. \\ &\quad \left. -K\Pr\left[\sum_{j=1}^{J} y_j(T)B_j(T,s) < \sum_{j=1}^{J} \ln A_j(T,s) - \ln K\right]\right) \end{aligned} \tag{13.98}$$

The next job is to solve for the two probabilities. These probabilities cannot be solved easily as they are linear combination of scaled non-central chi-squared random variables. If it were just the sum but not linear combination, then the solution would be easily solvable. Although the probabilities are not directly solvable, their characteristic functions are. The characteristic function of $\Sigma y_j B_j$ is:

$$\begin{aligned} f(u) &= \tilde{\mathbb{E}}_t\left[e^{-iu\Sigma y_j B_j}\right] \\ &= \prod_{j=1}^{J} \tilde{\mathbb{E}}_t\left[e^{-iuB_j y_j}\right] \\ &= \prod_{j=1}^{J} f_j(B_j u) \end{aligned} \tag{13.99}$$

where each characteristic function is

$$f_j(B_j u) = \left(1 - \frac{B_j(T,s)iu}{\phi_j + \psi_j}\right)^{-1/2\nu_j} \exp\left(\frac{1/2 B_j(T,s)\delta iu}{\phi_j + \psi_j - B_j(T,s)iu}\right) \tag{13.100}$$

and

$$\begin{aligned}
\nu_j &= \frac{4\alpha_j\mu_j}{\sigma_j^2} \\
\phi_j &= \frac{2\gamma_j}{\sigma_j^2\left(e^{\gamma_j(T-t)}-1\right)} \\
\psi_j &= \frac{\alpha_j+\lambda_j+\gamma_j}{\sigma_j^2} \\
\gamma_j &= \sqrt{(\alpha_j+\lambda_j)^2+\sigma_j^2} \\
\delta_j &= \frac{2\phi_j^2 e^{\gamma_j(T-t)}}{\phi_j+\psi_j}y_j(t)
\end{aligned} \tag{13.101}$$

Knowing the characteristic function, we can then easily obtain the probability function by the well known inversion formula:

$$\Pr[X<x] = \frac{1}{\pi}\int_{-\infty}^{\infty}\frac{\sin ux}{u}f(u)du \tag{13.102}$$

where $X = \Sigma y_j B_j$ and $x = \Sigma(\ln A_j(T,s) - \ln K)$. Similarly we can obtain:

$$\overset{*}{\Pr}[X<x] = \frac{1}{\pi}\int_{-\infty}^{\infty}\frac{\sin ux}{u}f^*(u)du \tag{13.103}$$

where the only difference is:

$$\delta_j^* = \frac{2\phi_j^2 e^{\gamma_j(T-t)}}{\phi_j+\psi_j+B_j(T,s)}y_j(t) \tag{13.104}$$

The Fourier inversion function can be computed very efficiently and Chen and Scott provide a short sketch of how to implement such a integration. Many mathematical libraries such as IMSL (International Mathematical and Statistical Libraries) provide easy calls to their libraries to execute such integrations accurately efficiently. Many software packages such as Matlab, Mathematica, and SAS also provide efficient routines for these calculations.

13.5 The Barrier Option Model and First Passage Time (Kwok, Ch.4)

A barrier option is an option whose payoff is activated (knocked in) and terminated (knocked out) when a barrier is reached by the stock price before maturity. Due the barrier can be higher or lower than the current stock price, barrier options are classified as down-and-in and down-and-out barrier options whose barrier lies below the current stock price; and up-and-in, and up-and-out barrier options whose barrier lies above the current stock price.

Down-and-in and up-and-in barrier options become call or put options when the barrier is reached. Similarly, down-and-out and up-and-out barrier options become worthless when the barrier is reached.

The barrier option problem is the first passage time problem. Assume the standard Black-Scholes model that the stock price follows the log normal process. Let the barrier be H and the strike be K. If the barrier is below the current stock price, i.e. $S_t > H$, then the risk neutral probability that the stock price stays above both H and K is either:

$$N\left(\frac{\ln S(t) - \ln K + (r - {}^1\!/_2\sigma^2)(T-t)}{\sigma\sqrt{T-t}}\right) \\ - \left(\frac{S(t)}{H}\right)^{1-\frac{2r}{\sigma^2}} N\left(\frac{2\ln H - \ln S(t) - \ln K + (r - {}^1\!/_2\sigma^2)(T-t)}{\sigma\sqrt{T-t}}\right) \tag{13.105}$$

if $H < K$ or

$$N\left(\frac{\ln S(t) - \ln H + (r - {}^1\!/_2\sigma^2)(T-t)}{\sigma\sqrt{T-t}}\right) \\ - \left(\frac{S(t)}{H}\right)^{1-\frac{2r}{\sigma^2}} N\left(\frac{\ln H - \ln S(t) + (r - {}^1\!/_2\sigma^2)(T-t)}{\sigma\sqrt{T-t}}\right) \tag{13.106}$$

if $H > K$. Note that the first part of equation (13.105) is just the probability of $S(T) > K$. The second part of equation (13.105) is the joint probability that $S(T) > K$ and $S(\tau) < H$. As a result, the difference is the joint probability that $S(T) > K$ and $S(\tau) > H$. When $H > K$, then staying above the barrier automatically guarantees in the money for the option and we simply replace K by H.

As a result, the down-and-out barrier call option model (i.e. $S(t) > H$) can be derived as:

$$\begin{aligned} C_{\text{do}} &= e^{-r(T-t)}\hat{\mathbb{E}}_t[(S(T)-K)1_{S(T)>K\cap S(\tau)>H}] \\ &= e^{-r(T-t)}\hat{\mathbb{E}}_t[S(T)\mathbb{I}_{S(T)>K\cap S(\tau)>H}] - e^{-r(T-t)}K\hat{\mathbb{E}}_t[\mathbb{I}_{S(T)>K\cap S(\tau)>H}] \\ &= S(t)\Pi_+ - e^{-r(T-t)}K\Pi_- \end{aligned} \tag{13.107}$$

where

$$\Pi_\pm = N\left(\frac{\ln S(t) - \ln K + (r \pm {}^1\!/\!_2\sigma^2)(T-t)}{\sigma\sqrt{T-t}}\right) - \left(\frac{S(t)}{H}\right)^{1-\frac{2r}{\sigma^2}} N\left(\frac{2\ln H - \ln S(t) - \ln K + (r \pm {}^1\!/\!_2\sigma^2)(T-t)}{\sigma\sqrt{T-t}}\right)$$

if $H < K$ or

$$\Pi_\pm = N\left(\frac{\ln S(t) - \ln H + (r \pm {}^1\!/\!_2\sigma^2)(T-t)}{\sigma\sqrt{T-t}}\right) - \left(\frac{S(t)}{H}\right)^{1-\frac{2r}{\sigma^2}} N\left(\frac{\ln H - \ln S(t) + (r \pm {}^1\!/\!_2\sigma^2)(T-t)}{\sigma\sqrt{T-t}}\right)$$

if $H > K$. Note that the two results for Π_- are just probability of first passage time. The results for Π_+ can be obtained via the change of measure.

The down-and-in barrier call option model can be derived by subtracting the down-and-out option model from the Black-Scholes model: $C_{\text{di}} = C_{\text{BS}} - C_{\text{do}}$.

If the barrier is above the current stock price, i.e. $S(t) < H$, then the risk neutral probability that the stock price stays above both H and K is either

$$\begin{aligned} &N\left(\frac{\ln S(t) - \ln K + (r - {}^1\!/\!_2\sigma^2)(T-t)}{\sigma\sqrt{T-t}}\right) \\ &- \left(\frac{S(t)}{H}\right)^{1-\frac{2r}{\sigma^2}} \left\{N\left(\frac{2\ln H - \ln S(t) - \ln K + (r - {}^1\!/\!_2\sigma^2)(T-t)}{\sigma\sqrt{T-t}}\right) - N\left(\frac{\ln S(t) - \ln K + (r - {}^1\!/\!_2\sigma^2)(T-t)}{\sigma\sqrt{T-t}}\right)\right\} \end{aligned} \tag{13.108}$$

if $H < K$ or

$$
\begin{aligned}
&N\left(\frac{\ln S(t)-\ln H+(r-{}^1\!/\!_2\sigma^2)(T-t)}{\sigma\sqrt{T-t}}\right)\\
&-\left(\frac{S(t)}{H}\right)^{1-\frac{2r}{\sigma^2}}\left\{N\left(\frac{\ln H-\ln S(t)+(r-{}^1\!/\!_2\sigma^2)(T-t)}{\sigma\sqrt{T-t}}\right)\right.\\
&\qquad\left.-N\left(\frac{\ln S(t)-\ln H+(r-{}^1\!/\!_2\sigma^2)(T-t)}{\sigma\sqrt{T-t}}\right)\right\}
\end{aligned}
\tag{13.109}
$$

if $H > K$. Note that the first part of equation (13.108) is just the probability of $S(t) > K$. The second part of equation (13.108) is the joint probability that $S(t) > K$ and $S_T < H$. As a result, the difference is the joint probability that $S(t) > K$ and $S_T > H$. When $H > K$, then staying above the barrier automatically guarantees in the money for the option and we simply replace K by H.

The up-and-in and up-and-out call option models are inverses of the above results, as Brownians are symmetrical. An up-and-out barrier call option becomes worthless if the barrier (which is higher than the current stock price) is reached.

13.6 Appendix

13.6.1 Black-Scholes model derived using Fourier Inversion

In this appendix, we derive the Black-Scholes model using the Fourier Inversion method. Knowing the Heston model, we can derive the Black-Scholes model easily as the Black-Scholes model is a special case of the Heston model. The PDE for the Black-Scholes model is a special case of (13.59):

$$rC = rSC_S + {}^1\!/\!_2\sigma^2 S^2 C_{SS} + C_t \tag{13.110}$$

After change of variable $x = \ln S$, it becomes a special case of (13.61):

$$rC = (r - {}^1\!/\!_2\sigma^2)C_x + {}^1\!/\!_2\sigma^2 C_{xx} + C_t \tag{13.111}$$

As in Heston, we guess the solution to be:

$$C = S\Pi_1 + e^{-r(T-t)}K\Pi_2 \tag{13.112}$$

where Π_1 and Π_2 are probabilities in the Black-Scholes model and satisfy the following PDEs:

$$0 = (r - {}^1\!/\!_2\sigma^2)\Pi_{jx} + {}^1\!/\!_2\sigma^2\Pi_{jxx} + \Pi_{jt} \tag{13.113}$$

where $j = 1, 2$. As in Heston, the characteristic functions are also solutions to the same PDEs, that is:

$$\begin{aligned} 0 &= (r + {}^1\!/\!_2\sigma^2)\phi_{1x} + {}^1\!/\!_2\sigma^2\phi_{1xx} + \phi_{1t} \\ 0 &= (r - {}^1\!/\!_2\sigma^2)\phi_{2x} + {}^1\!/\!_2\sigma^2\phi_{2xx} + \phi_{2t} \end{aligned} \tag{13.114}$$

Guess the solutions to be:

$$\phi_j(u) = e^{C_j(T-t)+iux} \tag{13.115}$$

where $j = 1, 2$ and then, following (13.68), the ODEs for $C_1(T-t)$ and $C_2(T-t)$ are respectively:

$$\begin{aligned} {}^1\!/\!_2\sigma^2(-u^2) + \left(r + {}^1\!/\!_2\sigma^2\right) iu + C_{1t} &= 0 \\ {}^1\!/\!_2\sigma^2(-u^2) + \left(r - {}^1\!/\!_2\sigma^2\right) iu + C_{2t} &= 0 \end{aligned} \tag{13.116}$$

Given the boundary condition (when $t \to T$) that $\phi_j(u) = e^{iux}$ or equivalently $C_j(0) = 0$, the solutions are straightforward:

$$\begin{aligned} C_{1t} &= \left[{}^1\!/\!_2\sigma^2\left(iu - u^2\right) + riu\right](T-t) \\ C_{2t} &= \left[{}^1\!/\!_2\sigma^2\left(-iu - u^2\right) + riu\right](T-t) \end{aligned} \tag{13.117}$$

Finally, equation (13.76) is used to compute Π_1 and Π_2.

13.6.2 Extending to d-dimensions

t	T_1	T_2
	$\underline{V_1 > K_1 + D_{12} + D_{13}}$	$\underline{V_2 > K_2 + D_{23}}$
D_{01}	$D_{11} = K_1$	$D_{22} = K_2$
	$D_{12} = e^{-rt}\mathbb{E}[\min V_2, K_2]$	$D_{23} = e^{-r\Delta t}\mathbb{E}[\min V_3, K_3]$
D_{02}	$D_{13} = e^{-2rt}\int_{\bar{V}_2}^{\infty}\mathbb{E}[\min V_3, K_3] + e^{-rt}\int_{\bar{E}_2}^{\bar{V}_2}(V_2 - K_2)$	$E_2 = e^{-r\Delta t}\mathbb{E}[\max V_3 - K_3, 0] - K_2$
D_{03}	$E_1 = e^{-2rt}\int_{\bar{V}_2}^{\infty}\mathbb{E}[\max V_3 - K_3, 0] - e^{-rt}\int_{\bar{V}_2}^{\infty} K_2 - K_1$	$\underline{K_2 < V_2 < K_2 + D_{23}(\equiv \bar{V}_{23})}$
E_0	$\underline{K_1 + D_{12} < V_1 < K_1 + D_{12} + D_{13}(\equiv \bar{V}_{13})}$	$D_{22} = K_2$
		$D_{23} = V_2 - K_2$
	$D_{11} = K_1$	$E_2 = 0$
	$D_{12} = e^{-rt}\mathbb{E}[\min V_2, K_2]$	$\underline{V_2 < K_2(\equiv \bar{V}_{22})}$
	$D_{13} = V_1 - D_{11} - D_{12}$	$D_{22} = V_2$
	$E_1 = 0$	$D_{23} = 0$
	$\underline{K_1 < V_1 < K_1 + D_{12}(\equiv \bar{V}_{12})}$	$E_2 = 0$
	$D_{11} = K_1$	
	$D_{12} = V_1 - K_1$	
	$D_{13} = 0$	
	$E_1 = 0$	
	$\underline{V_1 < K_1(\equiv \bar{V}_{11})}$	
	$D_{11} = V_1$	
	$D_{12} = 0$	
	$D_{13} = 0$	
	$E_1 = 0$	

Figure 13.1: Geske

where

$$
\begin{aligned}
D_{01} &= e^{-rt}\int_{K_1}^{\infty} K_1 + \int_{0}^{K_1} V_1 = e^{-rt}\int_{K_1}^{\infty} K_1 + e^{-rt}\int_{0}^{\infty} V_1 - e^{-rt}\int_{K_1}^{\infty} V_1 \\
&= e^{-rt}K_1 N(h_1^-(K_1)) + V_0(1 - N(h_1^+(K_1)))
\end{aligned}
$$

$$
\begin{aligned}
D_{02} &= e^{-2rt}\int_{\bar{V}_{12}}^{\infty} E[\min V_2, K_2] + e^{-rt}\int_{K_1}^{\bar{V}_{12}} (V_1 - K_1) \\
&= e^{-2rt}\int_{\bar{V}_{12}}^{\infty}\int_{0}^{K_2} V_2 + e^{-2rt}\int_{\bar{V}_{12}}^{\infty}\int_{K_2}^{\infty} K_2 + e^{-rt}\int_{K_2}^{\infty}(V_1 - K_1) - e^{-rt}\int_{\bar{V}_{12}}^{\infty}(V_1 - K_1) \\
&= e^{-2rt}\int_{\bar{V}_{12}}^{\infty}\int_{0}^{\infty} V_2 - \int_{\bar{V}_{12}}^{\infty}\int_{K_2}^{\infty} V_2 + e^{-2rt}K_2 N(h_1^-(V_{12}), h_2^-(K_2)) \\
&\quad + V_0[N(h_1^+(K_1)) - N(h_1^+(\bar{V}_{12}))] - e^{-rt}K_1[N(h_1^-(K_1)) - N(h_1^-(\bar{V}_{12}))] \\
&= V_0(N(h_1^+(K_1)) - N(h_1^+(\bar{V}_{12}), h_2^+(K_2))) + e^{-2rt}K_2 N(h_1^-(\bar{V}_{12}), h_2^-(K_2)) \\
&- \quad e^{-rt}K_1[N(h_1^-(K_1)) - N(h_1^-(\bar{V}_{12}))]
\end{aligned}
$$

$$
\begin{aligned}
D_{03} &= e^{-3rt}\int_{\bar{V}_{13}}^{\infty}\int_{\bar{V}_{23}}^{\infty} E[\min V_3, K_3] + e^{-2rt}\int_{\bar{V}_{13}}^{\infty}\int_{K_2}^{\bar{V}_{23}} V_2 - K_2 + e^{-rt}\int_{\bar{V}_{12}}^{\bar{V}_{13}} V_1 - K_1 - e^{-rt}E[\min V_2, K_2] \\
&= e^{-3rt}\int_{\bar{V}_{13}}^{\infty}\int_{\bar{V}_{23}}^{\infty}\int_{0}^{K_3} V_3 + e^{-3rt}\int_{\bar{V}_{13}}^{\infty}\int_{\bar{V}_{23}}^{\infty}\int_{K_3}^{\infty} K_3 + e^{-2rt}\int_{\bar{V}_{13}}^{\infty}\int_{K_{23}}^{\infty}(V_2 - K_2) - e^{-2rt}\int_{\bar{V}_{13}}^{\infty}\int_{\bar{V}_{23}}^{\infty}(V_2 - K_2) \\
&\quad + e^{-rt}\int_{\bar{V}_{12}}^{\infty}(V_1 - K_1) - e^{-rt}\int_{\bar{V}_{13}}^{\infty}(V_1 - K_1) - e^{-2rt}\int_{\bar{V}_{12}}^{\bar{V}_{13}}\int_{0}^{K_2} V_2 - e^{-2rt}\int_{\bar{V}_{12}}^{\bar{V}_{13}}\int_{K_2}^{\infty} K_2 \\
&= V_0[N(h_1^+(\bar{V}_{13}), h_2^+(V_{23})) - N(h_1^+(\bar{V}_{13}), h_2^+(V_{23}), h_3^+(K_3))] \\
&\quad + e^{-3rt}K_3 N(h_1^-(\bar{V}_{13}), h_2^-(V_{23}), h_3^-(K_3)) \\
&\quad + V_0[N(h_1^+(\bar{V}_{13}), h_2^+(K_2)) - N(h_1^+(\bar{V}_{13}), h_2^+(V_{23}))] \\
&\quad - e^{-2rt}K_2[N(h_1^-(\bar{V}_{13}), h_2^-(K_2)) - N(h_1^-(\bar{V}_{13}), h_2^-(V_{23}))] \\
&\quad + V_0[N(h_1^+(V_{12})) - N(h_1^+(\bar{V}_{13}))] - e^{-rt}K_1[N(h_1^-(V_{12})) - N(h_1^-(\bar{V}_{13}))] \\
&\quad - e^{-2rt}\int_{\bar{V}_{12}}^{\bar{V}_{13}}\int_{0}^{\infty} V_2 + e^{-2rt}\int_{\bar{V}_{12}}^{\bar{V}_{13}}\int_{K_2}^{\infty} V_2 - e^{-2rt}\int_{\bar{V}_{12}}^{\bar{V}_{13}}\int_{K_2}^{\infty} K_2 \\
&= V_0[N(h_1^+(\bar{V}_{13}), h_2^+(K_2)) - N(h_1^+(\bar{V}_{13}), h_2^+(\bar{V}_{23}), h_3^+(K_3))] \\
&\quad + e^{-3rt}K_3 N(h_1^-(\bar{V}_{13}), h_2^-(\bar{V}_{23}), h_3^-(K_3)) \\
&\quad - e^{-2rt}K_2[N(h_1^-(\bar{V}_{13}), h_2^-(K_2)) - N(h_1^-(\bar{V}_{13}), h_2^-(V_{23}))] \\
&\quad + V_0[N(h_1^+(\bar{V}_{12})) - N(h_1^+(\bar{V}_{13}))] - e^{-rt}K_1[N(h_1^-(\bar{V}_{12})) - N(h_1^-(\bar{V}_{13}))] \\
&\quad - V_0[N(h_1^+(\bar{V}_{12})) - N(h_1^+(\bar{V}_{13})) - N(h_1^+(\bar{V}_{12}), h_2^+(K_2)) + N(h_1^+(\bar{V}_{13}), h_2^+(K_2))] \\
&\quad - e^{-2rt}K_2[N(h_1^-(\bar{V}_{12}), h_2^-(K_2)) - N(h_1^-(\bar{V}_{13}), h_2^-(K_2))]
\end{aligned}
$$

$$
\begin{aligned}
&= V_0[N(h_1^+(\bar{V}_{12}), h_2^+(F_2)) - N(h_1^+(\bar{V}_{13}), h_2^+(\bar{V}_{23}), h_3^+(F_3))] \\
&\quad + e^{-3rt} K_3 N(h_1^-(\bar{V}_{13}), h_2^-(\bar{V}_{23}), h_3^-(K_3)) \\
&\quad + e^{-2rt} K_2[N(h_1^-(\bar{V}_{13}), h_2^-(\bar{V}_{23})) - N(h_1^-(\bar{V}_{12}), h_2^-(K_2))] \\
&\quad - e^{-rt} K_1[N(h_1^-(\bar{V}_{12})) - N(h_1^-(\bar{V}_{13}))]
\end{aligned}
$$

$$
\begin{aligned}
E_0 &= e^{-3rt} \int_{\bar{V}_{13}}^{\infty} \int_{\bar{V}_{23}}^{\infty} \{E[\max V_3 - K_3, 0] - K_2\} - e^{-rt} \int_{\bar{V}_{13}}^{\infty} K_1 \\
&= e^{-3rt} \int_{\bar{V}_{13}}^{\infty} \int_{\bar{V}_{23}}^{\infty} \int_{K_3}^{\infty} (V_3 - K_3) - e^{-2rt} \int_{\bar{V}_{13}}^{\infty} \int_{\bar{V}_{23}}^{\infty} K_2 - e^{-rt} K_1 N(h_1^-(V_1)) \\
&= V_0 N(h_1^+(\bar{V}_{13}), h_2^+(\bar{V}_{23}), h_3^+(K_3)) - e^{-3rt} K_3(h_1^-(\bar{V}_{13}), h_2^-(\bar{V}_{23}), h_3^-(K_3)) \\
&\quad - e^{-2rt} K_2 N(h_1^-(\bar{V}_{13}), h_2^-(\bar{V}_{23})) - e^{-rt} K_1 N(h_1^-(\bar{V}_{13}))
\end{aligned}
$$

$$
h_j^{\pm}(x) = \frac{\ln V_0 - \ln x + (r \pm \sigma^2/2)(T_j - t)}{\sigma\sqrt{T_j - t}}
$$

13.6.3 Extending to d-dimensions with an alternative recovery

t	T_1	T_2
	$\underline{V_1 > K_1}$	$\underline{V_2 > K_2}$
D_{01}	$D_{11} = K_1$	$D_{22} = F_2$
D_{02}	$D_{12} = e^{-rt}\mathbb{E}[\min V_2, K_2]$	$D_{23} = e^{-r\Delta t}\mathbb{E}[\min V_3, K_3]$
D_{03}	$D_{13} = e^{-2rt}\int_{K_2}^{\infty} \mathbb{E}[\min V_3, K_3]$	$E_2 = e^{-r\Delta t}\mathbb{E}[\max V_3 - K_3, 0] - K_2$
E_0	$E_1 = e^{-2rt}\int_{K_2}^{\infty} \mathbb{E}[\max V_3 - K_3, 0] - e^{-rt}\int_{K_2}^{\infty} K_2 - K_1$	note that equity may be < 0
	note that equity may be < 0	$\underline{V_2 < K_2}$
	$\underline{V_1 < K_1}$	$D_{22} = V_2$
	$D_{11} = V_1$	$D_{23} = 0$
	$D_{12} = 0$	$E_2 = 0$
	$D_{13} = 0$	
	$E_1 = 0$	

Figure 13.2: Geske

where

$$
\begin{aligned}
D_{01} &= e^{-rt}\int_{K_1}^{\infty} K_1 + \int_{0}^{K_1} K_1 \\
&= e^{-rt}\int_{K_1}^{\infty} K_1 + e^{-rt}\int_{0}^{\infty} V_1 - e^{-rt}\int_{K_1}^{\infty} V_1 \\
&= e^{-rt}K_1N(h_1^-(K_1)) + V_0 - V_0N(h_1^+(K_1)) \\
D_{02} &= e^{-2rt}\int_{K_1}^{\infty} E[\min V_2, K_2] \\
&= e^{-2rt}\int_{K_1}^{\infty}\int_{0}^{K_2} V_2 + e^{-2rt}\int_{K_1}^{\infty}\int_{K_2}^{\infty} K_2 \\
&= e^{-2rt}\int_{K_1}^{\infty}\int_{0}^{\infty} V_2 - \int_{K_1}^{\infty}\int_{K_2}^{\infty} V_2 + e^{-2rt}K_2N(h_1^-(K_1), h_2^-(K_2)) \\
&= -V_0[N(h_1^+(K_1), h_2^+(K_2)) - N(h_1^+(K_1))] + e^{-2rt}K_2N(h_1^-(K_1), h_2^-(K_2)) \\
D_{03} &= e^{-3rt}\int_{K_1}^{\infty}\int_{K_2}^{\infty} E[\min V_3, K_3] \\
&= e^{-3rt}\int_{K_1}^{\infty}\int_{K_2}^{\infty}\int_{0}^{K_3} V_3 + e^{-3rt}\int_{K_1}^{\infty}\int_{K_2}^{\infty}\int_{K_3}^{\infty} F_3 \\
&= V_0[N(h_1^+(K_1), h_2^+(K_2)) - N(h_1^+(K_1), h_2^+(K_2), h_3^+(K_3))] \\
&\quad + e^{-3rt}K_3N(h_1^-(V_1), h_2^-(V_2), h_3^-(V_3)) \\
E_0 &= e^{-3rt}\int_{K_1}^{\infty}\int_{K_2}^{\infty}\{E[\max V_3 - K_3, 0] - K_2\} - e^{-rt}\int_{\bar{V}_1}^{\infty} K_1 \\
&= e^{-3rt}\int_{K_1}^{\infty}\int_{K_2}^{\infty}\int_{K_3}^{\infty}(V_3 - K_3) - e^{-2rt}\int_{K_1}^{\infty}\int_{K_2}^{\infty} K_2 - e^{-rt}K_1N(h_1^-(K_1)) \\
&= V_0N(h_1^+(K_1), h_2^+(K_2), h_3^+(K_3)) - e^{-3rt}K_3(h_1^-(K_1), h_2^-(K_2), h_3^-(K_3)) \\
&\quad - e^{-2rt}K_2N(h_1^-(K_1), h_2^-(K_2)) - e^{-rt}K_1N(h_1^-(K_1))
\end{aligned}
$$

13.6.4 Derive the PDE of the Scott model

Similar to the PDE of the Heston model, the PDE for the Scott model without the jump is:

$$
\begin{aligned}
&{}^1\!/\!_2VC_{xx} + {}^1\!/\!_2\gamma^2VC_{VV} + {}^1\!/\!_2g^2rC_{rr} + \rho\gamma VC_{xV} + (r - {}^1\!/\!_2V)C_x + (\alpha\beta - (\alpha+\lambda)V)C_V \\
&\qquad + (ab - (a+l)r)C_r + C_t = rC
\end{aligned}
\tag{13.118}
$$

and the probabilities must satistify the following PDEs:

$$\begin{aligned}&\frac{V}{2}\Pi_{1xx}+\frac{\gamma^2V}{2}\Pi_{1VV}+\frac{g^2r}{2}\Pi_{1rr}+\rho\gamma V\Pi_{1xV}+[\alpha\beta-(\alpha+\lambda+\rho\gamma)V]\Pi_{1V}\\&+\left(r+\frac{V}{2}\right)\Pi_{1x}+[ab-(a+l)r]\Pi_{1r}+\Pi_{1t}=0\end{aligned}\tag{13.119}$$

$$\begin{aligned}&\frac{V}{2}\Pi_{1xx}+\frac{\gamma^2V}{2}\Pi_{1VV}+\frac{g^2r}{2}\Pi_{1rr}+\rho\gamma V\Pi_{1xV}+[\alpha\beta-(\alpha+\lambda)V]\Pi_{1V}\\&+\left(r-\frac{V}{2}\right)\Pi_{1x}+[ab-(a+l)r]\Pi_{1r}+\Pi_{1t}=0\end{aligned}\tag{13.120}$$

and the characteristic functions should satisfy the sam PDEs. The guessed form of the chacteristic functions is:

$$f_j=e^{C_j(T-t)+D_j(T-t)V+E_j(T-t)r+iux}\tag{13.121}$$

where $j=1,2$. This leads to two sets of ODEs for the characteristic functions:

$$\begin{aligned}&-\frac{1}{2}u^2+\frac{\gamma^2}{2}D_1^2-(\alpha+\lambda-\rho\gamma(1+iu))D_1+\frac{1}{2}iu+D_{1t}=0\\&iu+\frac{g^2}{2}E_1^2-(a+l)E_1+E_{1t}=0\\&\alpha\beta D_1+abE_1+C_{1t}=0\end{aligned}\tag{13.122}$$

and

$$\begin{aligned}&-\frac{1}{2}u^2+\frac{\gamma^2}{2}D_2^2-(\alpha+\lambda+\rho\gamma iu)D_2-\frac{1}{2}iu+D_{2t}=0\\&(iu-1)+\frac{g^2}{2}E_2^2-(a+l)E_2+E_{2t}=0\\&\alpha\beta D_2+abE_2-C_{2t}=0\end{aligned}\tag{13.123}$$

The solution to E is the Cox-Ingersoll-Ross model to the term structure:

$$\begin{aligned}E_1&=\frac{x}{g^2}\left[\frac{\frac{x-(a+l)}{x+(a+l)}e^{-x(T-t)}-1}{\frac{x-(a+l)}{x+(a+l)}e^{-x(T-t)}+1}+\frac{a+l}{x}\right]\\x&=\sqrt{(a+l)^2-2g^2iu}\end{aligned}\tag{13.124}$$

$$E_2 = \frac{x}{g^2}\left[\frac{\frac{x-(a+l)}{x+(a+l)}e^{-x(T-t)}-1}{\frac{x-(a+l)}{x+(a+l)}e^{-x(T-t)}+1}+\frac{a+l}{x}\right] \quad (13.125)$$
$$x = \sqrt{(a+l)^2 - 2g^2(iu-1)}$$

Adding jumps to the model is straightforward. We simply multiply the characteristic function (13.121) by (13.127) and 13.126) respectively as jumps are independent of the any other diffusion process. The characteristic functions of the Poisson process are given below:

$$f_{J2}(u) = \exp\begin{pmatrix} -iu\lambda_J e^{-\mu_J + 1/2\sigma_J^2}(T-t)\left(e^{\mu_J - 1/2\sigma_J^2} - 1\right) \\ +\lambda_J(T-t)e^{-\mu_J+1/2\sigma_J^2}\left(e^{iu(\mu_J - \sigma_J^2) - 1/2\sigma_J^2 u^2} - 1\right) \end{pmatrix} \quad (13.126)$$

$$f_{J1}(u) = \exp\begin{pmatrix} -iu\lambda_J e^{-\mu_J + 1/2\sigma_J^2}(T-t)\left(e^{\mu_J - 1/2\sigma_J^2} - 1\right) \\ +\lambda_J(T-t)\left(e^{iu\mu_J - 1/2\sigma_J^2 u^2} - 1\right) \end{pmatrix} \quad (13.127)$$

Then it follows from (13.76) to arrive at the probabilities.

Alternative Derivation of the Black-Scholes Model

Recall that the PDE for the Black-Scholes model is as follows:

$$rSC_S + 1/2\sigma^2 S^2 C_{SS} + C_t = rC \quad (13.128)$$

From (13.60), we re-write the above PDE as follows (in terms of the log price):

$$(r - 1/2\sigma^2)C_x + 1/2\sigma^2 C_{xx} + C_t = rC \quad (13.129)$$

As in the Heston case, we guess the solution (although we already knew what it is) to the above PDE as follows:

$$C = S\Pi_1 - e^{-r(T-t)}K\Pi_2 \quad (13.130)$$

and the probabilities follow the following PDEs:

$$\begin{aligned} (r + 1/2\sigma^2)\Pi_{1x} + 1/2\sigma^2\Pi_{1xx} + \Pi_{1t} = 0 \\ (r - 1/2\sigma^2)\Pi_{2x} + 1/2\sigma^2\Pi_{2xx} + \Pi_{2t} = 0 \end{aligned} \tag{13.131}$$

which are the same PDEs that the characteristic functions must also satisfy:

$$\begin{aligned} (r + 1/2\sigma^2)f_{1x} + 1/2\sigma^2 f_{1xx} + f_{1t} = 0 \\ (r - 1/2\sigma^2)f_{2x} + 1/2\sigma^2 f_{2xx} + f_{2t} = 0 \end{aligned} \tag{13.132}$$

As in Heston, we guess the solutions to the characteristic functions to be:

$$f_{\pm}(u) = e^{C_{\pm}(T-t)+iux} \tag{13.133}$$

As a result, the ODEs for the function $C_{\pm}$ are:

$$(r \pm 1/2\sigma^2)iu - 1/2\sigma^2 u^2 + C_{\pm t} = 0 \tag{13.134}$$

and hence the solution is a straightforward integration:

$$\begin{aligned} C_{\pm} &= \int (r \pm 1/2\sigma^2)iu - 1/2\sigma^2 u^2 \\ &= (r \pm 1/2\sigma^2)iu(T-t) - 1/2\sigma^2 u^2(T-t) \end{aligned} \tag{13.135}$$

Then it follows from (13.76) to arrive at the probabilities.

Chapter 14

Credit Risk Modeling

This chapter sketches the basic math for credit risk modeling. Many product details are omitted and readers should refer to other fixed income texts.

14.1 Introduction

Simply speaking, credit risk is default risk. However, credit risk can be transferred in different ways. That is, there are different transactable instruments that can transfer different parts of credit risk. Typically, there are three types of credit risk:

- bankruptcy
- migration
- spread

Bankruptcy is the ultimate default. The firm no longer operates and liquidates its assets to pay off its debts. Debt holders lose some of their investment value (known as loss given default or severity) and stock holders lose all of their investment value. Credit risk of this sort needs to evaluate the likelihood of bankruptcy (known as probability of default, or PD) and loss given default (LGD).

Usually bankruptcies do not happen suddenly. Companies lose money and ultimately lead to bankruptcy. Rating agencies monitor such transition and provide their assessment of the fundamental financial healthiness of the firm via credit ratings. Hence, rating migrations (especially from high to low) indicate how poorly the firm is operating. As a result, when a firm is downgraded by rating agencies, its bond and stock prices are lowered. Empirical evidence has shown that rating changes have significant impacts on stock and bond prices.

Last but not least is the spread credit risk. Ratings reflect firms' financial health. Yet, rating changes do not occur often. Furthermore, ratings are discrete measures. Stocks and bonds are traded continuously in the marketplace. As a result, we can transfer credit risk much more quickly by trading spreads.

In this chapter, we only discuss the basic math of credit risk modeling. That is, we only evaluate the bankruptcy credit risk. There are models for just migration and spread risks but are not covered here. Readers should be aware that many migration and spread models are ad-hoc and do not relate back to default. Readers should keep in mind that all three types of credit risk result from default because without default risk, there are no migration or spread changes.

14.2 Poisson Process of Defaults

A Poisson process, named after the French mathematician Simon-Denis Poisson (1781 - 1840), is the stochastic process in which events occur continuously and independently of one another. The Poisson process is a continuous-time process: its discrete-time counterpart is the Bernoulli process.

A homogeneous Poisson process is characterized by a rate parameter λ, also known as intensity, such that the number of events in time interval $(t, t+\tau]$ follows a Poisson distribution with the associated parameter $\lambda\tau$. This relation is given as:

$$\Pr[N(t+\tau) - N(t) = j] = \frac{(\lambda\tau)^j e^{-\lambda\tau}}{j!} \tag{14.1}$$

Hence,

$$\Pr[N(t+\tau) - N(t) = 0] = e^{-\lambda\tau} \tag{14.2}$$

The Poisson process is an ideal way to model unexpected defaults as an event occurs with no prior memory. (14.2) describes the probability of no default between $(t, t+\tau]$. The mean and variance of a Poisson distribution are:

$$\mathbb{E}[j] = \mathbb{V}[j] = \lambda\tau \tag{14.3}$$

14.2.1 Exponential Distribution

If an event occurs with a Poisson distribution, then it can be shown that the time of the event occurring follows an exponential distribution. The exponential distribution is (with parameter λ):

$$f(x) = \lambda e^{-\lambda x} \tag{14.4}$$

Hence, its cumulative density function is:

$$\begin{aligned}\Pr[x < \tau] &= F(\tau) \\ &= \int_0^t f(x)dx \\ &= 1 - e^{-\lambda\tau}\end{aligned} \tag{14.5}$$

or

$$\begin{aligned}\Pr[x > \tau] &= 1 - F(\tau) \\ &= e^{-\lambda\tau}\end{aligned} \tag{14.6}$$

which is the survival probability when t represents the default time. When the time interval is small, we can use Taylor's series expansion to approximate the survival probability as:

$$\Pr[x > \tau] = e^{-\lambda\tau} \approx 1 - \lambda\tau \tag{14.7}$$

or the default probability $\lambda\tau$. The mean and variance of the exponential distribution are:

$$\begin{cases}\mathbb{E}[x] = \dfrac{1}{\lambda} \\ \mathbb{V}[x] = \dfrac{1}{\lambda^2}\end{cases} \tag{14.8}$$

Note that the Poisson process has no memory. Hence, the exponential distribution has the following property:

$$\begin{aligned}\Pr[x > s + t | x > s] &= \frac{e^{-\lambda(s+t)}}{e^{-\lambda s}} \\ &= e^{-\lambda t} \\ &= \Pr[x > t]\end{aligned} \tag{14.9}$$

In plain words, what this means is that the expected wait time (till the event) has nothing to do with how long one has waited. The event is not expected to happen sooner because you have already waited for a long time.

14.2.2 Default Modeling

We compute survival probabilities when we model default. The survival probability between now and some future time T if we adopt the Poisson process (i.e. default time τ) for defaults is:

$$Q(t,T) = \Pr[\tau > T] = \hat{\mathbb{E}}_t[\mathbb{I}_{\{\tau>T\}}] = e^{-\lambda(T-t)} \tag{14.10}$$

As we can see, there is extreme similarity between the result of survival probability and risk-free discount. In fact, we shall show that they can be combined linearly if the recovery of a risky bond is 0. Due to this similarity, we shall proceed without proof (which is difficult in many situations) to borrow what we have known for the risk-free rate and use it for the intensity. In the insurance literature, the intensity parameter is called the hazard rate.

If the intensity parameter (hazard rate) is non-constant, then we can express the survival probability as:

$$Q(t,T) = \Pr[\tau > T] = \hat{\mathbb{E}}_t[\mathbb{I}_{\{\tau>T\}}] = \exp\left(-\int_t^T \lambda_u du\right) \tag{14.11}$$

Furthermore, if the hazard rate is random, then we simply compute the risk neutral expectation:

$$Q(t,T) = \Pr[\tau > T] = \hat{\mathbb{E}}_t[\mathbb{I}_{\{\tau>T\}}] = \hat{\mathbb{E}}_t\left[\exp\left(-\int_t^T \lambda_u du\right)\right] \tag{14.12}$$

Finally, if the interest rate is random, then we must use the forward measure (see Chapter 5):

$$Q(t,T) = \Pr[\tau > T] = \tilde{\mathbb{E}}_t^{(T)}[\mathbb{I}_{\{\tau>T\}}] = \tilde{\mathbb{E}}_t^{(T)}\left[\exp\left(-\int_t^T \lambda_u du\right)\right] \tag{14.13}$$

14.2.3 Survival and Default Probability Curves

Equations in 14.2.2 represent the survival probability till a certain time in the future. As a result, we can have a whole curve of survival probabilities.

Similarly, we can obtain a curve of default probabilities as:

$$\pi(t,T) = -\frac{dQ(t,T)}{dT} \tag{14.14}$$

Again, we remind the readers of the similarity between default probabilities and forward rates in the risk-free world.[1]

14.3 Zero Recovery Risky Bond

A zero recovery risky bond has the same analogy to the risk-free bond as the hazard rate to the risk-free rate. First we shall look at the risky discount factor.

14.3.1 Risky Discount Factor

A risky discount factor, similar to the risk-free discount factor, discounts \$1 paid in the future. Assuming the same notation for the risk-free discount factor, $P(t,T)$, that represents the present value of \$1 paid in time T, we can denote the survival probability as $Q(t,T)$. A risky discount factor is the present value of \$1 paid in T only if default does not occur. As a result, the present value of \$1 paid in time T is $P(t,T)Q(t,T)$.

14.3.2 Zero Recovery Risky Bond

A zero recovery risky coupon bond pays a periodic coupon c till maturity T_n. If the bond has no recovery, then its price must be:

$$B(t,\underline{T}) = \sum_{j=1}^{n} cP(t,T_j)Q(t,T_j) + P(t,T_n)Q(t,T_n) \tag{14.15}$$

where $\underline{T} = < T_1 \cdots T_n >$.

[1] See Chapter 12, equation (12.19).

14.4 Positive Recovery Risky Bond

With recovery, the valuation of the risky bond becomes much more complex. Deciding a recovery value of a defaulted bond is a complex process. When a firm defaults, its assets are under a liquidation process and when it ends, bond holders know what they can recover. This process can sometimes take multiple years to finish. For some bond holders who do not wish to wait, they can sell their bonds to the marketplace (distressed bond market) to gain cash earlier. This is similar to Account Receivable factorization.

When such a market exists, then investors estimate a fair present value for the ultimate recovery. This then represents the fair market value of recovery of the bond. When such a market does not exist, then the recovery value must be estimated. It is common for practitioners to use a historical average. Rating agencies provide historical averages for various categories of bonds. For example, senior unsecured bonds recover on average 35% to 45% and junior unsecured bonds recover on average 15% to 25%.

In the literature, there are two major approaches to model positive recovery. Jarrow and Turnbull (1995) assume recovery to be a fixed percentage of the face value of the bond and Duffie and Singleton (1997) assume the recovery to be a fixed percentage of the market value of the bond immediately prior to default. As we shall show later, the Jarrow-Turnbull model is particularly useful in building the credit curve (i.e. bootstrapping) and the Duffie-Singleton model is useful in integrating with the term structure models.

14.4.1 Recovery of Face Value – The Jarrow-Turnbull Model

When the recovery rate is a fixed amount, we can modify the pricing formula of (14.15) as follows:

$$\begin{aligned} B(t,\underline{T}) = \sum\nolimits_{j=1}^{n} cP(t,T_j)Q(t,T_j) + P(t,T_n)Q(t,T_n) \\ +R\sum\nolimits_{j=1}^{n} P(t,T_j)[Q(t,T_{j-1}) - Q(t,T_j)] \end{aligned} \tag{14.16}$$

where c is coupon (or cash flow), $P(t,T)$ is risk-free discount factor between now and time T, $Q(t,T)$ is survival probability between now and time T, and R is the recovery rate that is assumed constant. The last term is added due to recovery. Note that $Q(t,T_{j-1}) - Q(t,T_j)$ is the default probability between T_{j-1} and T_j. Note that in continuous time, this is $-dQ(t,T)$ which is equal to $\pi(t,T)dT$. As a result, the above formula can be written in continuous time as:

$$
\begin{aligned}
B(t,\underline{T}) = \sum\nolimits_{j=1}^{n} cP(t,T_j)Q(t,T_j) + P(t,T_n)Q(t,T_n) \\
+R\int_t^{T_n} P(t,u)\pi(t,u)du
\end{aligned}
\tag{14.17}
$$

The above equation is not a closed-form solution as it requires integration over the default probability measure. One particularly easy way to keep the closed-form solution is to assume the recovery to be received at a fixed time (and not upon default). Then the above equation can be simplified assuming the recovery is received at T_n:

$$
B(t,\underline{T}) = \sum\nolimits_{j=1}^{n} cP(t,T_j)Q(t,T_j) + P(t,T_n)Q(t,T_n) + RP(t,T_n)[1-Q(t,T_n)]
\tag{14.18}
$$

The last term $1 - Q(t,T_n)$ is the cumulative default probability. This is the Jarrow-Turnbull model.

14.4.2 Recovery of Market Value – The Duffie-Singleton Model

Another easy way to arrive at a closed-form solution is to let the recovery be proportional to the otherwise undefaulted value. That is, upon default (at default time u), the recovery value is $R_t = \delta Z(t,T)$ where $Z(t,T)$ is the price of a zero coupon risky bond as if it has not defaulted.

Under the Poisson process, for a very small time interval Δt, we can write the bond equation as:

$$
Z(t,T) = \frac{Z(t+\Delta t,T)\delta\lambda\Delta t + Z(t+\Delta t,T)(1-\lambda\Delta t)}{1+r\Delta t}
\tag{14.19}
$$

which then can be simplified to, assuming n periods between now t and maturity T:

$$
\begin{aligned}
Z(t,T) &= \frac{Z(t+\Delta t,T)(1-\lambda\Delta t(1-\delta))}{1+r\Delta t} \\
&= Z(T,T)\left[\frac{1-\lambda\Delta t(1-\delta)}{1+r\Delta t}\right]^n \\
&\sim Z(T,T)\left[\frac{e^{-\lambda\Delta t(1-\delta)}}{e^{r\Delta t}}\right]^n \\
&\sim Z(T,T)e^{-(r+s)(T-t)}
\end{aligned}
\tag{14.20}
$$

where $s = \lambda(1 - \delta)$ can be viewed as a spread over the risk-free rate. This is the Duffie-Singleton model. $Z(T,T)$ is the terminal value of the bond which is usually the face value.

A nice feature of the Duffie-Singleton model is that a coupon bond can then a portfolio of such zeros, as each coupon is treated as a zero bond it recovers market value. That is:

$$B(t,\underline{T}) = \sum_{j=1}^{n} cZ(t,T_j) + Z(t,T_n) \tag{14.21}$$

This model is practically appealing in that it reflects the usual industry practice that credit risk is reflected in spreads. This model is also convenient to be combined with existing term structure models. It simply adds a second state variable. The drawback of the model is that the recovery parameter and the intensity parameter always are inseparable. This adds to difficulty in calibration this model to the market.

14.5 Credit Default Swap

A Credit Default Swap, or CDS, is a bilateral contract which allows an investor to buy protection against the risk of default of a specified reference credit. The fee may be paid up front, but more often is paid in a "swapped" form as a regular, accruing cashflow. A CDS is a negotiated contract and there are a number of important features that need to be agreed between the counterparties and clearly defined in the contract documentation.

First and foremost is the definition of the credit event itself. This is obviously closely linked to the choice of the reference credit and will include such events as bankruptcy, insolvency, receivership, restructuring of debt and a material change in the credit spread. This last materiality clause ensures that the triggering event has indeed affected the price of the reference asset. It is generally defined in spread terms since a fall in the price of the reference asset could also be due to an increase in the level of interest rates.

Many CDS contracts define the triggering of a credit event using a reference asset. However, in many cases the importance of the reference asset is secondary as the credit event may also be defined with respect to a class of debt issued by a reference entity. In this case the importance of the reference asset arises solely from its use in the determination of the recovery price used to calculate the payment following the credit event.

The contract must specify what happens if the credit event occurs. Typically, the protection buyer will usually agree to do one of the following:

- Deliver the defaulted security to the protection seller in return for Par in cash. Note that the contract usually specifies a basket of securities which are ranked *pari passu* which may be delivered in place of the reference asset. In effect the protection seller is long a "cheapest to deliver" option.

- Receive Par minus the default price of the reference asset settled in cash. The price of the defaulted asset is typically determined via a dealer poll conducted within a few weeks to months of the credit event, the purpose of the delay being to let the recovery price stabilize.

Some CDS have a different payoff from the standard Par minus recovery price. The main alternative is to have a fixed pre-determined amount which is paid out immediately after the credit event. This is known as a binary default swap. In other cases, where the reference asset is trading at a significant premium or discount to Par, the payoff may be tailored to be the difference between the initial price of the reference asset and the recovery price.

The protection buyer automatically stops paying the premium once the credit event has occurred, and this property has to be factored into the cost of the swap payments. It has the benefit of enabling both parties to close out their positions soon after the credit event and so eliminates the ongoing administrative costs which would otherwise occur.

A CDS can be viewed as a form of insurance with one important advantage – efficiency. Provided the credit event in the default swap documentation is defined clearly, the payment due from the triggering of the credit event will be made quickly. Contrast this with the potentially long and drawn out process of investigation and negotiation which may occur with more traditional insurance.

However it is possible to get a very good idea of the price of the CDS using a simple "static replication" argument. This involves recognizing that buying a CDS on a risky par floating rate asset which only defaults on coupon dates is exactly equivalent to going long a default-free floating rate note and short a risky floating rate note of the same credit quality. If no default occurs, the holder of the position makes a net payment equal to the asset swap spread of the asset on each coupon date until maturity. This spread represents the credit quality of the risky floater at issuance. If default does occur, and we assume that it can only occur on coupon payment dates, the position can be closed out by buying back the defaulted asset in return for the recovery rate and selling the par floater. The net value of the position is equal to the payoff from the default swap. The following table summarizes.

CDS vs. Floater			
Event	Riskless FRN	Risky FRN	CDS
At inception	Pay par	Pay par	0
No default	Receive LIBOR	Receive LIBOR + spread	Pay spread
Upon default	Receive par	Receive recovery	Receive par, pay recovery
Maturity	Receive par	Receive par	0

From the above table, it is clear that the spread of a CDS must equal to the spread of the equivalent risky FRN to avoid arbitrage.

14.5.1 Restructuring Definitions by ISDA

CDS contracts provide default protection. When a default occurs, CDS buyers stop paying the premium (spread), deliver the defaulted bond (cheapest if possible), and collect full face value as payment. However, default is hard to define. It is extremely rare for a company to file bankruptcy. What is usually happening is that losses happen over the years and reduce the asset value of the company, to a point where the company is at the brink of bankruptcy. Then the management of the company will start looking for alternatives to save the firm. One popular alternative to save the firm is to ask debt holders to change their debt contracts to the company known as debt restructuring. Debt restructuring often means that debt holders convert parts of their debts into equity and participate in the operation of the firm. To protect their own interests, debt holders, especially large ones, will be willing to agree to debt restructuring.

Hence debt restructuring is commonly regarded as a form of default. However, each restructuring can be very different. Some restructurings are major and equivalent to defaults. But some could be minor as precautionary actions to avoid further deterioration of the firm, which are not equivalent to default.

To regulate if a CDS is triggered, ISDA (International Swaps and Derivatives Association) defines various restructuring standards:

- Full restructuring (FR), based on the ISDA 1999 Definition
- Modified restructuring (MR), based on the ISDA 2001 Supplement Definition
- Modified-modified restructuring (MMR), based on the ISDA 2003 Definition,
- No restructuring (NR).

- The definitions are as follows:

  FR Any bond of maturity up to 30 years

  MR $T \leqslant \bar{T} < T + 30$ months

  MMR Allow additional 30 months for the restructured bond.

  For other obligations, same as MR.

CDS contracts traded in different regions follow different ISDA conventions.

14.5.2 Why Has the CDS Market Developed So Rapidly?

CDS is the most popular credit derivatives contract and has grown rapidly in late 90's and early 00's. The following is a direct quote of Rene Stulz's article on CDS (2010)[2]

Back in the mid-1990s, one of the first credit default swaps provided protection on Exxon by the European Bank for Reconstruction and Development to JP Morgan (Tett, 2009). It took months to negotiate. By 1998, the total size of the credit default swap market was a relatively small $180 billion (Acharya, Engle, Figlewski, Lynch, and Subrahmanyam, 2009). The credit default swap market has grown enormously since then, although there is no definite measure of how much. Based on survey data from the Bank for International Settlements (BIS) at http://www.bis.org/statistics/derstats.htm, the total notional amount of the credit default swap market was $6 trillion in 2004, $57 trillion by June 2008, and $41 trillion by the end of 2008. Credit-default swap contracts that insure default risk of a single firm are called single-name contracts; in contrast, contracts that provide protection against the default of many firms are called multi-name contracts. [3]

In addition to the efficiency in hedging and transferring credit risk, the potential benefits of CDS include:

- A short positioning vehicle that does not require an initial cash outlay
- Access to maturity exposures not available in the cash market
- Access to credit risk not available in the cash market due to a limited supply of the underlying bonds
- The ability to effectively exit credit positions in periods of low liquidity

[2] "Credit Default Swaps and the Credit Crisis," Journal of Economic Perspectives, Volume 24, Number 1, Winter 2010, pp. 73-92.

[3] Stulz noted that DTCC statics are a lot smaller.

- Off-balance sheet instruments which offer flexibility in terms of leverage

- To provide important anonymity when shorting an underlying credit

14.5.3 Relationship between Default Probabilities and CDS Spreads

There is a simple formula that relates the CDS spread, the risk-free rate, default/survival probabilities, and the fixed recovery rate. Due to the swap nature, CDS, similar to IRS (interest rate swap), has two legs – floating and fixed. The floating leg of a CDS contract is called the protection leg as it pays only if default occurs. The fixed leg of a CDS contract is called the premium leg because the fixed payments (i.e. spreads) are like insurance premiums. As in a standard swap contract, at inception, the values of the two legs must equal each other. This is how CDS spreads are calculated. Recently affected by the crisis, CDS premiums have been split into an upfront and a spread (which is the index, such as CDX, trading convention). As we shall see later, this extra calculation does not add any complexity to the model. We simply deduct the upfront amount from the protection value of the CDS. For now, we shall proceed with no upfront.

Using the formulation given earlier, the protection and premium values of a CDS are as follows:

$$\begin{aligned} V_{\text{prot}}(t,T) &= (1-R)\int_t^T P(t,u)[-dQ(t,u)] \\ V_{\text{prem}}(t,T) &= s(t,T)\sum\nolimits_{i=1}^{n} P(t,T_i)Q(t,T_i) \end{aligned} \tag{14.22}$$

where $T_n = T$. As a result, the spread (known as par spread) can be computed as:

$$s(t,T) = \frac{(1-R)\int_t^T P(t,u)[-dQ(t,u)]}{\sum_{i=1}^{n} P(t,T_i)Q(t,T_i)} \tag{14.23}$$

Often times, the continuous forward default probability $-dQ(t,u)$ can be approximated by $Q(t,T_{i-1}) - Q(t,T_i)$ if default can only occur on a cashflow day. Note that (14.23) assumes no accrued interest if default occurs in between coupons. In reality there are accrued interests on both legs and they may not be equal. If default is assumed to happen on cashflow days only, then there is no accrued interest. Note that if there is an upfront, we simply deduct it from the protection value, V_{prot}.

14.5.4 Back-of-the-envelope Formula

In a one-period model where default is a Bernulli event, as the following picture demonstrates,

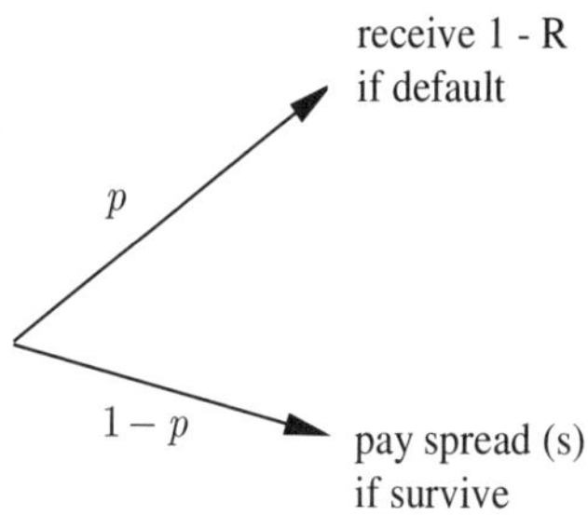

Figure 14.1: One-period Default Diagram

We know that for the CDS to have 0 value it must be true that:

$$p(1-R) = (1-p)s \tag{14.24}$$

Note that risk-free discount cancel from both sides. Hence, we arrive at the famous back-of-the-envelope formula for the default probability:

$$p = \frac{s}{1-R} \tag{14.25}$$

This formula, while simple, provides a powerful intuition of spreads and default probabilities. If the recovery is 0, then spread is (forward) default probability. This is not only true in (14.25) but also true in continuous time. Spreads are not equal to (i.e. smaller) forward default probabilities in that they are compensated by recovery. Note that CDS buyers acquire default protection by paying spreads as an insurance premium. If recovery is high, the protection value is low, and so should be the spread. In an extreme case where the recovery is 100%, the spread should be 0, which is suggested by (14.25).

14.6 Bootstrapping (Curve Cooking)

We need credit curves to price credit derivatives. Credit curves are obtained from liquid cash products such as CDS or corporate bonds. Due to the liquidity concern,

CDS is a better choice for curve cooking.

The basic bootstrapping idea of constructing a risky curve is the same as the risk-free curve. We use a pricing formula to back out the parameter(s). In the traditional fixed income world (Treasuries and IRS), we back out spot and forward rates from the market prices of bonds and swap rates. Here, we back out survival probabilities from a series of CDS contracts. As in the world of traditional fixed income, we need a term structure of CDS spreads in order to back out the entire survival probability curve. A popular smoothing technique in LIBOR curves is piece-wise flat.

The CDS market has been standardized over the years to have the following on-the-run maturities: 1, 2, 3, 5, 7, and 10 years to maturities. As in the IRS market, these contracts are on-the-run which are issued periodically. In the early years, only 5-year CDS contracts were issued. A few years ago, the market started to trade 10-year CDS contracts. The other maturities have gradually been introduced to the market but their liquidity is still a concern. Assume for now that we observe market prices of these CDS spreads.

14.6.1 Poisson Assumption

From (14.11) $\sim$ (14.13), we know that if we assume piece-wise flat intensity values, then the survival probability can be approximated as follows:

$$Q(t, T_n) = \exp\left\{-\sum_{i=1}^{n} \lambda_i (T_i - T_{i-1})\right\} \tag{14.26}$$

where $t = T_0$.

14.6.2 Simple Demonstration (annual frequency)

To make matters simple, we assume CDS spreads are paid annually. There are 6 CDS spreads observed in the market (1, 2, 3, 5, 7, 10). Take Disney as an example. On 12/23/2005, we observe the following spreads (in basis points):

CDS quotes	
term	sprd
1	9
2	13
3	20
5	33
7	47
10	61

Continue to assume fixed 0.4 recovery ratio under MR for Disney. The following binomial chart presents possible cash flows. Lets assume 5% risk-free rate.

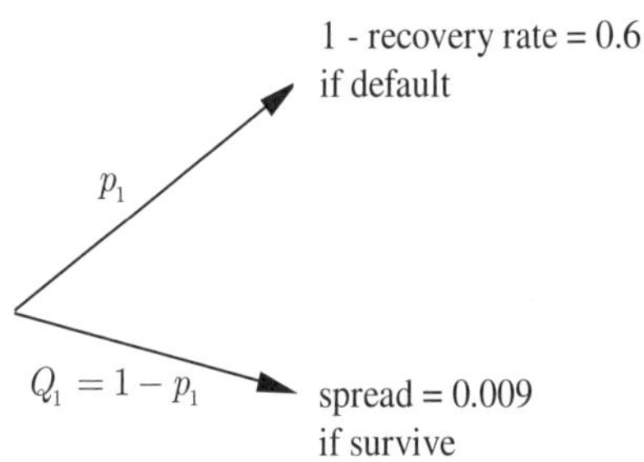

Figure 14.2: First period

CDS is a swap contract so there is no cash changed hands on day 1. Hence, it must be the case that the expected payment (9 basis points) equals the expected compensation (60%). In a single period, since both payment and compensation are discounted, the risk-free does not matter. Note that the survival probability for one year is $e^{-\lambda_1} = 1 - p_1$. Hence, using 5% interest rate, we have:

$$\frac{0.6 \times \left(1 - e^{-\lambda_1}\right) - 0.0009 \times e^{-\lambda_1}}{1.05} = 0$$

which is equivalent to:

$$0.6 \times \left(1 - e^{-\lambda_1}\right) = 0.0009 \times e^{-\lambda_1}$$

Solve for λ_1 to be 0.001499, or 14.99 basis points. Hence, the survival and default probabilities are 99.85% and 0.15% respectively.

Now we proceed to bootstrap the second period.

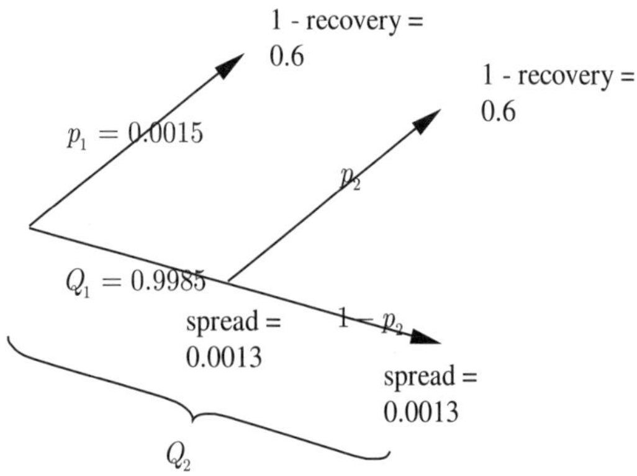

Figure 14.3: Two-period Default Diagram

There are three scenarios. Either Disney defaults in period 1, or default in period 2, or survive in period 2 (note that to survive till period 2, Disney must first survive period 1). We know the first default probability, which is $p_1 = 0.0015$. But we do not know the other two probabilities. Using the same Poisson algorithm, we can compute the second year present value as:

$$\frac{0.6 \times \left(1 - e^{-\lambda_2}\right) - 0.0013 \times e^{-\lambda_2}}{1.05}$$

Note that this quantity itself is not 0; but combining it with the first year cash flows is:

$$\frac{0.6 \times 0.0015 + 0.9985 \left(-0.0013 + \frac{0.6 \times \left(1 - e^{-\lambda_2}\right) - 0.0013 \times e^{-\lambda_2}}{1.05}\right)}{1.05} = 0$$

Solve for λ_2 to get 0.002865, or 28.65 basis points. The conditional default and survival probabilities are $p_2 = 1 - e^{-\lambda_2}$ which is 28.62 bps and $1 - p_2 = e^{-\lambda_2}$ which is 99.7139% respectively. The unconditional survival probability equals $(1 - p_1)(1 - p_2)$ which is 99.5645%. The unconditional default probability is $1 - 99.5645\% = 0.4355\%$.

Similar process applies to all periods as the following figure depicts.

The results are given as follows. The first three columns are taken from the market. Columns A and B are the same spread input given earlier. Column C is

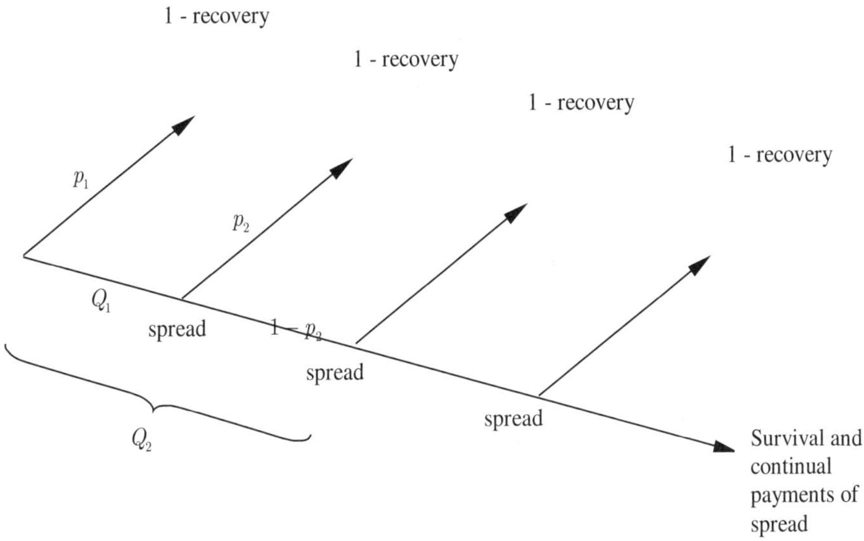

Figure 14.4: Multi-period Default Diagram

the risk-free discount factors that are computed using 5% flat in the example.

CDS Bootstrapping					
A	B	C	D	E	F
Term	Market Spread	Risk-free $P(t)$	Fwd. $\lambda(t)$	Surv.Pr. $Q(t)$	Fwd.Def.Pr. $-dQ(t)$
1	0.0009	0.9512	0.0015	0.9985	0.0015
2	0.0013	0.9048	0.0029	0.9956	0.0029
3	0.0020	0.8607	0.0059	0.9898	0.0058
5	0.0033	0.7788	0.0066	0.9769	0.0129
7	0.0047	0.7047	0.0098	0.9580	0.0190
10	0.0061	0.6065	0.0090	0.9326	0.0254

Column D is the main result which is the piece-wise flat values. We need to solve for these values.

CDS Bootstrapping					
G	H	I	J	K	L
$P(t)Q(t)$	$P(t)[-dQ(t)]$	Prem. Leg	Prot. Leg	Model Spread	Cond. Surv.Pr.
0.9498	0.0014	0.9498	0.0009	0.0009	
0.9009	0.0026	1.8507	0.0024	0.0013	0.9971
0.8520	0.0050	2.7027	0.0054	0.0020	0.9942
0.7608	0.0100	3.4635	0.0114	0.0033	0.9870
0.6751	0.0134	4.1386	0.0195	0.0047	0.9806
0.5656	0.0154	4.7042	0.0287	0.0061	0.9735

Column E presents the survival probabilities that are computed sequentially as in equqation (14.26):

$$\begin{aligned} Q(t,T_n) &= \exp\left\{-\sum\nolimits_{i=1}^{n}\lambda_i(T_i-T_{i-1})\right\} \\ &= Q(t,T_{n-1})\exp\left\{-\lambda_n(T_n-T_{n-1})\right\} \end{aligned} \tag{14.27}$$

For example, $Q(0,2) = 0.9956 = 0.9985 \times e^{-0.0029\times(2-1)}$. Column F is the forward default probabilities which is the differences are two consecutive survival probabilities. For example, $0.0029 = 0.9985 - 0.9956$. Column G is known as the risky discount factor (introduecd earlier), or \$1 present value with default risk. These values are needed in order to compute the default swap spread, i.e. the denominator of (14.23). Similarly, column H provides the values for the numerator of (14.23). Columns I and J are accumulations of columns G and H respectively. Column K is the division of column J by column I, which is the spread of CDS. The values of this column must match the market quotes in column B. In fact, we solve for column D so that values in column K are identical to values in column B. Finally, column L, conditional survival probabilities, is 1 – column F (that is conditional default probabilities).

The table presented here provides a nice algorithm for further automate the calculations for more complex situations in reality, which we shall demonstrate later. Once all the λ_j values are bootstrapped out, we can then compute any survival probability of any future time, as follows:

$$Q(t,v) = Q(t,T_{n-1})\exp\left\{-\lambda_n(v-T_{n-1})\right\} \tag{14.28}$$

where $T_{n-1} < v < T_n$. For example, the survival probability of 6.25 years is $0.9769 \times e^{-0.0098\times(6.5-5)} = 0.9627$.

14.6.3 In Reality (quarterly frequency)

In the above example, we assume spreads are paid annually. As a result, the calculation is quite simple. In reality, this is not the case. Spreads are paid by the swap market convention which is quarterly. In this case, default can occur at any quarter. We then need to alter the one period calculation shown above.

Note that within a year (for the first few spreads), all per-quarter default probabilities are equal. This is because we have only one spread (e.g. 0.0009 in year 1) to cover four quarters. The basic formula is still the same. Mainly we solve the following equation for p (note that at each period, the interest rate is 1.25%):

$$\begin{aligned} x_i &= x_0 + 0.6p_1 + (1 - p_1)x_{i-1} \\ x_n &= x_0 \end{aligned} \tag{14.29}$$

where n represents the number of periods that shares the probability. For the first year, $n = 4$ and $x_0 = 0.000225$. We then solve for $p_1 = 0.000375$ or 3.75 basis points and $\lambda_1 = 0.0015$. The full expansion of this equation is shown in the Appendix.

Note that while this equation is solvable by hand if proper re-arrangement of terms is performed, it is much faster if we set up the equation and use the Solver in Excel. This equation can be set up recursively as the discounting and expected values are nested. We can set up an Excel sheet to compute all the results. As a demonstration, we provide the results up the 3 years.

The layout of the table is the same as before. The frequency of the CDS premium payments is now quarterly (see column A). Column B is still market CDS spreads that are available every four quarters. Column C is quarterly risk-free discount factors (at 5%). Columns E $\sim$ H are computed similarly to the previous section, only with quarterly frequency.

Columns, I, J, and K are computed similarly to the previous section but only every year. Note that column K is column J $\div$ column I $\times$ 4 in order to annualize to match annual CDS market quotes in column B.

Readers should complete the table for the full 10 years.

14.7 Duffie-Singleton Model

Both the Jarrow-Turnbull and the Duffie-Singleton models assume defaults to be unexpected like Poisson events. Different from the Jarrow-Turnbull model that assumes fixed amount recovery (or known as recovery of face value), the Duffie-Singleton model assumes the recovery to be proportional to the market value of the

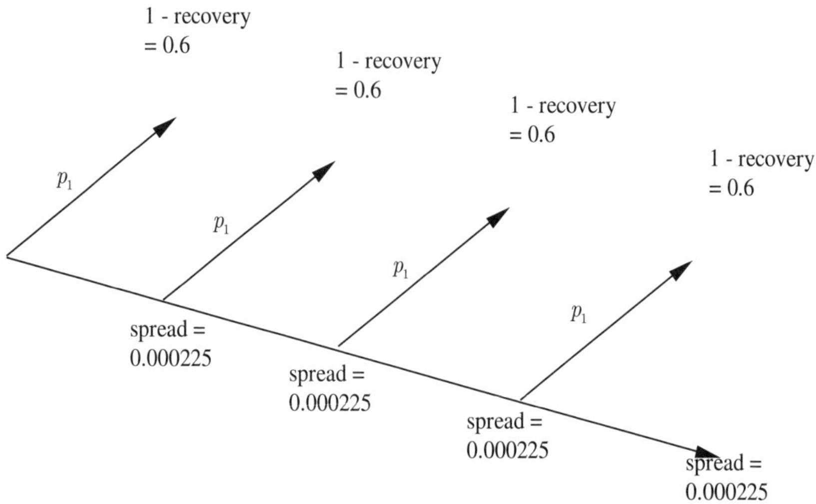

Figure 14.5: Default Spine

CDS Bootstrapping					
A	B	C	D	E	F
Term	Market Spread	Risk-free $P(t)$	Fwd. $\lambda(t)$	Surv.Pr. $Q(t)$	Fwd.Def.Pr. $-dQ(t)$
0.25		0.9876	0.0015	0.9996	0.000375
0.5		0.9753	0.0015	0.9993	0.000375
0.75		0.9632	0.0015	0.9989	0.000375
1	0.0009	0.9512	0.0015	0.9985	0.000374
1.25		0.9394	0.002868	0.9978	0.000716
1.5		0.9277	0.002868	0.9971	0.000715
1.75		0.9162	0.002868	0.9964	0.000715
2	0.0013	0.9048	0.002868	0.9956	0.000714
2.25		0.8936	0.00586	0.9942	0.001457
2.5		0.8825	0.00586	0.9927	0.001455
2.75		0.8715	0.00586	0.9913	0.001453
3	0.002	0.8607	0.00586	0.9898	0.001451

CDS Bootstrapping					
G	H	I	J	K	L
$P(t)Q(t)$	$P(t)[-dQ(t)]$	premium leg	protection leg	computed spread	cond. surv.prob.
0.9872	0.0004				
0.9746	0.0004				0.9996
0.9621	0.0004				0.9996
0.9498	0.0004	3.8737	0.0009	0.0009	0.9996
0.9373	0.0007				0.9993
0.925	0.0007				0.9993
0.9129	0.0007				0.9993
0.9009	0.0006	7.5498	0.0025	0.0013	0.9993
0.8884	0.0013				0.9985
0.8761	0.0013				0.9985
0.8639	0.0013				0.9985
0.8519	0.0012	11.0302	0.0055	0.002	0.9985

debt (known as recovery of market value) immediately prior to default.

The Jarrow-Turnbull model is suitable for bootstrapping and the Duffie-Singleton model is convenient to combine with term structure models. Following the similar analysis for equations (14.11) $\sim$ (14.13), we can write (14.20) as:

$$Z(t,T) = \hat{\mathbb{E}}_t \left[\exp\left(-\int_t^T (r(u) + s(u))du \right) \right] \tag{14.30}$$

which allows us to directly model spread as another state variable. Note that $s(t) = \lambda(t)(1-\delta)$ according to (14.20) and hence the spread process is similar to the intensity process. If the intensity and the risk-free rate are independent, then the Duffie-Singleton model of (14.21) is similar to (14.15) but with positive recovery.

We can easily conduct a CIR model with the Duffie-Singleton approach. We can have the following joint square-root process:

$$\begin{aligned} dr(t) &= \hat{\alpha}_r(\hat{\mu}_r - r(t))dt + \sigma_r\sqrt{r(t)}d\hat{W}_r(t) \\ ds(t) &= \hat{\alpha}_s(\hat{\mu}_s - s(t))dt + \sigma_s\sqrt{s(t)}d\hat{W}_s(t) \end{aligned} \tag{14.31}$$

where $d\hat{W}_r(t)d\hat{W}_s(t) = 0$. Then (14.30) has a closed-form solution as each expectation in the following equation is a CIR solution.

$$Z(t,T) = \hat{\mathbb{E}}_t \left[\exp\left(-\int_t^T r(u)du \right) \right] \hat{\mathbb{E}}_t \left[\exp\left(-\int_t^T s(u)du \right) \right] \tag{14.32}$$

The following example is to demonstrate how the Duffie-Singleton model can be easily combined with any interest rate model. In the following a simple binomial model for the risk-free rate is given. The probabilities of the up and down branches are $\frac{1}{2}$ and $\frac{1}{2}$.

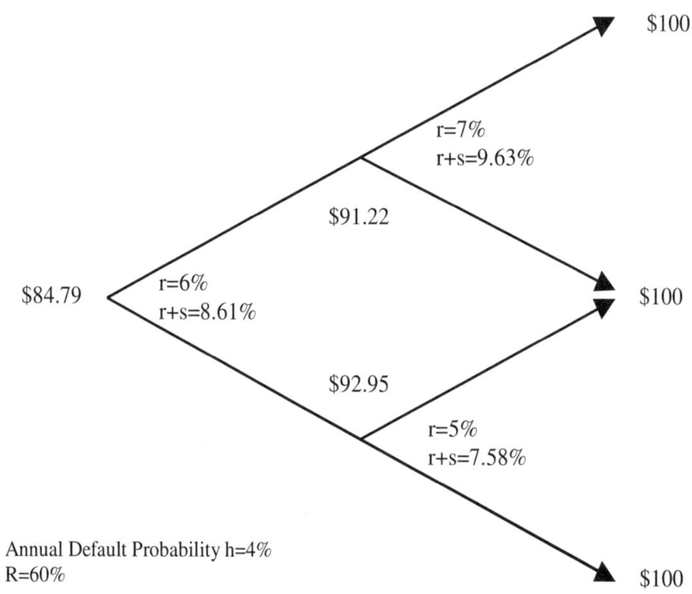

Figure 14.6: Duffie-Single Model

From the input information provided, we can compute the bond prices in the diagram, as follows:

$$91.215 = \frac{1}{1.07}[(1 - 4\%) \times 100 + 4\% \times 40]$$

$$92.95 = \frac{1}{1.05}[(1 - 4\%) \times 100 + 4\% \times 40]$$

$$84.79 = \frac{1}{1.06}\left[(1 - 4\%) \times \frac{91.215 + 92.95}{2} + 4\% \times \frac{91.215 + 92.95}{2} \times 0.4\right]$$

The spreads of these bonds are computed as follows:

$$\frac{100}{91.215} - 1 - 7\% = 9.63\% - 7\% = 2.63\%$$
$$\frac{100}{92.95} - 1 - 5\% = 7.58\% - 5\% = 2.58\%$$
$$\frac{92.0825}{84.79} - 1 - 6\% = 8.61\% - 6\% = 2.61\%$$

where $92.0825 = \frac{1}{2}(91.215 + 92.95)$. Note that these spreads are not (even though close to) the continuous spread in the Duffie-Singleton model, which is computed as follows:

$$\begin{aligned} &\frac{1}{\text{Default Prob} \times \text{Recovery} + \text{Survival Prob}} - 1 \\ &= \frac{1}{4\% \times 0.4 + 96\%} - 1 \\ &= 2.46\% \end{aligned}$$

14.8 Corporate Finance Approach of Modeling Default

This approach of modeling default is called structural approach. It does not assume defaults as an unexpected event but rather a firm's failure to pay its debts. The earliest model is Black-Scholes (1973) and Merton (1974) who regard the equity of the firm as a call option. This is because the equity of the firm has a residual claim of the firm's assets. In a simple example where the firm has only one zero coupon debt, the face value of the debt is the strike of the call option. When the asset value at the maturity of the debt is above the face value, the firm is able to pay off its debt and the equity receives the residual value of $A(T) - K$ where A is the asset value of the firm at the maturity date, T, of the debt and K is the face value of the debt. Otherwise (the asset value is smaller than the face value), the equity is worthless.

As a result, the equity value in this simple case can be priced by the Black-Scholes model:

$$E(t) = A(t)N(d_1) - e^{-r(T-t)}KN(d_2) \tag{14.33}$$

where

$$\begin{aligned} d_2 &= \tfrac{\ln A(t) - \ln K + (r - \sigma^2/2)(T-t)}{\sigma\sqrt{T-t}} \\ d_1 &= d_2 + \sigma\sqrt{T-t} \end{aligned}$$

r is the risk-free rate, and σ is the asset return volatility. The debt value which is the difference between asset and equity is then equal to:

$$\begin{aligned} D(t) &= A(t) - E(t) \\ &= \underbrace{A(t)[1 - N(d_1)]}_{\text{Recovery value}} + \underbrace{e^{-r(T-t)} K N(d_2)}_{\text{Survival value}} \end{aligned} \tag{14.34}$$

On the right hand side of the equation, the debt value is decomposed into two parts – surivial value and recovery value. The survival value is the amount received by the debt holders if the firm survives. In this case it is the face value weighted by the survival (in-the-money) probability $N(d_2)$ and discounted at the risk-free rate. The recovery value is the amount received by the debt holders if the firm defaults. In this case it is the asset value weighted by the default probability and discounted at the risk-free rate. But since the asset value is random, we must compute the following expected value: $e^{-r(T-t)}\hat{\mathbb{E}}_t \left[A(T)\mathbb{I}_{A<K}\right]$ which is equal to the recovery value given above.

Note that $D(t,T) < e^{-r(T-t)}K$ which is the risk-free debt value. The (risk-neutral) default probability is $p = 1 - N(d_2)$.

The following diagram is usually used to visualize how the Black-Scholes-Merton model works. The plotted are two possible sample paths of the asset value, one drifts upward and the other drifts downward. The asset value starts at 100 with the default barrier at 50. One sample path stays above the default barrier and is a survival path and one crosses the barrier and is a default path. To be explored fully later, the log difference between the asset value and the default barrier is known as the distance to default, or DD.

On the very right of the diagram is the log normal distribution that describes the asset price. The area below the default barrier is the probability of default (commonly abbreviated as PD).

14.8.1 The Geske Model

Although the Black-Scholes-Merton model reflects the reality that the only reason firms default is due to failure to pay, it is a one period model which is too limited to be used for a regular firm that has multiple debts.

Geske extended the model to include multiple debts. The Geske model can be seen in Chapter 13.

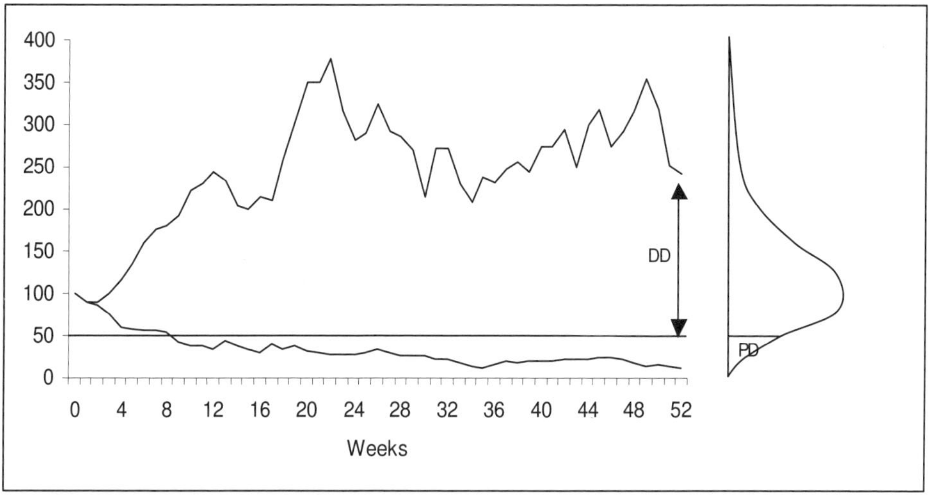

Figure 14.7: Merton Model

14.8.2 The Leland-Toft Model

The Leland-Toft model (which is an extension of the Leland model) is another structural model that extended the Black-Scholes-Merton model to include multiple debts. Different from the Geske model, the Leland-Toft model assumes continuous coupons and contiuous issuance and redemption of the debts. As a result, the firm has a steady-state debt level that will not change, as the amount of debt issued exactly cancels the amount of debt redeemed.

14.8.3 Hybrid Models

The Geske model is difficult to use in reality due to its computational difficulties. Hence, a number of hybrid models emerged to bridge the gap between the true structural model such as Geskes (where the meaning of default is failure to pay) and the reduced form models (that focus on calibration ease). These hybrid models assume a default barrier that is exogenously specified. The default barrier is then calibrated to the market.

Once there is an external barrier, the modeling complexity drops substantially. There are several barrier option models that are popular. The first one is the flat barrier option model introduced in Chapter 13, Section 13.5. The second one is by Black and Cox (1976) where the barrier is not flat but exponentially decayed. The first passage time (survival probability) by Black and Cox is given as follows:

$$N\left(\frac{\ln V - \ln Ke^{-\gamma(T-\tau)} + (r - a - 1/2\sigma^2)(\tau - t)}{\sqrt{\sigma^2(T-t)}}\right) - \left(\frac{V}{Ke^{-\gamma(T-t)}}\right)^{1-\frac{2(r-a-\gamma)}{\sigma^2}} \times$$
$$N\left(\frac{2\ln Ke^{-\gamma(T-t)} - \ln V - \ln Ke^{-\gamma(T-\tau)} + (r - a - 1/2\sigma^2)(\tau - t)}{\sqrt{\sigma^2(T-t)}}\right) \tag{14.35}$$

where K is the barrier amount, γ is the decay rate of the barrier (in other words, the barrier is smaller currently than close to maturity), τ is the default time (i.e., the first passage time to the boundary), and V is the asset value. To put it plainly, if t^* is the first passage time to the boundary, then the probability that $t^* \geq \tau$ is given by the above equation.

The last one is the CreditGrades model (2002) that was developed by Goldman Sachs, JP Morgan, Deutsche Bank, and RiskMetrics. This is essentially a random barrier model. The technical document can be downloaded from creditgrades.com. The main equation is its approximation formula for the survival probability, as follows:

$$Q(t) = N\left(-\frac{A}{2} + \frac{\ln d}{A}\right) - dN\left(-\frac{A}{2} - \frac{\ln d}{A}\right) \tag{14.36}$$

where

$$A = \sqrt{\sigma_S^2 \xi^2 t + \lambda^2}$$
$$d = \frac{1}{1-\xi} e^{\lambda^2}$$
$$\xi = \frac{E}{E+D}$$

and λ is the volatility parameter for the stochastic barrier, E is the equity value and D is the "target" debt value. Hence, it is logical to view ξ as the leverage ratio. With the survival probability function, we can proceed with CDS valuation.

14.9 Understanding Credit Portfolio and Credit Correlation

14.9.1 Basics

The following Venn diagram helps explain the basic setup of portfolio credit risk valuation.

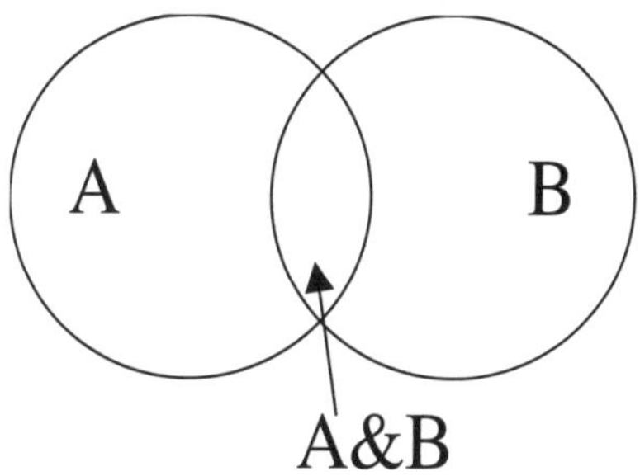

Figure 14.8: Venn Diagram for Default

where A and B represent the two default events for company A and company B respectively. The area in the middle represents joint default.

In order to properly correlate default events, maintain the flexibility of freely specifying individual default probabilities, and be able to uniquely define a joint distribution, we use conditional default probabilities to describe the dependency of two default events (rather than specifying the correlation). In other words, we specify the default probability of the second bond given that the first bond has already defaulted.

For a single period, the default event should follow a Bernoulli distribution where 0 represents no default and 1 represents default. Under the Bernoulli distribution, the specification of the conditional probability (given the marginal probabilities) uniquely defines the joint default probability. The joint default probability can be formally delineated as:

$$p(A \cap B) = \begin{cases} p(A|B)p(B) & \text{or} \\ p(B|A)p(A) \end{cases} \tag{14.37}$$

where A and B represent default events, i.e. a short hand notation for A = 1 and B = 1.

This specification of the joint distribution is desirable in that it is realistic in defining the default relationship between two bonds. If two bonds have different default probabilities (marginals) but they are extremely highly correlated, it means (in our model) that if the less risky one defaults, the riskier one will surely default but not the other way around. Imagine a small auto part supplier whose sole client is a large automobile manufacturer. If the manufacturer defaults, it is highly likely that the small part supplier will eventually default as well. The small part supplier may also default on its own, independent of what happens to the automobile manufacturer.

For example, we let A be default on the part of the manufacturer and B be the default on the part of the small supplier. Each has 10% and 20% default probability for the period, respectively. The part supplier is completely dependent upon the manufacturer, i.e. the conditional default probability of the supplier on the manufacturer is one. Then the joint probability of both firms defaulting is:

$$\begin{aligned} p(A \cap B) &= p(B|A)p(A) \\ &= p(A) \\ &= 10\% \end{aligned} \tag{14.38}$$

Hence, we obtain the following joint distribution for the two companies:

Bivariate Bernulli Distribution			
	0	1	A
0	80%	0%	80%
1	10%	10%	20%
B	90%	10%	100%

The survival of the small supplier depends completely on the large auto manufacturer:

$$p(B|A) = \frac{p(A \cap B)}{p(A)} = 100\% \tag{14.39}$$

The dependency of the manufacturer on the supplier can be calculated to be:

$$p(A|B) = \frac{p(A \cap B)}{p(B)} = \frac{10\%}{20\%} = 50\% \tag{14.40}$$

which is equal to the ratio of the two individual default probabilities. The unconditional correlation of the two companies is 0.6667. The two companies can also have perfectly negative dependency: default of one company implies the survival of the other. In this case, the conditional probability is:

$$p(B^C|A) = 1 \quad \text{or} \quad p(B|A) = 0 \tag{14.41}$$

Using the numbers in the previous example yields the following joint distribution:

Bivariate Bernulli Distribution			
	0	1	A
0	70%	10%	80%
1	20%	0%	20%
B	90%	10%	100%

Notice that, even though there is perfect negative dependency, the unconditional correlation is only −0.1667.

The examples here show that perfect dependency does not translate to perfect correlation. This is because dependency is directional but correlation is not. In the first case, A defaults causes B defaults but not vice versa. Hence the dependency of B on A is 100% but A on B is only 50%. The correlation is in between, 67%. In the second case, A can survive only if B defaults and vice versa. This is perfectly negative dependency but the correlation is only −17%. The unconditional default correlation is calculated as follows:

$$\begin{aligned}\rho(A,B) &\overset{\text{either}}{=} \frac{p(B|A)p(A) - p(A)p(B)}{\sqrt{p(A)(1-p(A))p(B)(1-p(B))}} \\ &\overset{\text{or}}{=} \frac{p(A|B)p(B) - p(A)p(B)}{\sqrt{p(A)(1-p(A))p(B)(1-p(B))}}\end{aligned} \tag{14.42}$$

So it is seen that it is not possible to reach 100% or -100% correlation unless the following conditions are satisfied:

- $p(A) = p(B)$ for 100% correlation
- $p(A) + p(B) = 1$ for -100% correlation

When there is perfect dependency (i.e. $p(B|A) = p(A|B) = 1$), for $\rho(A,B) = 1$, it must be true that:

$$
\begin{aligned}
1 &= \frac{p(A)(1-p(B))}{\sqrt{p(A)(1-p(A))p(B)(1-p(B))}} \\
&= \frac{\sqrt{p(A)(1-p(B))}}{\sqrt{(1-p(A))p(B)}}
\end{aligned}
\tag{14.43}
$$

in which the only solution is $p(A) = p(B)$. When there is perfect negative dependency (i.e. $p(B^C|A) = 1$, or equivalently $p(B|A) = 0$, and vice versa), for $\rho(A,B) = -1$, it must be true that:

$$
\begin{aligned}
-1 &= \frac{-p(A)p(B)}{\sqrt{p(A)(1-p(A))p(B)(1-p(B))}} \\
1 &= \frac{\sqrt{p(A)p(B)}}{\sqrt{(1-p(A))(1-p(B))}}
\end{aligned}
\tag{14.44}
$$

in which the only solution is $p(A) + p(B) = 1$.

14.9.2 Default Baskets (First to default)

Default baskets usually contain only a few credit names (individual bonds). The first-to-default basket pays principal and accrued interest minus the recovery value of the first defaulted bond in the basket.

For the sake of easy exposition, we examine a two-asset basket with no counterparty risk. The model can easily be extended to a multi-asset basket, but the value must be solved numerically. The closed-form solution for the basket's value is:

$$
\begin{aligned}
V &= \frac{1}{1+r} p(A \cup B) \\
&= \frac{1}{1+r} [p(A) + p(B) - p(B|A)p(A)]
\end{aligned}
\tag{14.45}
$$

where r is the risk-free discount rate. As can be seen, the basket value is linear in the conditional probability. It is easy to demonstrate that the basket value is a (negative) linear function of the default correlation. When the default correlation is small (or even negative), the issuers tend to default alternately: this increases the basket risk. When the correlation is large, the issuers tend to default together: this decreases the basket risk. As will be demonstrated below, the value of the default basket will vary between that of a single default swap (when there is perfect dependency) and the sum of the two individual default swaps (when there is zero dependency).

Note that in the event of perfect dependency, i.e. $p(B|A) = 1$, the basket valuation formula in (14.45) becomes:

$$\begin{aligned} V &= \frac{1}{1+r}[p(A) + p(B) - p(B|A)p(A)] \\ &= \frac{1}{1+r}p(B) \end{aligned} \tag{14.46}$$

which implies that the value of the basket is equal to the value of a single name default swap. On the other hand, if the dependency is perfectly negative, i.e.$p(B^C|A) = 0$ or $p(B|A) = 1$, the basket valuation becomes:

$$\begin{aligned} V &= \frac{1}{1+r}[p(A) + p(B) - p(B|A)p(A)] \\ &= \frac{1}{1+r}[p(A) + p(B)] \end{aligned} \tag{14.47}$$

which implies that the value of the basket is equal to the sum of the two individual swaps. Note that if there is no dependency and the two companies were to have default probabilities that sum to exactly 100%, then the joint distribution would degenerate and result in an unconditional correlation of -1.

The results can be extended to multiple assets, though the calculations of the probabilities become multi-dimensional. We can use Monte-Carlo methods to calculate the joint normal probabilities in high dimensions. We can write down the valuation equation as follows:

$$V = \hat{\mathbb{E}}_t \left[P(t, \min\{u_j\}) \mathbb{I}_{\min\{u_j\}<T} (1 - R_j(u_j)) N_j \right] \tag{14.48}$$

where $\mathbb{I}$ is the indicator function, u_j is the default time of the j-th bond, R_j is recovery rate of the j-th bond, and N_j is the notional amount of the j-th bond. The basket pays when it experiences the first default, i.e. $\min\{u_j\}$.

Obviously, the above equation has no easy solution when the default events (or default times, u_j) are correlated. Hence, in the following, we only solve the model under indepedent defaults. Under independence, we can write (14.48) as:

$$V = \int_t^T \sum\nolimits_{i=1}^n P(t, u) \left[-d\Pi_{j=1}^i Q_j(t, u) + d\Pi_{j=0}^{i-1} Q_j(t, u) \right] (1 - R_i(u)) \tag{14.49}$$

where $Q_0(t, u) = 1$ and hence $dQ_0(t, u) = 0$. The above formula assumes that the last bond (i.e. bond n) has the highest priority in compensation, i.e. if the last bond jointly defaults with any other bond, the payoff is determined by the last bond. The second to last bond has the next highest priority in a sense that if it jointly defaults with any other bond but the last, the payoff is determined by the second to last bond. This priority prevails recursively to the first bond in the basket.

14.9.3 Including Counterparty Risk

Investment banks that sell or underwrite default baskets are themselves subject to default risks. If a basket's underlying issuers have a higher credit quality than their underwriting investment bank, then it is possible that the bank may default before any of the issuers. In this case, the buyer of the default basket is subject to not only the default risk of the issuers of the bonds in the basket, but also to that of the bank as well. This is called the counterparty risk. If the counterparty defaults before any of the issuers in the basket does, the buyer suffers a total loss of the whole protection (and the spreads that had been paid up to that point in time). We modify the model to incorporate the counterparty risk, by adding a new asset with 0 payoff to the equation:

$$V = \int_t^T \sum_{i=1}^{n+1} P(t,u)\left[-d\Pi_{j=1}^{i} Q_j(t,u) + d\Pi_{j=0}^{i-1} Q_j(t,u)\right](1 - R_i(u)) \qquad (14.50)$$

where the $(n+1)$-th asset represents the counterparty whose payoff is 0, i.e.:

$$1 - R_{n+1}(t) = 0 \qquad (14.51)$$

for all t.

Note that the counterparty payoff has the lowest priority because the buyer will be paid if the counterparty jointly defaults with any issuer.

The default swap is a special case of the default basket with $N = 1$. However, with a default swap, the counterparty risk is more pronounced than that with a basket deal. With only one issuer, the model can be simplified to:

$$\begin{aligned} V &= \int_t^T P(t,u)\left[-dQ_1(t,u)(1 - R_1(t,u)) + (-dQ_1(t,u)Q_2(t,u) + dQ_1(t,u))(1 - R_2(t))\right] \\ &= \int_t^T P(t,u)\left[-dQ_1(t,u)Q_2(t,u) + dQ_1(t,u))(1 - R_2(t))\right] \end{aligned} \qquad (14.52)$$

Above equation implies that the investor who buys a default swap on the underlying issuer effectively sells a default swap of joint default back to the counterparty.

When the defaults of the issuers and the counterparty are correlated, the solution to (14.52) becomes very complex. When the correlations are high, issuers in the basket tend to default together. In this case, the riskiest bond will dominate the default of the basket. Hence, the basket default probability will approach the default probability of the riskiest bond. On the other hand, when the correlations are low, individual bonds in the basket shall default in different situations. No bond will dominate the default in this case. Hence, the basket default probability will be closer to the sum of individual default probabilities.

14.10 Copula and CDO Pricing

14.10.1 Basics

A CDO is a collection of bonds (or loans and in that case it is called CLO) that are resold to the secondary market in tranches. The actual tranche structure is highly flexible and takes no definite form but in general, it has a waterfall structure where tranches are paid off in a sequential order. This structure is known as the n-th loss tranche structure. For example, a CDX CDO is a CDO with 100 credit default swaps with US$ 10 million notional. The CDO is sliced up into the following tranches: 0-3%, 3-7%, 7-10%, 10-15%, and 15-30%. Losses over 30% are extremely unlikely (due to collaterals) and hence are not analyzed. The structure is the standard waterfall structure that the first tranche (equity tranche) takes upon loss of default up to 3%, or $300,000, Then the following mezzanine tranche will take over the next 4% loss (from 3 to 7%), or $400,000, and so on.

As we see above, the basic structure of a CDO is similar to that of a CMO (Collateral Mortgage Obligations) in that the waterfall defines the sequence of the payoff to each tranche. The difference lies in tranche payoffs. A typical CMO uses all its revenues (mortgage payments) to pay its senior tranche. After the senior tranche is completely paid, the mezzanine tranche can take over. A CDO structure is to pay every tranche if there is enough revenue (coupons from bonds). Any reduction in revenue due to defaults is taken from the junior tranche, and then the mezzanine tranche, and then the senior tranche.

The primary pricing objective of pricing CDOs is to compute the spreads of the tranches. However, an equally important objective is to study the loss distributions of the tranches. As it will be clear later, even though spreads can be computed for the investors to trade, risk managing these tranches require much more than just the spreads. Due to the extremely exotic loss distributions, without the whole distribution, it is nearly impossible to do a good job in risk management.

A typical number of a CDOs constituents is in hundreds. And all constituents are highly correlated. As a result, to understand the loss distribution of a CDO is difficult. Monte Carlo simulation is the usual technique to accomplish such a goal. However, for 100 bonds in a CDO, there will be $\frac{100\times 99}{2} = 4950$ correlations to estimate and simulations themselves take a large amount of time. As a result, a fast approximation method is developed. The use of FI quickly reduces the multi-dimensional problem to uni-dimensional. However, some simplifications must be made. While details are to be seen later, the major intuition is as follows. First, we must assume a factor model (single factor to begin with) through which all bonds are correlated. This limits the flexibility of how assets are correlated and yet it provides consistency (of correlation estimates) over time and a parametric structure that makes FI possible. Secondly, we must assume a barrier structural model that explain how defaults occur.

Tranches are regarded as a series of unsecured debts from senior to junior. The equity tranche of a CDO is like the regular equity of a company, which is a call option. The problem here is to model the underlying asset which is portfolio loss.

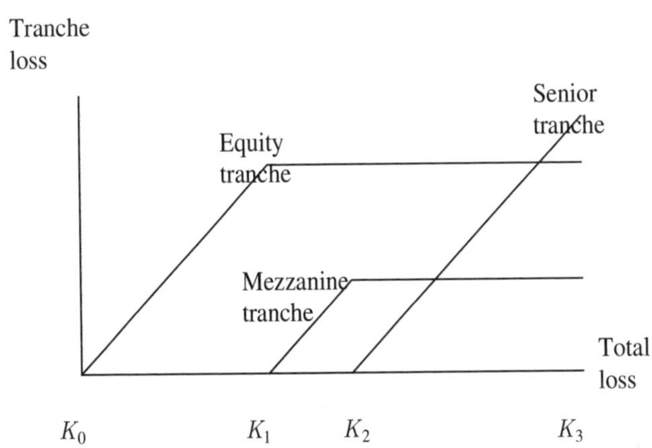

Figure 14.9: CDO Waterfall

In Figure 14.9, $K_0 = 0$ and $K_j - K_{j-1}$ is the size of the jth tranche. For example, K_1 (or $K_1 - K_0$) is the notional of the equity tranche, $K_2 - K_1$ is the notional of the mezzanine tranche, and $K_3 - K_2$ is the notional of the senior tranche.

Note that the total loss of the pool cannot exceed the total size of the pool as long as there is recovery. In a simple case where all bonds are equal size (e.g. $ 100,000 notional for each bond) and equal recovery (e.g. 40%), a pool of 100 bonds has a total size of $10 million and the total loss will not exceed $4 million. One can size up the pool to three tranches as given in the above diagram to be $2 million equity, $5 million mezzanine, and $3 million senior tranches. Identically, this implies $K_1 = 2$ million, $K_2 = 7$ million, and $K_3 = 10$ million. Note that the maximum loss in this example is $6 million (recovery rate is 40% for all bonds) and therefore the senior tranche will never suffer any loss. As a result, the senior tranche is risk-free.

A typical contract of a CDO tranche is a swap. That is, the protection buyer pays a series of spreads and in return receives payments identical to default losses. In other words, a CDO tranche is itself a CDS. However, different from a simple corporate CDS where default can happen only once and the contract stops, a CDO pool will have multiple defaults (up to the total number of bonds in the pool, which is 100 in the above example). Upon each default, a payment is made to a proper tranche investor. According to the loss payoff diagram above, for the first few defaults experienced by the pool, the loss amounts are paid to equity investors up to K_1 ($2 million in the example). Then losses of the next defaults are paid to mezzanine investors up to K_2. The process continues until the either all bonds in the pool default or the CDO contract expires, whichever earlier.

To facilitate the understanding of a CDO structure and how each tranche is priced, we start with the following notation. We first assume that there are a total of m bonds in the CDO pool and the pool will liquidate by time T. And then,

- N_i is the notional of the ith bond in the pool for $i = 1, \cdots, m$.

  $N_0 = L(t) + W(t) + N(t) = \sum_{i=1}^m N_i$ is initial notional

- R_i is the recovery rate of the ith bond in the pool for $i = 1, \cdots, m$ so that:

  $N(t) = \sum_{i=1}^m N_i \mathbb{I}_{\tau_i > t}$ is survival pool notional at time t, where τ_i is default time for the ith bond,

  $L(t) = \sum_{i=1}^m (1 - R_i) N_i \mathbb{I}_{\tau_i < t}$ is the total pool loss at time t, where τ_i is default time for the ith bond,

  $W(t) = \sum_{i=1}^m R_i N_i \mathbb{I}_{\tau_i < t}$ is write-down-from-above at time t, where τ_i is default time for the ith bond,

  $L(t) + W(t) = \sum_{i=1}^m N_i \mathbb{I}_{\tau_i < t} = 1 - N(t)$ is the loss in pool notional,

- $L^{\max} = \sum_{i=1}^m (1 - R_i) N_i$ is the maximum loss of the pool (all bonds default) and $L(t) < L^{\max}$ for all $t < T$.

- $W^{\max} = \sum_{i=1}^m R_i N_i$ is the maximum write-down-from-above of the pool (all bonds default) and $W(t) < W^{\max}$

- $N_0 = \sum_{i=1}^{m} N_i = L^{\max} + W^{\max}$ is the total notional for the entire pool at inception. It is the sum of notionals of individual bonds in the pool. Note that the pool notional will decrease over time as losses occur.

Note that, due to the indicator function, $N(t)$, $L(t)$, and $W(t)$ are random variables.

As a bond defaults, its loss is paid to a proper tranche. Assume that the maturity time of the CDO contract is T and that there are an unknown number of (n) defaults occur before maturity. Then let L_t^X be the loss paid to tranche X for the default at time t $(< T)$. As defaults occur at discrete points in time, we label default times as t_j where $j = 1, \cdots, n < m$ and n is a random variable. As a result, at any future default time t_n, the total loss and total write-down-from-able can be written as:

$$
\begin{aligned}
L(t_n) &= \sum_X \sum_{j=1}^{n} L_{t_j}^X = \sum_X \int_0^{t_n} L_t^X \, dt \\
W(t_n) &= \sum_X \sum_{j=1}^{n} W_{t_j}^X = \sum_X \int_0^{t_n} W_t^X \, dt
\end{aligned} \tag{14.53}
$$

Note that the actual loss must be less than the maximum loss and so must be write-down-from-above.

$$
\begin{aligned}
L^{\max} &\leqslant L(T) \\
W^{\max} &\leqslant W(T)
\end{aligned} \tag{14.54}
$$

where equalities hold when all bonds default prior to the maturity time T. Also note that the total pool size should be equal to all the tranches combined at inception:

$$
N_0 = \sum_X N_0^X \tag{14.55}
$$

As defaults happen, the tranche notional decreases:

$$
N_{t_n}^X = N_0^X - \sum_{j=1}^{n} L_{t_j}^X - \sum_{j=1}^{n} W_{t_j}^X \tag{14.56}
$$

The spread of each tranche must be set such that the total spread payment (present value of all spreads) is equal to the total discounted expected loss for the tranche. The present value of all spreads is the value of the premium leg of the swap:

$$
\begin{aligned}
V_{\text{prem}}^{X} &= s^{X}\hat{\mathbb{E}}\left[\sum\nolimits_{j=1}^{n} \exp\left(-\int_{0}^{t_j} r(u)du\right) N_{t_j}^{X}\right] \\
&= s^{X}\sum\nolimits_{j=1}^{n} P(0,t_j)\left(N_0^X - \tilde{\mathbb{E}}^{(t_j)}\left[\sum\nolimits_{i=1}^{j} L_{t_i}^{X}\right] - \tilde{\mathbb{E}}^{(t_j)}\left[\sum\nolimits_{i=1}^{j} W_{t_i}^{X}\right]\right)
\end{aligned}
\tag{14.57}
$$

where s^X is the tranche-X spread, L_t^X is the loss occured at time t to tranche X, $W_{t_j}^X$ is write-down-from-above at same time t. The protection leg of the tranche can be written as:

$$
\begin{aligned}
V_{\text{prot}}^{X} &= \hat{\mathbb{E}}\left[\sum\nolimits_{j=1}^{n} \exp\left(-\int_{0}^{t_j} r(u)du\right) L_{t_j}^{X}\right] \\
&= \sum\nolimits_{j=1}^{n} P(0,t_j)\tilde{\mathbb{E}}^{(t_j)}[L_{t_j}^{X}]
\end{aligned}
\tag{14.58}
$$

Define $C_{t_n}^X = \sum_{j=1}^{n} L_{t_j}^X$ as the cumulative tranche loss up to time t_n. Consequently $L_{t_n}^X = C_{t_n}^X - C_{t_{n-1}}^X$ and hence we can re-write the protection leg value as:

$$
V_{\text{prot}}^{X} = \sum\nolimits_{j}^{n} P(0,t_j)\left\{\hat{\mathbb{E}}[C_{t_j}^X] - \hat{\mathbb{E}}[C_{t_{j-1}}^X]\right\} \tag{14.59}
$$

Although the two results are the same, it is more convenient to use the cumulative tranche loss as it is a call spread of the total pool loss, as the loss payoff diagram depicts:

$$
C_t^X = \max\left\{\int_0^t L(u)du - K_X, 0\right\} - \max\left\{\int_0^t L(u)du - K_{X+1}, 0\right\} \tag{14.60}
$$

As a result, as long as we can model the total loss, we can easily compute the cumulative tranche loss and in turn the value of the protection leg and the value of the premium leg of each tranche.

14.10.2 Factor Copula

In the standard factor copula model used now as the industry standard, we write the following:

$$
x_i = \sqrt{\rho}\hat{W}_M + \sqrt{1-\rho}\hat{W}_i \tag{14.61}
$$

where x_i is the factor that computes the default probability for firm i and ρ is the base correlation. Both W_M and W_i are normally distributed with 0 mean. Hence x_i is normally distributed. For any given period (say a year), we can write the conditional default probability as follows:

$$\begin{aligned}\hat{p}_{i|f} &= \widehat{\Pr}\left(x_i < K_i | W_M = f\right) = \widehat{\Pr}\left(\sqrt{\rho} f + \sqrt{1-\rho} W_i < K_i\right) \\ &= \widehat{\Pr}\left(W_i < \tfrac{\sqrt{\rho} f - K_i}{\sqrt{1-\rho}}\right) \\ &= N\left(\tfrac{K_i - \sqrt{\rho} f}{\sqrt{1-\rho}}\right) \\ &= N\left(\tfrac{N^{-1}(\hat{p}_i) - \sqrt{\rho} f}{\sqrt{1-\rho}}\right)\end{aligned} \tag{14.62}$$

where $\widehat{\Pr}(x_i < K_i) = N(K_i) = \hat{p}_i$ is the unconditional default probability. Naturally,

$$\int_{\Omega(f)} \hat{p}_{i|f} \phi(f) df = \hat{p}_i \tag{14.63}$$

where $\phi(f)$ is the standard normal distribution for the common factor, f. Conditional on the market factor f, firms default independently. Hence, under the conditional probability $\hat{p}_{i|f}$, defaults are independent. There are several methods to model conditional loss distribution. Here we introduce the Vasicek method, the Fourier inversion method, and the recursive algorithm.

The Vasicek mothod is a special case of the Fourier inversion method and the recursive algorithm. It assumes that the bond notionals and recoveries in the CDO pool are the same. The Fourier inversion method and the recursive algorithm have relative advantages and disadvantages. Recursive algorithm is faster but it works best when the recoveries of the bonds are the same. Fourier inversion method is more general but it is generally slower and it gives continuous loss distribution where the true loss distribution is discrete.

14.10.3 The Vasicek Model

If the exposures (notionals) are identical, then the percentage loss can be found by carrying out the following equation,

$$\begin{aligned}\Pr(\ L=i/m\ ) &= \binom{m}{i}\Pr(A_1<K_1,\cdots,A_i<K_i,A_{i+1}>K_{i+1},\cdots,A_m>K_m)\\ &= \binom{m}{i}\int_{-\infty}^{\infty}\Pr(A_1<K_1,\cdots,A_i<K_i,A_{i+1}>K_{i+1},\cdots,A_m\\ &\quad >K_m|W_M=f)dF(W_M<f)\end{aligned} \tag{14.64}$$

Independence conditional on $W_M = f$ gives:

$$\begin{aligned}\Pr(L=i/m) &= \binom{m}{i}\int_{-\infty}^{\infty}\Pr(A_1<K_1|W_M=f)\cdots\Pr(A_i<K_i|W_M=f)\\ &\quad \Pr(A_{i+1}>K_{i+1}|W_M=f)\cdots\Pr(A_n>K_n|W_M=f)dF(W_M<f)\\ &= \binom{m}{i}\int_{-\infty}^{\infty}\prod_{j=1}^{i}N\left(-\tfrac{N^{-1}(p_j)+\sqrt{\rho}f}{\sqrt{1-\rho}}\right)\prod_{j=i+1}^{m}N\left(\tfrac{N^{-1}(p_j)+\sqrt{\rho}f}{\sqrt{1-\rho}}\right)\\ &\quad dF(W_M<f)\end{aligned} \tag{14.65}$$

Vasicek further simplifies by setting all probabilities equal, i.e. $p_j = p$:

$$\Pr(L=i/m) = \binom{m}{i}\int_{-\infty}^{\infty}N\left(-\tfrac{N^{-1}(p)+\sqrt{\rho}f}{\sqrt{1-\rho}}\right)^{i}N\left(\tfrac{N^{-1}(p)+\sqrt{\rho}f}{\sqrt{1-\rho}}\right)^{m-i}dF(W_M<f) \tag{14.66}$$

Cumulative loss is:

$$\Pr(L\leqslant i/m) = \sum_{\ell=0}^{i}\Pr(L=\ell/m) \tag{14.67}$$

This can be very computationally expensive if probabilities are not equal, and still quite expensive if probabilities are equal (sum over m-choose-i terms). Vasicek (Limiting Loan Loss Probability Distribution) proposes an asymptotic formula and Schonbucher presents the FFT method. The FFT method is more plausible because it is computationally possible and less limitation than the Vasicek method.

14.10.4 Fourier Inversion

Fourier Inversion method has been introduced to finance for more than 10 years now. Unofficially, Dilip Madan has spoken about using Fourier Inversion to compute multi-factor interest rate options back in 1991. The first published work in finance using such a technique is dated back to Steve Heston in 1993 when he published his seminal work in option pricing with random volatility in the Journal of Finance. Since then an explosive trend of option pricing with FI has emerged. Encouraged by the conversation with Madan, Scott discovered (1997), independently of Heston, the European option pricing model with not only random volatility, but random interest rates and jumps. In other words, Scotts model is a 4-factor model but can be simply computed with uni-variate integration by using the FI technique. Scott and Chen (1995) applied the technique in our interest rate option model, and Yeh and Chen (2003) applied the technique to provide bounds for American option prices under Scotts model. Peter Carr and Dilip Madan himself (1999) have also provided nice work in the variance gamma process. Recently Leippold and Wu (2002) use it for modeling the term structure of interest rates and Bakshi, Madan, and Zhang (2002) use it to price credit default swaps.

Fourier transform is used when the pricing problem is multi-dimensional. Fourier transform turns the multi-dimensional problem into a single dimensional problem and hence closed-form solution is readily available. Recently, the method has been applied to valuing large credit portfolios. Fourier inversion is particular efficient in credit portfolios because there is a large number of bonds in a portfolio. Monte Carlo is too slow to analyze large portfolios. With some reasonable simplifications, one can compute prices or risk analytics of large portfolios extremely fast with FI. Interestingly, there is no original reference to be found as to by whom and where FI has been used for credit portfolios, and yet the method has been widely circulated and used on Wall Street nowadays. However, lack of implementation details prevents wider use of this powerful method. The purpose of this paper is to derive in full details the CDO (Collateral Debt Obligations) pricing model using FI and correct some mistakes in the current materials that are being circulated.

In addition to Vasicek's approximation, we can implement Fourier Transform to compute the cumulative loss. Continue with the previous notation and let L_i where $i = 1, \cdots, m$ be individual losses (independent conditional on market factor, f). Under the same one factor model defined above, the characteristic function for each loss, given the market (i.e. $\hat{W}_M = f$) is:

$$\hat{\psi}_{i|f}(u) = \hat{\mathbb{E}}[e^{-\iota u L_i} | W_M = f] \tag{14.68}$$

where $\iota = \sqrt{-1}$ is the imaginary number. In the reduced form framework (binomial event), the individual characteristic function is straightforward:

$$\begin{aligned}\hat{\psi}_{i|f}(u) &= \hat{\mathbb{E}}[e^{-\iota u L_i}|\hat{W}_M = f] \\ &= (1 - \hat{p}_{i|f}) + \hat{p}_{i|f} e^{-\iota u N_i(1-R_i)}\end{aligned} \tag{14.69}$$

when R_i (recovery) and N_i (exposure) are fixed. If R is random, then the characteristic function becomes:

$$\begin{aligned}\hat{\psi}_{i|f}(u) &= \hat{\mathbb{E}}[e^{-\iota u L_i}|\hat{W}_M = f] \\ &= (1 - \hat{p}_{i|f}) + \hat{p}_{i|f} \hat{\mathbb{E}}[e^{-\iota u N_i(1-R_i)}|\hat{W}_M = f]\end{aligned} \tag{14.70}$$

where $p_{i|f}$ is conditional on knowing the market, given by (14.62). The second term in the above equation is more complex but can still be computed in the Merton model.

Let

$$L = \sum\nolimits_{i=1}^{m} L_i \tag{14.71}$$

be the portfolio loss. Then:

$$\begin{aligned}\hat{\psi}_{L|f}(u) &= \hat{\mathbb{E}}[e^{-\iota u L}|\hat{W}_M = f] \\ &= \prod\nolimits_{i=1}^{m} \hat{\psi}_{i|f}(u)\end{aligned} \tag{14.72}$$

Substituting (14.68) back into (14.72) and integrate over the distribution of market, f, gives the unconditional characteristic function:

$$\hat{\psi}_L(u) = \int_{-\infty}^{\infty} \prod\nolimits_{i=1}^{m} \hat{\psi}_{i|f} \phi(f) df \tag{14.73}$$

where $\phi(f) = \frac{1}{\sqrt{2\pi}} \exp\left(-\frac{f^2}{2}\right)$ is the standard normal density function. The inversion of the characteristic function gives the probability:

$$\widehat{\Pr}(L > z) = 0.5 + \frac{1}{\pi} \int_0^{\infty} \text{Re}\left[\frac{e^{-\iota u z} \hat{\psi}_L(u)}{\iota u}\right] du \tag{14.74}$$

Under identical exposure and recovery, i.e. $K_i = K$ and $R_i = R$ for all i, as in the case of synthetic CDOs, (14.72) can be simplified as:

$$
\begin{aligned}
\hat{\psi}_{L|f}(u) \quad & = \prod_{i=1}^{m}(1-\hat{p}_{i|f}) + \hat{p}_{i|f}e^{-\iota uK(1-R)} \\
& = \prod_{i=1}^{m}(1-\hat{p}_{i|f}) \\
& + e^{-\iota uN(1-R)}\begin{Bmatrix} (\hat{p}_{1|f}\hat{p}_{2|f}\cdots\hat{p}_{m-1|f}(1-\hat{p}_{m|f}) \\ +\cdots+(1-\hat{p}_{1|f})\hat{p}_{2|f}\cdots\hat{p}_{m-1|f}\hat{p}_{m|f}) \end{Bmatrix} \qquad \cdots(1) \\
& + e^{-2\iota uN(1-R)}\begin{Bmatrix} \hat{p}_{1|f}\hat{p}_{2|f}\cdots\hat{p}_{m-2|f}(1-\hat{p}_{m-1|f})(1-\hat{p}_{m|f}) \\ +\cdots+(1-\hat{p}_{1|f})(1-\hat{p}_{2|f})\hat{p}_{3}\cdots\hat{p}_{m-1}\hat{p}_{m} \end{Bmatrix} \qquad \cdots(2) \\
& + e^{-3\iota uN(1-R)}\begin{Bmatrix} (\hat{p}_{1}\hat{p}_{2}\cdots(1-\hat{p}_{m-2|f})(1-\hat{p}_{m-1|f})(1-\hat{p}_{m|f}) \\ +\cdots+(1-\hat{p}_{1|f})(1-\hat{p}_{2|f})(1-\hat{p}_{3|f})\cdots\hat{p}_{n-1}\hat{p}_{n}) \end{Bmatrix} \qquad \cdots(3) \\
& + \cdots \\
& + \prod_{i=1}^{m}\hat{p}_{j}e^{-m\iota uN(1-R)}
\end{aligned}
\tag{14.75}
$$

where (1) has m terms, (2) has C_2^m terms, and (3) has C_3^m terms and so on. Integrating over the common factor, as shown in (14.73), we get:

$$
\hat{\psi}_L(u) = A_0 + A_1e^{-\iota uN(1-R)} + A_2e^{-2\iota uN(1-R)} + \cdots + A_me^{-m\iota uN(1-R)} \tag{14.76}
$$

where each A is the integral of ps and $(1-p)$'s over the domain of f. Then,

$$
e^{\iota uL}\hat{\psi}_L(u) = A_0e^{\iota uL} + A_1e^{\iota u(L-N(1-R))} + A_2e^{\iota u(L-2N(1-R))} + \cdots + A_me^{\iota u(L-nN(1-R))} \tag{14.77}
$$

Note that A's are not functions of u. Hence, the Fourier integral is:

$$
\begin{aligned}
G &= \int_0^\infty \text{Re}\left[\frac{e^{\iota uz}\hat{\psi}_L(u)}{\iota u}\right]du \\
&= A_0\int_0^\infty \text{Re}\left[\frac{e^{\iota uz}}{\iota u}\right]du + A_1\int_0^\infty \text{Re}\left[\frac{e^{\iota u(z-N(1-R))}}{\iota u}\right]du \\
&+ A_2\int_0^\infty \text{Re}\left[\frac{e^{\iota u(z-2N(1-R))}}{\iota u}\right]du + \cdots + A_n\int_0^\infty \text{Re}\left[\frac{e^{\iota u(z-nN(1-R))}}{\iota u}\right]du
\end{aligned}
\tag{14.78}
$$

Now, we focus on each integral. First, note that

$$
\begin{aligned}
\int_0^\infty \text{Re}\left[\frac{e^{\iota ux}}{\iota u}\right]du &= \lim_{k\to 0}\int_0^\infty \text{Re}\left[\frac{e^{\iota ux}}{\iota u}e^{-k^2u^2}\right]du \\
&= \lim_{k\to 0} f(k,x)
\end{aligned}
\tag{14.79}
$$

then,

$$\begin{aligned}\frac{\partial f(k,x)}{\partial x} &= \int_0^\infty \text{Re}\left[e^{\iota ux - k^2u^2}\right] du \\ &= \int_0^\infty \text{Re}\left[\exp\left\{-k^2\left(u - \tfrac{\iota x}{2k^2}\right)^2 - \left(\tfrac{x}{2k}\right)^2\right\}\right] du \\ &= \exp\left\{-\left(\tfrac{x}{2k}\right)^2\right\} \int_0^\infty \text{Re}\left[\exp\left\{-\left(ku - \tfrac{\iota kx}{2k^2}\right)^2\right\}\right] du\end{aligned} \tag{14.80}$$

Let $v = ku$. Then

$$\begin{aligned}\frac{\partial f(k,x)}{\partial x} &= \frac{1}{k}\exp\left\{-\left(\tfrac{x}{2k}\right)^2\right\} \int_0^\infty \text{Re}\left[\exp\left\{-\left(v - \tfrac{ikx}{2k^2}\right)^2\right\}\right] dv \\ &= \frac{1}{k}\exp\left\{-\left(\tfrac{x}{2k}\right)^2\right\} \frac{\sqrt{\pi}}{2}\end{aligned} \tag{14.81}$$

Since $f(k,0) = 0$,

$$\begin{aligned}f(k,x) &= \frac{\sqrt{\pi}}{2k}\int_0^x \exp\left\{-\left(\tfrac{y}{2k}\right)^2\right\} dy \\ &= \frac{\sqrt{\pi}}{2k}(2k\sqrt{\pi})\int_0^x \frac{1}{\sqrt{2}k\sqrt{2\pi}}\exp\left\{-\left(\tfrac{y}{2k}\right)^2\right\} dy \\ &= \pi N\left(\frac{x}{2k}\right)\end{aligned} \tag{14.82}$$

hence,

$$\int_0^\infty \text{Re}\left[\frac{e^{\iota ux}}{\iota u}\right] du = \lim_{k\to 0} f(k,x) = \pi \tag{14.83}$$

Note that this result is independent of x. As a result,

$$\begin{aligned}\widehat{\Pr}(L > z) &= 0.5 + \frac{1}{\pi}\int_0^\infty \text{Re}\left[\frac{e^{-\iota uz}\hat{\psi}_L(u)}{\iota u}\right] du \\ &= 0.5 + \frac{1}{\pi}G \\ &= 0.5 + \frac{1}{\pi}\left(\pi\sum_{j=1}^N A_j\right) \\ &= 0.5 + \sum_{j=1}^N A_j\end{aligned} \tag{14.84}$$

While both (14.74) and (14.84) can calculate the loss distribution (the former is Fourier Inversion and the latter is closed-form), we note that the latter is not necessarily more efficient in that the number of terms to sum up is growing in a combinatorial manner.

14.10.5 Recursive Algorithm

Note that the loss amounts are discrete. If we fix the recovery rate at R (e.g. 40%), then the loss amounts are: 0, $1-R$, $2(1-R)$, L, $m(1-R)$ where m is the total number of bonds in the CDO. The recursive algorithm is similar to the binomial expansion. Given that the recursive algorithm assumes defaults to be independent, it must be used under the factor copula model in that conditional on the common factor, f, defaults are independent.

To begin the algorithm, we must set two initial values:

$$\begin{aligned} q_0(L=0) &= q_0(0) = 1 \\ q_0(L>0) &= 0 \end{aligned} \tag{14.85}$$

where L is loss. For each bond $i = 1, \cdots, m$, we have $L_1, \cdots, L_m$ representing the loss of the bond respectively. The algorithm is described by the following formula:

$$\begin{aligned} q_{i+1}(L,t) &= \{1-\hat{p}_{i+1|f}(t)\}q_i(L,t) + \hat{p}_{i+1|f}(t)q_i(L-L_{i+1},t) & L \geqslant L_{i+1} \\ q_{i+1}(L,t) &= \{1-\hat{p}_{i+1|f}(t)\}q_i(L,t) & L < L_{i+1} \end{aligned} \tag{14.86}$$

where $\hat{p}_{i|f}$ and R_i are the conditional (on common factor f) probability of default and recovery rate of the ith bond.

For simplicity and easy exposition, we first assume that $\hat{p}_{i|f} = p$ and $R_i = 0.4$ for all bonds (in which case $L_i = 0.6$ for all i). We start with one bond. In a CDO pool of only one bond with a face value N, the only possible outcomes are no default with no loss and default with a loss of \$0.6N. Then we can write the probability distribution as:

Loss	0	$0.6N$	$> 0.6N$
Prob	$q_1(0)$	$q_1(0.6)$	0

where

$$\begin{aligned} q_1(0) &= q_0(0)(1-p) = 1-p \\ q_1(0.6) &= q_0(0.6)(1-p) + q_0(0)p = p \end{aligned}$$

If we assume that there are two bonds each with a notional of N, the possible losses are \$0 and \$$0.6N$ and \$$1.2N$.

Loss	0	$0.6N$	$1.2N$	$> 1.2N$
Prob	$q_2(0)$	$q_2(0.6)$	$q_2(1.2)$	0

where

$$q_2(0) = q_1(0)(1-p) = (1-p)^2$$
$$q_2(0.6) = q_1(0.6)(1-p) + q_1(0)p = 2p(1-p)$$
$$q_2(1.2) = q_1(1.2)(1-p) + q_1(0.6)p = p^2$$

By induction, we can write the general result as:

Loss	0	$0.6N$	$1.2N$				$0.6mN$	$> 0.6mN$
Prob	$q_m(0)$	$q_m(0.6)$	$q_m(1.2)$	$\cdots$	$\cdots$	$\cdots$	$q_m(mN)$	0

where for $i = 0, 1, \cdots, m$,

$$q_m(i(1-R)) = \binom{m}{i} p^i (1-p)^{m-i}$$

which is precisely the binomial distribution. Apparently the recursive algorithm is more general than the binomial distribution as neither the base probability p nor the recovery needs to be fixed, as (14.86) shows.

The conditional (on the common factor f) loss distribution is a discrete distribution. The unconditional loss distribution is to integrate this conditional loss distribution over the market factor f:

$$\int_{\Omega(f)} q_m(L,t)\phi(f)df = Q_m(L,t) \tag{14.87}$$

Now we have obtained the unconditional loss distribution at any given time t. Once we find the total loss distribution, we can then price each tranche as specified in the loss payoff diagram described in Section A.

14.10.6 An Example

In the following, we show an example where there are 100 bonds in the pool. This probability is exaggerated so to magnify the behavior of the loss distribution. The recovery rate is 40% for all bonds. Hence, the maximum loss of the pool is 60% of the total pool size.

The shape of the loss distribution is very sensitive to the correlation. In the following several diagrams, we shall see how the loss distribution changes as the correlation increases. On the left is plotted the loss distribution for various correlation numbers when the default probability is 6%. On the right is plotted the same correlation scenarios when the default probability is 2%.

As we can see obviously, the lower is the default probability; the more right-skewed is the loss distribution, which implies the safer is the pool. Also we see that distribution is multi-modal when the correlation is high. The multi-modal phenomenon is more pronounced when the default probability is high. This can be easily seen from the two graphs when the correlation is 0.5.

Once we obtain the loss distribution, we can then proceed to price various CDO tranches. For example, for the equity,

$$\begin{aligned} V_{\text{prot}}^{\text{eq}} &= e^{-r(T-t)}\hat{\mathbb{E}}_t[\max\{L(T)-K_0,0\} - \max\{L(T)-K_1,0\}] \\ &= e^{-r(T-t)}\{\hat{\mathbb{E}}_t[L(T)] - \hat{\mathbb{E}}_t[\max\{L(T)-K_1,0\}]\} \\ &= e^{-r(T-t)}\{\hat{\mathbb{E}}_t[L(T)] - \hat{\mathbb{E}}_t[L(T)\mathbb{I}_{L>K_1}] - K_1\widehat{\Pr}[L(T)>K_1]\} \end{aligned} \tag{14.88}$$

where $\widehat{\Pr}[\cdot]$ is the risk-neutral probability. The risk-neutral probability $\widehat{\Pr}[\cdot]$ can be directly computed by using (14.74) via Fourier Inversion or (14.87) via Recursive Algorithm. The quantity $\hat{\mathbb{E}}_t[L(T)\mathbb{I}_{L>K_1}]$ can be computed numerically as we know the probability function of the loss.

14.10.7 Extensions

One significant drawback of the copula model is that it is a single period model. It successfully incorporates multiple bonds and generates the loss distribution but it is not able to take into account defaults and losses over time. For this, we need to combine the Merton model with the copula model. Also, in the factor copula model, $\sqrt{\rho}$ is the correlation everybody has with the market. However, this is not necessarily the case.

As a result, we can modify the factor copula model as follows. Assume that

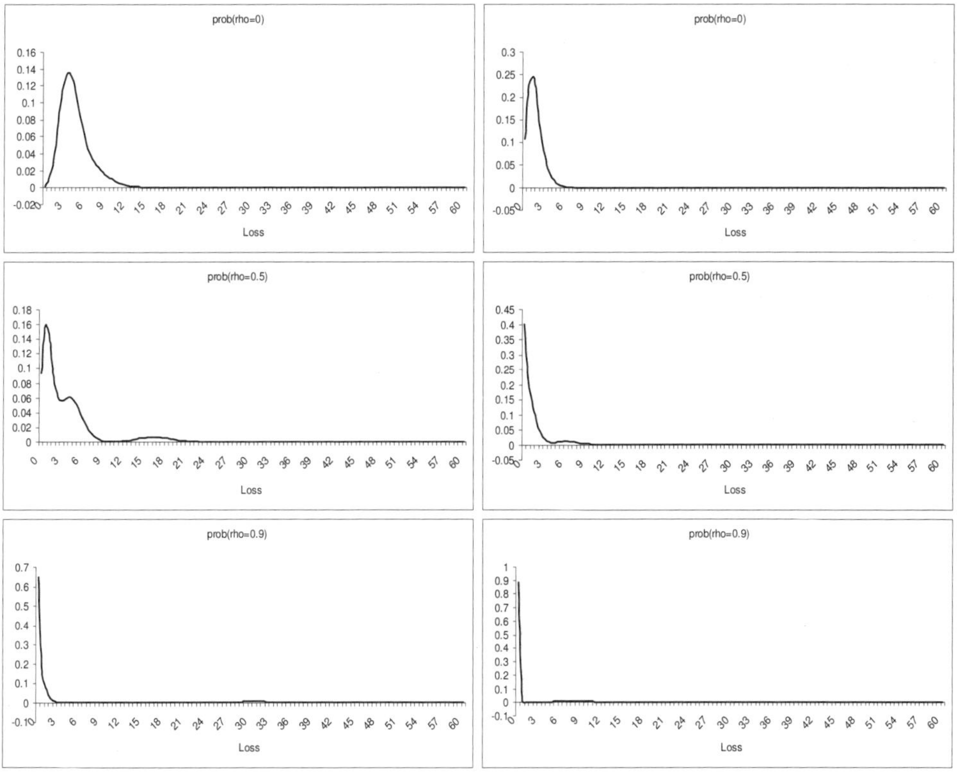

6% default probability 2% default probability

Figure 14.10: Fourier Inversion Results

each bond in the CDO pool has a publicly traded stock whose dynamics over time follows the Black-Scholes model:

$$\begin{aligned} d\ln S_i &= (\mu_i - 1/2\sigma^2)dt + \sigma dz_i \\ &= (r - 1/2\sigma^2)dt + \sigma d\hat{z}_i \end{aligned} \tag{14.89}$$

where

$$\begin{aligned} dz_i &= \sqrt{\rho_{iM}} dW_M + \sqrt{1-\rho_{iM}} dW_i \\ d\hat{z}_i &= \sqrt{\rho_{iM}} d\hat{W}_M + \sqrt{1-\rho_{iM}} d\hat{W}_i \end{aligned} \tag{14.90}$$

follows a copula model so that dW_i, dW_M, $d\hat{W}_i$, and $d\hat{W}_M$ are Brownian motions under the real and risk-neutral measures for the company and market respectively. It follows that $\sqrt{\rho_{iM}}$ is the correlation between the market and asset i and $dz_i dz_j = d\ln S_i d\ln S_j = \sqrt{\rho_{iM}\rho_{jM}}dt$. This is similar to the market model:

$$\ln S_i(T) - \ln S_i(t) = a_i + b_i r_M + e_i \tag{14.91}$$

where r_M is the return on market, and e_i is the idiosyncratic risk so that $E[e_i] = 0$ for all i and $E[e_i e_j] = 0$ for all $i \neq j$. Linking back to (14.89), we arrive at:

$$\begin{aligned} \ln S_i(T) - \ln S_i(t) &= \left(\mu_j - 1/2\sigma_i^2\right)(T-t) + \sigma_i\left(z_i(T) - z_i(t)\right) \\ &= \left(\mu_j - 1/2\sigma_i^2\right)(T-t) \\ &+ \sigma_i\left[\sqrt{\rho_{iM}}(W_M(T) - W_M(t)) + \sqrt{1-\rho_{iM}}(W_i(T) - W_i(t))\right] \end{aligned} \tag{14.92}$$

and

$$r_M = \mu_M(T-t) + W_M(T) - W_M(t) \tag{14.93}$$

As a result, we have:

- $\mathbb{E}[r_M] = \mu_M(T-t)$
- $\mathbb{V}[r_M] = \sigma_M^2(T-t)$
- $\mathbb{E}[r_i] = (\mu_i - 1/2\sigma_i^2)(T-t) = a_i + b_i\mathbb{E}[r_M] = a_i + b_i\mu_M$
- $b_i = \frac{\mathbb{K}[r_i, r_M]}{\mathbb{V}[r_M]} = \frac{\sigma_i(T-t)\sqrt{\rho}}{\mathbb{V}[r_M]}$

- $\mathbb{V}[r_i] = \sigma_i^2(T-t) = b_i^2\mathbb{V}[r_M] + \mathbb{V}[e_i]$
- $\sqrt{\rho_{iM}} \equiv \text{corr}[r_j, r_M] = \frac{\mathbb{K}[r_i, r_M]}{\sqrt{mathbbV[r_i]\mathbb{V}[r_M]}} = \frac{b_i\mathbb{V}[r_M]}{\sqrt{\mathbb{V}[r_i]\mathbb{V}[r_M]}} = b_i\frac{\sigma_M}{\sigma_i}$

Note that this is precisely the regression result. Furthermore, the correlation between r_i and r_j becomes:

$$\rho_{ij} = \frac{\text{cov}[r_i, r_j]}{\sigma_i\sigma_j(T-t)} = \frac{b_i b_j \sigma_M^2}{\sigma_i\sigma_j} = \sqrt{\rho_{iM}\rho_{jM}} \tag{14.94}$$

Note that conditional on r_M, $\rho_{ij|M} = 0$. Note that p_i is NOT a function of r_M and ρ_{iM}, but only inputs that affect the asset price process but the conditional probability $p_{i|M}$ is, hence it is random unless it is conditioned knowing r_M. Define $U_M = (r_M - \mu_M(T-t))/\sigma_M\sqrt{T-t}$ which is standard normal. For convenience, the conditional default probability (conditional on $U_M = f/\sqrt{T-t}$) is restated here as:

$$\hat{p}_{i|f} = N\left(\frac{N^{-1}(\hat{p}_i) - \rho_{iM}f/\sqrt{T-t}}{\sqrt{1-\rho_{iM}^2}}\right) \tag{14.95}$$

As we integrate over the domain of U_M, it returns the unconditional probability. Apply Fourier inversion to arrive at:

$$\widehat{\Pr}(L(t) > z) = 0.5 + \frac{1}{\pi}\int_0^\infty \text{Re}\left[\frac{e^{-\iota uz}\hat{\psi}_{L(t)}(u)}{\iota u}\right] du \tag{14.96}$$

$$f(z) = \frac{1}{2\pi}\int_{-\infty}^{\infty} e^{\iota uz}\hat{\psi}_L(u)du \tag{14.97}$$

Simplification of $\hat{\mathbb{E}}[e^{-\iota uL(t)}] = \psi_L(u) = \int_{-\infty}^{\infty} dU_M\phi(U_M)\Pi_{i=1}^m(1-\hat{p}_i)+\hat{p}_i e^{-\iota uN_i(1-R_i)}$.

14.10.8 Monte Carlo

In the simulation, we simulate a market factor plus idiosyncratic factors, all normals. The simulation for each reference entity is a bivariate normal (or multi-variate normal for multiple factors) and we run this normal for each reference entity through the normal probability function to transform the simulations for each x_i into 0-1 random variables, which will be correlated. Then the 0-1 random variables are applied to the survival probability functions to determine default times.

- Compute intensity parameter from default probability: $\lambda_i = -\frac{\ln[1-q_i]}{T}$ (flat lambda)
- Simulate $N+1$ independent normals (f and $\varepsilon_1 \cdots \varepsilon_N$)
- Compute $\underline{x}_i = \sqrt{\rho}\underline{f} + \sqrt{1-\rho}\underline{\varepsilon}_i$ using the assumed ρ
- Compute $\underline{p}_i = N(\underline{x}_i)$, which are then correlated
- Simulate (correlated) default time $\underline{\tau}_i = -\frac{\ln \underline{p}_i}{\lambda_i}$.

If there is only one asset, for any given path, j, we have default time, τ_1^j. The default leg value for that path is therefore $P_{\tau_1^j}(1-R_1)$ and the expected value is $\frac{1}{N}(1-R_1)\sum_j P_{\tau_1^j}$. For that path, the premium leg pays:

$$s\sum_{t=1}^{\tau_1^j \wedge T} P_t Q_t \tag{14.98}$$

and the expected value is

$$s\frac{1}{N}\sum_j \sum_{t=1}^{\tau_1^j \wedge T} P_t Q_t \tag{14.99}$$

As a result, the spread is:

$$s = \frac{\sum_j P_{\tau_i^j}(1-R_i)}{\sum_j \sum_{t=1}^{\tau_i^j \wedge T} P_t Q_t} \tag{14.100}$$

14.11 Default Prediction

Predicting bankruptcy is an important task for accountants. Annually, accountants issue Going Concern status on the firms they audit. A Going Concern status is to certify if a firm can survive the coming year. Accounting ratios are used to help such audits. Recently improvements have been proposed by Edward Altman who use discriminate analysis (Z score) and James Ohlson who use logistic regression (O score).

Financial economists also try to predict bankruptcy. They use corporate finance models as opposed to financial ratios. Fischer Black, Myron Scholes, and Robert Merton were the first to describe bankruptcy as the firms inability to pay its debt. The first commercial product of this idea was KMV. However, the Black-Scholes-Merton model is a single period model. To accommodate the multi-debt problem, KMV empirically estimated the one-year equivalent debt value to be:

$$
\begin{aligned}
\text{One Year Equivalent Debt} &= \text{Short Term Debt} + \frac{\text{Long Term Debt}}{2} \\
&= \frac{\text{Short Term Debt} + \text{Total Debt}}{2}
\end{aligned}
$$

and then use the Black-Scholes-Merton model. The original KMV model is to solve the simultaneous equation system:

$$
\begin{cases}
E(t) = A(t)N(d_+) - P(t,T)KN(d_-) \\
\sigma_E = \sigma_A \dfrac{A(t)}{E(t)} N(d_+)
\end{cases}
\tag{14.101}
$$

where K is the One Year Equivalent Debt, $A(t)$ is the asset value, $E(t)$ is the equity value (or market capitalization), $P(t,T)$ is the risk-free discount factor between now t and some future time T, σ_A is the asset volatility, σ_E is the equity volatility, and

$$
d_\pm = \frac{\ln \frac{A(t)}{KP(t,T)} \pm {}^1\!/\!_2\,\sigma_A^2 (T-t)}{\sigma_A \sqrt{T-t}}
$$

Both equity value (i.e. market capitalization) and equity volatility are observable. Hence we can solve for the asset value and the asset volatility.

KMV further defines a variable, distance to default (or abbreviated as DD), as follows:

$$
\begin{aligned}
\text{DD} &= \frac{E[\ln A(T)] - \ln K}{\sigma_A} \\
&= \frac{\ln A(t)/P(t,T) - {}^1\!/\!_2\,\sigma_A^2 (T-t) - \ln K}{\sigma_A} \\
&= d_-
\end{aligned}
\tag{14.102}
$$

Note that if we adopt a more complex model than the Black-Scholes-Merton, then we simply replace $N(d_-)$ in equation (14.101) by the delta of the model. For example, for the Geske model, the delta is defined in Chapter 13.

Recently, Stein (2002, 2005) proposed a non-parametric method to test models default prediction power. The Receiver Operating Characteristic (ROC) curves were first developed in the 1950s for radio signals contaminated by noise. Later it was introduced to medical research for testing the power of various diagnosis tests. A powerful test in disease detection should produce the least possible false positives and false negatives and the ROC curve is ideal for depicting these errors. False positive and false negative of a medical test is similar to the Type-I and Type-II errors in statistics. Hence formal statistical testing is possible when we use the ROC methodology. Similar to the medical test that tries to differentiate healthy individuals from unhealthy ones, a powerful default model should be able to successfully differentiate financially healthy firms from unhealthy firms. Yet only recently, this methodology has been introduced to evaluating credit risk. The first application of such methodology in finance appeared in a series of Moodys quantitative credit risk modeling group (e.g. Stein (2002, 2005)).

Stein (2002, 2005) argues that the power of a default model to predict defaults is its ability to detect True Positive, or true default, and the capability of a default model to calibrate to the data is its ability to detect True Negative, or true survivals. A good model should be both powerful and capable to calibrate.

The details of the ROC methodology can be found, for example, in Hanley and McNeil (1982) and the default application can be found in Stein (2002, 2005). Briefly, there are two populations, default and survival, each of which is plotted against distance to default, or DD as follows:

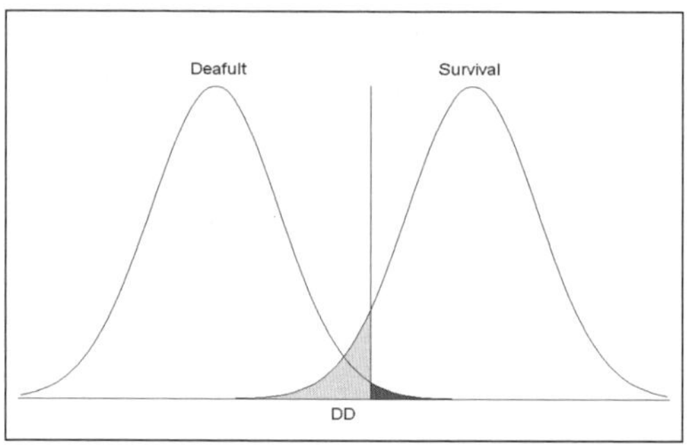

Figure 14.11: DD Distributions

Since no model is absolutely perfect and powerful, it is always possible that some small quantity of defaulted firms are erroneously evaluated to have higher DDs than some (also small quantity) survival firms. As a result, no matter where we set the cut-off point of DD to differentiate defaulted firms from the survived ones, there will always be false defaults (i.e. firms that are classified as default actually survived) and false survivals (firms that are classified as survival actually defaulted). The light-shaded (yellow-shaded) area on the left of the cut-off line represents false default and the dark-shade (gray-shaded) area on the right of the cut-off line represents false survival. Setting the null hypothesis to be H0: DD ¡ x represents default where x is an arbitrary cut-off threshold, we are able to relate the Type-I error (α) to false survival and the Type-II error (β) to false default. To compare different models, we can control the Type-II error at a certain level and compare the Type-I errors generated by the models.

The following diagrams depict what a good and a bad model are. The model on the left generates DDs that have a large overlap between default and survival samples. On the other hand, the model on the right generates DDs that are a lot more discriminative. As a result, it is easy to conclude that the model on the right is better than the model on the left. For the same Type-II error, the model on the right has a smaller Type-I error.

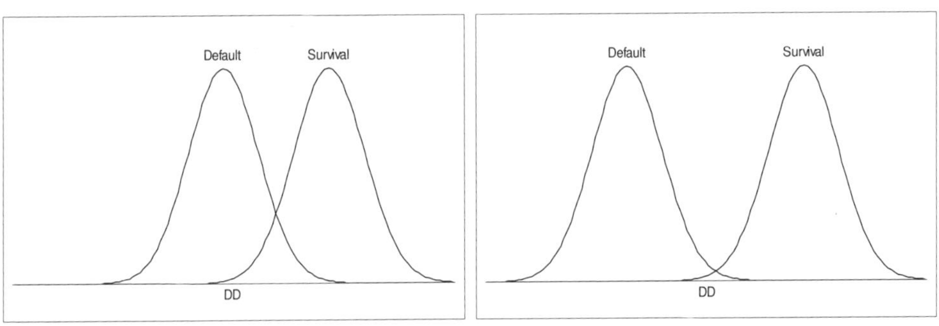

Figure 14.12: DD Distributions

The ROC curve is a plot of probability of False Positive, or false default, and the probability of True Positive, or true default. Note that the survival sample contains true survivals and false defaults and the default sample contains true defaults and false survivals. Hence the probabilities of true survival and false default sum to unity (and similarly the probabilities of true default and false survival sum to unity.) The diagram on the left explains graphically the ROC curve.

The diagram on the right plots three different models. The more powerful model should be more concave because it predicts more defaults successfully than the less powerful model. A model that is not able to differentiate default from survival is a model that produces DDs that overlap the default sample and the survival sample completely (i.e. the two distributions are one distribution). In this case, the ROC curve is a 45 degree line. Certainly, a model can be wrong in predicting default in that it gives defaulted firms higher DDs than survived firms. Such a model will have a convex ROC curve. In sum, the more powerful is the model that successfully set apart the two distributions, and then the more concave is the ROC curve.

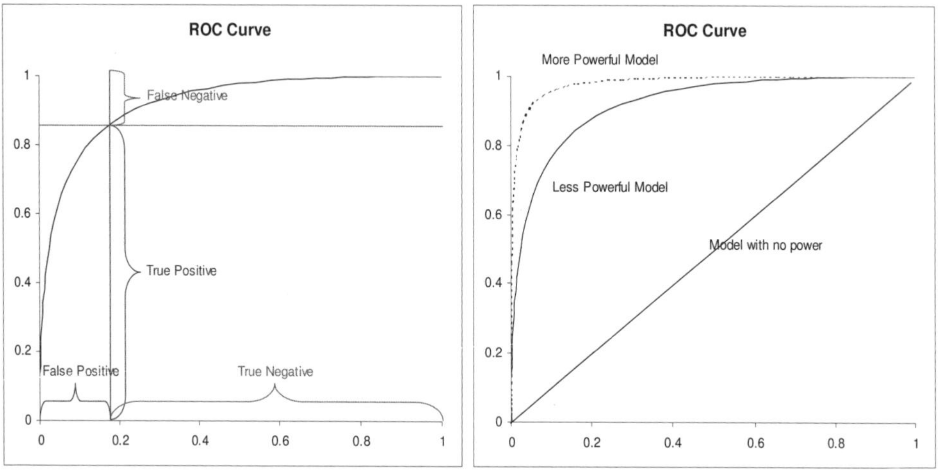

Figure 14.13: ROC Curve

The key statistic in the ROC methodology, which is also known as the Cumulative Accuracy Profile (CAP) or the Power Curve, is the Accuracy Ratio (AR) that is the area under the ROC curve. Since the ROC curve is generally concave, the more powerful is the model, the higher is the AR. At the end, we use the ARs of various models to test their prediction powers of default. We test 6 different models covering distinctly different assumptions in the paper. Note that the ROC curve is non-parametrical. Hence, it is ideal for the two distributions when they are non-parametrical.

Hanley and McNeil (1982) derive the standard error of AR. Note that their estimates for the standard error depend to a degree on the shapes of the distributions, but are conservative so even if the distributions are not normal, estimates of the standard error will tend to be a bit too large, rather than too small.

$$\sigma(A) = \sqrt{\frac{A(1-A) + (n_1 - 1)(Q_1 - A^2) + (n_2 - 1)(Q_2 - A^2)}{n_1 n_2}} \tag{14.103}$$

where A is the area under the curve, n_1 is the number of defaulted firms, n_2 is the number of survival firms, and:

$$
\begin{aligned}
Q_1 &= \tfrac{A}{2-A} \\
Q_2 &= \tfrac{2A^2}{1+A}
\end{aligned}
\tag{14.104}
$$

To test two models if they are different from each other, the standard error of the difference of the two areas, $A_1 - A_2$, is:

$$\sigma(A_1 - A_2) = \sqrt{\sigma(A_1)^2 + \sigma(A_2)^2 - 2\rho\sigma(A_1)\sigma(A_2)} \tag{14.105}$$

where ρ is the quantity that represents the correlation induced between the two areas by the same study of the same set of cases. Assuming we have two models M1 and M2, that classify all firms under study into either default and survival and we have already calculated the areas under the ROC curves (A_1 and A_2). The procedure is as follows:

- Look at survived firms. Find how the two sets of DDs correlate for these firms, and obtain a value ρ_1 for this correlation;
- Look at the defaulted firms, and similarly derive ρ_2, the correlation between the two tests for these patients;
- Average out ρ_1 and ρ_2;
- Average out the areas A_1 and A_2,
- Use Hanley and McNeil's (1983) Table I to look up a value of ρ, given the average areas, and average of ρ_1 and ρ_2.

Finally, the test statistic is Gaussian:

$$z = \frac{A_1 - A_2}{\sigma(A_1 - A_2)} \tag{14.106}$$

The following describes the needed steps to compute the ROC curve.

- Rank all DDs from smallest to greatest.
- To calculate DD, we need the most recent data on the stock price, stock volatility, and accounting data. For example, if we want to calculate DD at the end of December 1987, we use the market capitalization at the end of December and the annualized standard deviation of stock return calculated from 52 weekly stock returns ending at the end of December. The accounting

data are from the latest quarterly financial statements available to investors. We assume that financial statements will be available to investors three months after the end of a quarter.

- Divide the entire region into 100 groups (z from 1 $\sim$ 100). This choice is arbitrary. With more data, the granularity can be finer. With less data the granularity can be coarser (e.g. 10 groups).
- Divide the sample into default and survival groups. If the forecast horizon is 1 month, then firms that go bankruptcy in the following month (it will be January 1988 if we use DD calculated at the end of December 1987) will fall in the default group. If the forecast horizon is 12 month, then firms that goes bankruptcy within the following 12 month (it will be January 1988 if we use DD calculated at the end of December 1987) will be in the default group.
- In the default group, z is divided into 100 percentiles and compute cumulative probability (y). Again, the granularity depends on amount of data.
- In the survival group, DD is divided into 100 percentiles and compute cumulative probability (x). Again, the granularity depends on amount of data.
- Plot (x, y) curve.
- Calculate AR.

14.12 Appendix

14.12.1 Equation 14.29

For the first year, which contains four quarters, $n = 4$ and we have:

$$\begin{aligned} x_1 &= x_0 + 0.6p_1 + (1 - p_1)x_0 \\ x_2 &= x_0 + 0.6p_1 + (1 - p_1)x_1 \\ x_3 &= x_0 + 0.6p_1 + (1 - p_1)x_2 \\ x_4 &= x_0 = -0.000225 \end{aligned}$$

Substitute recursively into x_4 to get:

$$0 = \frac{0.6p_1 + (1 - p_1)\left[-0.000225 + \frac{0.6p_1 + (1-p_1)\left[-0.000225 + \frac{0.6p_1 + (1-p_1)\left[-0.000225 + \frac{0.6p_1 - 0.000225(1-p_1)}{1.0125}\right]}{1.0125}\right]}{1.0125}\right]}{1.0125}$$

and solve for $p_1 = 0.000375$ or 3.75 basis points.

14.12.2 Derive U_M

$$\begin{aligned} r_j &= a_j + b_j r_M + e_j \\ r_j - \mu_j &= b_j(r_M - \mu_M) + e_j \\ Z_j &= b_j Z_M + e_j \\ \frac{Z_j}{\sigma_M} &= b_j U_M + \frac{e_j}{\sigma_M} \end{aligned} \tag{14.107}$$

where $E[Z_j] = E[Z_M] = E[e_j] = 0$ and $\text{var}[Z_j] = \text{var}[r_j] = b_j^2\sigma_M^2 + \delta_j^2$. But previously, we have $d\ln A_j = (m_j - 0.5s_j^2)dt + s_j\left(\sqrt{\rho}dW_M + \sqrt{1-\rho}dW_j\right)$. To reconcile, we make the following identities:

$$\begin{aligned} \frac{Z_j}{\sigma_M} &= \ln\frac{A_{jT}}{A_{j0}} - (m_j - 0.5s_j^2)T \\ U_M &= W_M \\ \frac{e_j}{\sigma_M} &= s_j\sqrt{1-\rho}W_{jT} \end{aligned} \tag{14.108}$$

which result in the following relationships:

$$\begin{aligned} b_j &= s_j\sqrt{T}\sqrt{\rho} \\ \rho_{jM}\frac{\sigma_j}{\sigma_M} &= s_j\sqrt{T}\sqrt{\rho} \end{aligned} \tag{14.109}$$

which leads to:

$$\begin{cases} \rho_{jM} = \sqrt{\rho} \\ \dfrac{\sigma_j}{\sigma_M} = s_j\sqrt{T} \end{cases} \tag{14.110}$$

Checking for consistency:

$$\text{var}\left[\frac{Z_j}{\sigma_M}\right] = \frac{1}{\sigma_M^2}\sigma_j^2 \Leftrightarrow s_j^2 T \tag{14.111}$$

$$\text{var}\left[\frac{e_j}{\sigma_M}\right] = \frac{1}{\sigma_M^2}\delta_j^2 \Leftrightarrow s_j^2T(1-\rho) = \frac{\sigma_j^2}{\sigma_M^2}(1-\rho) \Leftrightarrow \delta_j^2 = \sigma_j^2(1-\rho) = \sigma_j^2(1-\rho_{jM}^2) \tag{14.112}$$

Chapter 15

Greeks, P&L Attribution and Value-at-Risk

Risk management has two very important calculations – Value at Risk and P&L Attribution. In this chapter, in addition to VaR and P&L, we also discuss briefly credit risk management and liquidity risk management. Risk management is an industry topic as well as an academic topic.

15.1 P&L Attribution

One of the most important areas in the banking industry where the derivative models are used is risk management. In particular, when a derivative model is well calibrated to the market price, its Greeks can then be used for risk management. The first risk management application is P&L attribution.

The P&L attribution (also known as P&L explain) is to explain where the profits and losses come from. Due to the fact that luck plays a critical role in trading, managers need to make sure that their traders make money not due to luck but due to skills (or talents). As a result, P&L attribution has become essential in trading and fund management. The following chart which is taken from http://www.pnlexplained.com/ on 12/15/2009, while quite subjective and incomplete, describes very well the concept of P&L attribution.

At the bottom layer, we see various first order Greeks such as Delta and Vega, and various higher order Greeks such as Gamma and cross Greeks (cross Gamma and Vega-Gamma). At the top layer, various risks are aggregated by different criteria defined by the management.

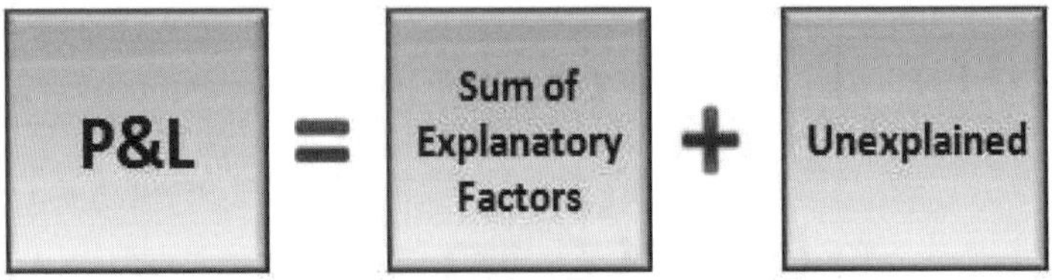

Figure 15.1: P&L Attribution

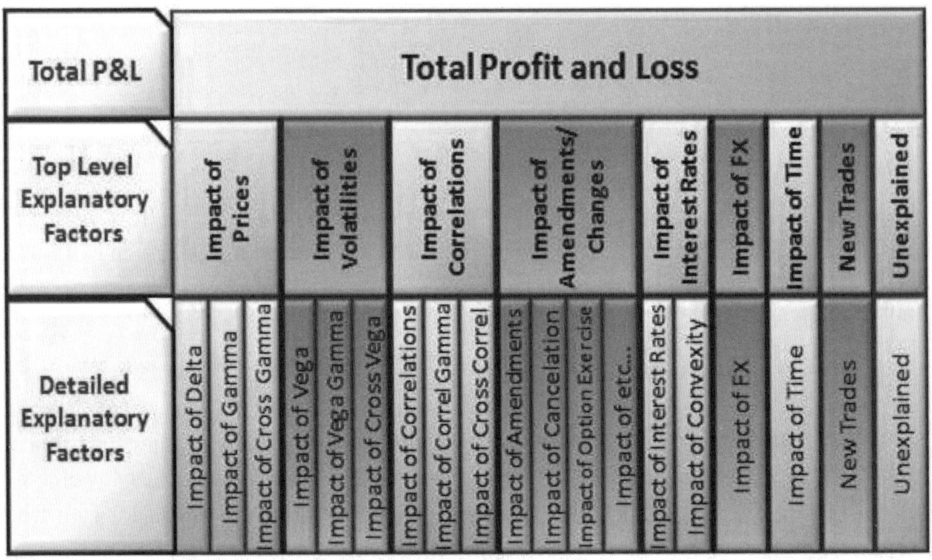

Figure 15.2: P&L Attribution

We define the basic Greeks as follows:

Delta $= \Delta = \frac{\partial C}{\partial S}$ = partial derivative of target (e.g. call) with respective to the underlying (e.g. stock)

Vega $= \mathcal{V} = \frac{\partial C}{\partial \sigma}$ = partial derivative of target (e.g. call) with respective to the volatility [1]

Rho $= \mathcal{R} = \frac{\partial C}{\partial r}$ = partial derivative of target (e.g. call) with respective to the the interest rate [2]

Theta $= \Theta = \frac{\partial C}{\partial t}$ = partial derivative of target (e.g. call) with respective to time (known as time decay)

Gamma $= \Gamma = \frac{\partial^2 C}{\partial S^2}$ = partial derivative of target (e.g. call) twice with respective to the underlying

Cross Greeks can be defined as well. The following are two popular cross Greeks used in FX trading:

Vanna $= \Delta_{\mathcal{V}} = \mathcal{V}_{\Delta} = \frac{\partial^2 C}{\partial S \partial \sigma}$

Volga $= \mathcal{V}_{\mathcal{V}} = \frac{\partial^2 C}{\partial \sigma^2}$

Take a single product as an example (e.g. call option). The explanatory factors are price, volatility, and time which translate to Delta, Gamma, Vega, and Theta. If the Black-Scholes model is the correct model, then we know that:

$$
\begin{aligned}
dC &= \left(\mu S C_S + \sigma^2 S^2 C_{SS} + C_t\right) dt + \sigma S C_S dW \\
&= \underbrace{\left(\mu S \Delta + \sigma^2 S^2 \Gamma + \Theta\right)}_{\text{explanatory factors}} dt + \underbrace{\sigma S \Delta dW}_{\text{unexplained}}
\end{aligned} \tag{15.1}
$$

and there is no Vega. This equation is derived using Ito's lemma. dC represents the change of call value, which is profit or loss. On the right-hand-side of the equation there are two components – explained and unexplained. The explained part is deterministic so there is no uncertainty. If the Black-Scholes model were correct, this component is expected. The unexplained part is pure luck. It is unpredictable. To measure against a trader or fund manager's performance, we use the explained part. If a trader or fund manager fails to deliver something that is expected, then he or she is a bad trader or manager. On the other hand, if he or she "beats the market", then he or she is a good trader or manager. The firm can use any model for performance measurement. The above result is Black-Scholes and can be replaced by any model desired by the management.

[1]Note that vega is not a Greek letter so we use a script $\mathcal{V}$ for this volatility sensitivity.

[2]Note that the capital Greek letter Rho is "P" which is like an English letter P. As a result, to avoid confusion, we use a script $\mathcal{R}$ for this Greek.

If the volatility and the risk-free rate are both random, then we must first define the volatility and interest rate processes. Usually we write both of them as mean-reverting square root processes:

$$
\begin{aligned}
dS &= \mu S dt + \sqrt{V} S dW_1 \\
dV &= \alpha(\beta - V)dt + \sqrt{V}\gamma dW_2 \\
dr &= a(b - r)dt + \sqrt{r} g dW_3
\end{aligned}
\tag{15.2}
$$

where $dW_i dW_j = \rho_{ij} dt$. Then the P&L attribution becomes:

$$
\begin{aligned}
dC &= C_S dS + C_V dV + C_r dr + C_{SS}(dS)^2 + C_{VV}(dV)^2 \\
&\quad + C_{rr}(dr)^2 + C_{SV}(dS)(dV) + C_{Sr}(dS)(dr) + C_{Vr}(dV)(dr) + C_t \\
&= \begin{bmatrix} \mu S C_S + \alpha(\beta - V)C_V + a(b-r)C_r + C_t + \\ VS^2 C_{SS} + \gamma^2 V C_{VV} + g^2 r C_{rr} + VS\gamma\rho_{12} C_{SV} + rSg\rho_{13} C_{Sr} + \sqrt{rV}\gamma g \rho_{23} C_{Vr} \end{bmatrix} dt \\
&\quad + \sqrt{V} S C_S dW_1 + \sigma\sqrt{V} C_V dW_2 + g\sqrt{r} \mathrm{P} dW_3 \\
&= \underbrace{\begin{bmatrix} \mu S\Delta + \alpha(\beta - V)\mathcal{V} + a(b-r)\mathcal{R} + \Theta + \\ VS^2\Gamma + \gamma^2 V \mathcal{V}_{\mathcal{V}} + g^2 r \mathcal{R}_{\mathcal{R}} + VS\gamma\rho_{12}\Delta_{\mathcal{V}} + \\ rSg\rho_{13}\Delta_{\mathcal{R}} + \sqrt{rV}\gamma g\rho_{23}\mathcal{V}_{\mathcal{R}} \end{bmatrix} dt}_{\text{explanatory factors}} \\
&\quad + \underbrace{\sqrt{V} S\Delta dW_1 + \gamma\sqrt{V}\mathcal{V} dW_2 + g\sqrt{r}\mathcal{R} dW_3}_{\text{unexplained}}
\end{aligned}
\tag{15.3}
$$

As we can see, the model can expand to as many random factors as we wish. As shown in the exhibit earlier, the random factors can include FX, multiple key interest rates, credit, prepayment, and any other risk factors. At the end, the equation can be very long and the number of parameters can become quite large. Some simlifications must be necessary.

The first simplification is as easy as just ignoring higher order (and cross) Greeks. We can assume that higher order Greeks such as Vega-Vega and Delta-Vega effects are small and ignore them. The second simplification is to adopt the VaR methodology and choose several key benchmark indices.

15.2 VaR

The concept of Value at Risk (or VaR) began with a thought that had been developed in JP Morgan in early 1990s about how their money managers can sleep well at night and not be surprised the next morning. In other words, how could their managers know how much money was AT RISK over night. In 1998, Bank for International Settlements (BIS) used it for standard banking risk management (known as Basel Accord I).

15.2.1 Historical VaR

An early concept of VaR was one standard deviation value. In other words, if the market swings by one standard deviation, how much money would be lost. In a normal distribution, one standard deviation is the 16th percentile. In other words, one standard deviation drop in value can happen with 16% probability. Similarly, a two-standard deviation loss is 2.3% probability. If the manager wants to know what is 1% likely loss, under normal distribution, it is 2.33 standard deviations. If one standard deviation is $100,000, then with 1% probability the portfolio will lose more than %233,000. Under normal distribution of returns, VaR is scalable by square-root of time.

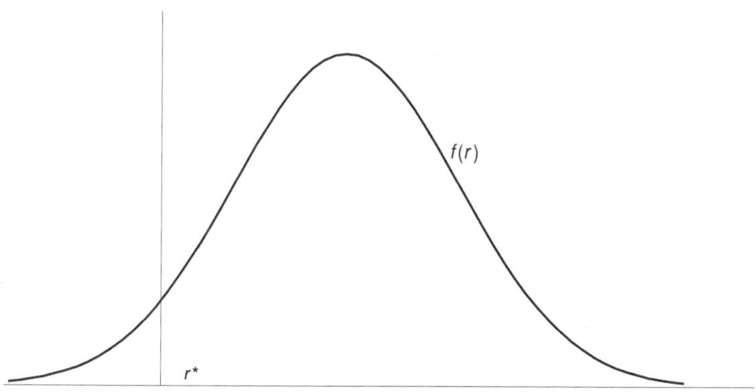

Figure 15.3: Value at Risk

In addition to the calculation of the standard deviation calculation, VaR also includes stress tests, scenario analyses, and worst case analyses.

15.2.2 Using Factor Models

Historical VaR is subject to many shortcomings. When the asset mix of a firm is simple (e.g. only stocks), it is easier to estimate a historical probability distribution. But if the asset mix is not so simple (e.g. stocks and bonds), then a historical probability distribution is not so easy to estimate. Especially if some assets are not liquidly traded, then there is no history of these assets. Then the historical VaR method will fail.

Historical VaR also suffers from a serious criticism. That is, historical VaR is not forward looking. Past distributions may not be accurate representations of the future. Hence, it is important that we forecast what future distribution should look like. Other criticisms of historical VaR include:

- security returns are usually not normal
- assuming a correlation (again, this is normal-based) is inaccurate
- difficult to integrate VaRs of different securities (e.g. bond and stock)
- derivative assets have certain arbitrage-free relationship with its underlying asset
- empirically very unstable

Given the above obvious reasons, VaR models have become parametric. That is, we employ a model and no longer rely on history to compute VaR. The standard VaR model used by the financial industry is a linear factor model. Although the details vary, a VaR model is typically a linear model in widely observed economic variables. As a result, the VaR models used today are connected with P&L attributions.

From P&L attributions, we know that there are a number of explanatory factors, such as FX rates, interest rates, equity indices, various volatilities, $\cdots$, etc. To put all assets under one VaR umbrella, we need a flexible and yet manageable model. First of all the risks must be additive. Standard deviations are NOT additive. So we must use something else. We use delta, or more generally, the first order sensitivities with respect to the underlying risk factors.

We assume a large factor model:

$$dP_i = a_i + \sum\nolimits_{k=1}^{K} b_{ik} f_k + e_i \text{for} i = 1, \cdots, N \tag{15.4}$$

where dP_i is the price change of asset i. There are $K < N$ factors in the economy. To make the model manageable, we first divide the factors into n groups, usually industry sectors. This simplifies the model in that the variance-covariance matrix can be simplified as follows:

$$\Omega_F = \begin{bmatrix} \boxed{\begin{matrix} \sigma_{11} & \cdots & \sigma_{1n_1} \\ & \ddots & \\ & & \sigma_{n_1 n_1} \end{matrix}} & 0 & 0 & 0 \\ 0 & \boxed{\begin{matrix} \sigma_{n_1+1,n_1+1} & \cdots & \sigma_{n_1+1,n_1+n_2} \\ & \ddots & \\ & & \sigma_{n_1+n_2,n_1+n_2} \end{matrix}} & 0 & 0 \\ 0 & & \ddots & 0 \\ 0 & 0 & 0 & \boxed{\begin{matrix} \sigma_{n_1+\cdots+n_{n-1},1} & \cdots & \sigma_{n_1+\cdots+n_{n-1},n_n} \\ & \ddots & \\ & & \sigma_{n_1+\cdots+n_n,n_1+\cdots+n_n} \end{matrix}} \end{bmatrix} \tag{15.5}$$

where $n_1 + \cdots + n_n = K$ and Ω_F is a $K \times K$ matrix. The above matrix implies that inter-industry correlations are assumed 0. Note that the standard deviations here are for the changes in price and not for return. The correlation between any two financial assets (change in price) can be computed as follows:

$$\text{cov}[dP_i, dP_j] = \sum\nolimits_{k=1}^{K} b_{ik} b_{jk} \, \text{var}[f_k] \tag{15.6}$$

In matrix notation, we can write the above equation as:

$$\Omega_P = \underline{b}' \Omega_F \underline{b} \tag{15.7}$$

where $\underline{b}$ is a $K \times N$ matrix. Let the portfolio weights be $\underline{w}$ which is an $N \times 1$ matrix. As a result, the VaR number of the portfolio is the following calculation:

$$\begin{aligned} \text{VaR} &= \sqrt{\underline{w}' \Omega_P \underline{w}} \\ &= \sqrt{\underline{w}' \underline{b}' \Omega_F \underline{b} \underline{w}} \end{aligned} \tag{15.8}$$

This method reduces the computation complexity of an $N \times N$ variance-covariance matrix to a $K \times K$ variance-covariance matrix. Furthermore, the factor model can be estimated easily (to obtain $\underline{b}$).

We should note that the volatilities computed here are "dollar" volatilities (i.e. the volatilities on value changes) as opposed to return volatilities. The relationship is as easy as follows:

$$\begin{aligned} \sigma_{\%} &= \sqrt{\text{var}\left[\frac{\Delta P}{P}\right]} \\ &= \sqrt{\frac{1}{P_t^2} \text{var}[\Delta P]} \\ &= \frac{1}{P_t} \sigma_{\$} \end{aligned} \tag{15.9}$$

and

$$\begin{aligned} \sigma_{\%}^{XY} &= \text{cov}\left[\frac{\Delta X}{X}, \frac{\Delta Y}{Y}\right] \\ &= X_t Y_t \, \text{cov}[\Delta X, \Delta Y] \\ &= X_t Y_t \sigma_{\$}^{XY} \end{aligned} \tag{15.10}$$

In many ways, we prefer dollar volatilities to return volatilities, especially in risk management and P&L attribution, as dollar volatilities are additive. We know that VaR is 99% (or any percentile) and in a normal distribution, this is equal to 2.326 standard deviations.

For example, we can use a three factor model for the interest rate:

- short rate
- slope
- curvature (convexity)

and any bond is expanded by the three factors. Or we can use 11 key rates for the model. We can use two factors for the stocks:

- market index (e.g., S&P 500)
- industry indices (e.g., SIC code 9 industries)

15.2.3 An Example

In this example, we limit ourselves to only interest rate risk. Let the instantaneous rate follow a three-factor model:

$$r = y_1 + y_2 + y_3 \tag{15.11}$$

It would be nice if these factors are directly usable for all bonds. Unfortunately these common factors can be used only through mathematical models. The price of a zero coupon bond (discount factor) is the following risk neutral expectation:

$$\begin{aligned}
P(t,T) &= \hat{\mathbb{E}}_t\left[\exp\left(-\int_t^T r(u)du\right)\right] \\
&= \hat{\mathbb{E}}_t\left[\exp\left(-\int_t^T y_1(u)du\right)\right]\hat{\mathbb{E}}_t\left[\exp\left(-\int_t^T y_2(u)du\right)\right]\hat{\mathbb{E}}_t\left[\exp\left(-\int_t^T y_3(u)du\right)\right] \\
&= Y_1(t,T)Y_2(t,T)Y_3(t,T)
\end{aligned} \tag{15.12}$$

if the factors are independent. Note that all default risk-free bonds can be a linear function of the discount factors:

$$\Pi(t, \underline{T}) = \sum_{\underline{T}} P(t, T_j)c(T_j) \tag{15.13}$$

where c is cash flow and $\underline{T}$ is a vector of coupon payment dates. Default risky bonds will be discussed briefly later.

Now we can assume any process for the factors. The two post popular choices are Vasicek and Cox-Ingersoll-Ross. For each factor y_k, we have (we drop the subscript for convenience):

$$\begin{aligned} dy &= \alpha(\mu - \tfrac{\sigma\lambda}{\alpha} - y)dt + \sigma d\hat{W} \\ dy &= (\alpha + \lambda)(\tfrac{\alpha\mu}{\alpha+\lambda} - y)dt + \sigma\sqrt{y}d\hat{W} \end{aligned} \tag{15.14}$$

where λ is the market price of risk. Note that the parameters are under the risk neutral measure which is obtained by including the risk premium of the factor. Also note that in the Vasicek model only the reverting level is affected by risk premium and in the CIR model both speed and level are affected. For the details of the risk premium of each model please see Ingersoll.

Under either model, we have a closed-form solution to the discount factor:

$$Y(t, T) = A(t, T)e^{-y(t)B(t,T)} \tag{15.15}$$

where under the Vasicek model we have:

$$\begin{aligned} -\ln A(t, T) &= \left(\mu - \frac{\sigma\lambda}{\alpha} - \frac{\sigma^2}{2\alpha^2}\right)(T - t - F(t, T)) + \frac{\sigma^2 F(t, T)^2}{4\alpha} \\ B(t, T) &= \frac{1 - e^{-\alpha(T-t)}}{\alpha} \end{aligned} \tag{15.16}$$

and under the CIR model we have:

$$\begin{aligned} A(t, T) &= \left[\frac{2\gamma e^{(\alpha+\lambda+\gamma)(T-t)/2}}{(\alpha + \lambda + \gamma)\left(e^{\gamma(T-t)} - 1\right) + 2\gamma}\right]^{2\alpha\mu/\sigma^2} \\ B(t, T) &= \frac{2\left(e^{\gamma(T-t)} - 1\right)}{(\alpha + \lambda + \gamma)\left(e^{\gamma(T-t)} - 1\right) + 2\gamma} \\ \gamma &= \sqrt{(\alpha + \lambda)^2 + 2\sigma^2} \end{aligned} \tag{15.17}$$

The fundamental factors, y_ks, span the entire risk-free fixed income universe. Hence, the job now is to estimate these factors, namely we need to estimate α_k,

μ_k, σ_k, and λ_k for $k = 1, \cdots, K$. To estimate this, we need to implement the MLE (maximum likelihood estimation). In a portfolio of many different types of Treasuries and derivatives, each security is modeled properly and their correlations are determined endogenously by the model. For example, a zero coupon bond and coupon bond are, respectively:

$$\begin{aligned} dP(t,T) &= \sum_{k=1}^{K} Y_{y_k}(t,T)dy_k + e_T \\ &= \sum_{k=1}^{K} b_{Pk}dy_k + e_T \end{aligned} \tag{15.18}$$

and,

$$\begin{aligned} d\Pi(t,\underline{T}) &= \sum_{\underline{T}} dP(t,T_j)c(T_j) \\ &= \sum_{\underline{T}} \sum_{k=1}^{K} Y_{y_j}(t,T_j)dy_k c(T_j) + c(T_j)e(T_j) \\ &= \sum_{k=1}^{K} \left[\sum_{\underline{T}} Y_{y_k}(t,T_j)c(T_j) \right] dy_k + \sum_{\underline{T}} c(T_j)e(T_j) \\ &= \sum_{k=1}^{K} \left[\sum_{\underline{T}} Y_{y_k}(t,T_j)c(T_j) \right] dy_k + u \\ &= \sum_{k=1}^{K} b_{\Pi k}dy_k + u \end{aligned} \tag{15.19}$$

It can be seen that the change of any securitys value is a weighted linear combination of the factors. The variance-covariance is:

$$\begin{aligned} \text{cov}[dP, d\Pi] &= \sum_{k=1}^{K} b_{Pk}b_{\Pi k}\,\text{var}[dy_k] \\ &= \begin{cases} \sum_{k=1}^{K} b_{Pk}b_{\Pi k}\sigma_k^2 & \text{Vasicek} \\ \sum_{k=1}^{K} b_{Pk}b_{\Pi k}\sigma_k^2 y_k & \text{CIR} \end{cases} \\ \text{cov}[dP] &= \sum_{k=1}^{K} b_{Pk}^2\,\text{var}[dy_k] \\ \text{cov}[d\Pi] &= \sum_{k=1}^{K} b_{\Pi k}^2\,\text{var}[dy_k] \end{aligned} \tag{15.20}$$

If we invest w in P and $1 - w$ in Π, then the VaR is:

$$\text{VaR} = \sqrt{[w, 1-w] \begin{bmatrix} \text{var}[dP] & \text{cov}[dP, d\Pi] \\ \text{cov}[dP, d\Pi] & \text{var}[d\Pi] \end{bmatrix} \begin{bmatrix} w \\ 1-w \end{bmatrix}} \quad (15.21)$$

We can easily compute any other covariance and fill the variance-covariance matrix. As we can see, once all the parameters are estimated, we complete the variance-covariance matrix.

If we do not want to use a full model (CIR or Vasicek), we can let the fundamental factors to be 11 key Treasury rates. Instead of using the CIR or Vasicek model, we adopt the following yield to maturity formula:

$$P(t,T) = e^{-y(t,T)(T-t)} \quad (15.22)$$

and the coupon bond formula remains the same. The market only observes 11 key rates but we need all the rates for $\underline{T}$. So we need some interpolation. For the time being, lets assume piece-wise flat. That is $y(t,T) = y(t,T_j)$ for $T_j < T < T_{j+1}$.

Simulation. It is quite straightforward to simulate the full model, since the stochastic processes are fully parametized and all factors are independent of one another. As a result it is a series of univariate simulations.

The Vasicek model is straighforward:

$$y(t) = \mu\left(1 - e^{-\alpha h}\right) + y(t-h)e^{-\alpha h} + \sqrt{\frac{\sigma^2\left(1-e^{-2\alpha h}\right)}{2\alpha}}e(t) \quad (15.23)$$

where e is a standard normal variable. We simulate each factor y with this process. For the CIR model, it is complex. Readers please refer to Chapter 9 for how to simulate a non-central chi-square distribution for y.

15.3 Risk Management for Credit Risk

While there has been more a long history in VaR, the risk management technology in credit risk management is very limited. People often are confused between market credit risk and default credit risk. Market credit risk refers to spread changes over time. Spreads reflect how default probabilities and expected recoveries are priced. Hence changes in spreads reflect how default probabilities and expected recoveries change over time. As default probabilities and expected recoveries are expectations, changes in spreads hence have nothing to do with actual defaults. As a result, spread changes are similar to any price changes of any financial securities.

Default credit risk is different. It measures losses caused by actual defaults. Banks have been computing counterparty risk exposures for a long time. This is usually understood as a call option. That is, if the counterparty defaults, then all its debts to the bank will vanish (assuming no recovery). As a result, the loss is identical to the postive net exposure to the counterparty. If the net exposure is negative (i.e. the bank owes to its counterpary), then there is no risk. Hence, the loss mimics a call option. Banks usually simulate all possible scenarios of net exposures and sum those that have positive net exposure paths.

15.3.1 Unexpected Loss

Recovery of a debt is a lengthy process and the amount is highly unpredictable. For those distressed debts that are traded in the secondary market, recovery is often assumed to be the fair market price of the debt. But for those who do not have secondary market prices, rough estimates are applied. Rating agencies such as Moodys and S & P provide rough statistics for senior unsecured and subordinated debts. Often these constant statistics are used as the recoveries for these debts. Under constant recovery, we may compute the expected loss and unexpected loss for a reduced form Bernulli model:

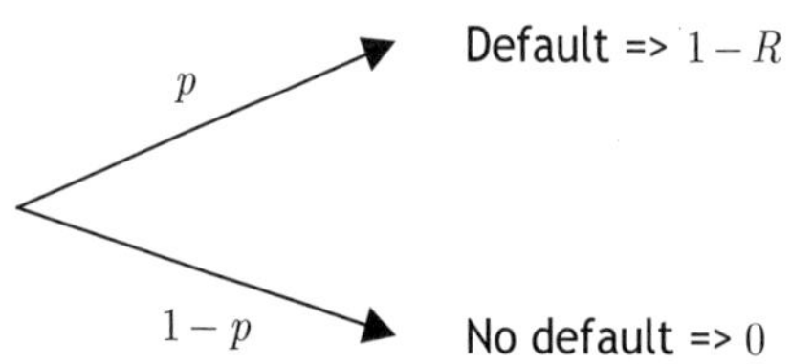

Figure 15.4: Expected Loss and Unexpected Loss

Hence, we compute the expected loss as,

$$\text{EL} = p(1 - R) \tag{15.24}$$

and unexpected loss as,

$$\text{UL} = \sqrt{p(1-p)}(1 - R) \tag{15.25}$$

Note that the unexpected loss is the standard deviation of loss under the Bernulli distribution. Also note that the quantity of UL is at the maximum when $p = \frac{1}{2}$. Clearly this is when the uncertainty is the highest. When p is either high (extremely likely to default) or low (extremely unlikely to default), the uncertainty is low. Hence, there is little unexpected loss.

15.3.2 Term Structure of Credit VaR

ISDA in March 1998 published a manuscript entitled Credit Risk and Regulatory Capital in which it calls for a term structure of Credit VaR. Returns that embed credit risk are highly skewed. The standard nomality assumption for returns cannot apply. As a result, Credit VaR is not scalable by the square root of time. Credit-risky returns are affected by three major sources of risk:

- Market risk – over a short horizon where migration and defaults are not likely to occur,
- Migration risk – for medium horizon where actual defaults are not likely to occur but the perception of default likelihood has changed (binned into the transition matrix, Moodys 2001 one year average default rate for a BBB rating is 15 basis points but the likelihood to be downgraded is 13%)
- Default loss risk – loss due to actual default over the long run.

which is described in the following diagram:

Term Structure of Credit Risk		
Short horizon ($< 1/4$ year)	Medium horizon (between 1/4 and 1 year)	Long horizon (over 1 year)
Market risk	Migration risk	Default risk

Hence, Credit VaR is expressed as:

$$\begin{aligned}\text{credit VaR} &= \text{market risk weight} \times \text{market risk VaR}\\ &+ \text{migration risk weight} \times \text{migration risk metrics}\\ &+ \text{default loss weight} \times \text{unexpected loss}\\ &+ \text{idiosyncratic risk}\end{aligned} \tag{15.26}$$

As a result, a full term structure of Credit VaR must be considered.

15.3.3 CVA – Credit Value Adjustment

CVA, or credit value adjustment, is a method to compute proper value if a trade takes on subtantial counterparty risk. If the trade has subsequent net cash inflows (so called in-the-money) from the counterparty but the counterparty is in default, then there is loss of unrealized gains. Under the requirement of marking to market (as investment banks), such unrealized gains have already been recognized at the time of the trade (so called day-1 P&L). As a result, the loss of the unrealized gains that have been recognized must be credited back and show in the current financial statements as a loss. In order not to show lumpy P&L fluctuations due to counterparty defaults, it has been a recent Wall Street trend that trading desks must put up reserves over time to accomondate losses due to counterparty defaults. This is called Credit Value Adjustment, or CVA.

To operationalize such a reserve-charging process, investment banks start to establish a CVA trading desk that sells protection on counterparty defaults. As opposed to burdening each and every trading desk to compute its counterparty default reserve, the CVA trading desk computes the fair price of the protection and sell CDS on each counterparty. CVA traders are then responsible for their own P&L trading CDS on counterparties. This is another successful securitization move on Wall Street that it securitizes default charges. Trading with CVA desk, each and every other trading desk gains protection on its counterparties. As the protection is provided in a form of swap, it is like an installment plan that requires a small payment periodically that is exactly like a reserve-charging system.

In a way, CVA is a mechanism of internal transfer pricing used widely in regular corporations. Yet through securitization, it lets market decide what the correct transfer price should be. This is consistent with the strong belief by Wall Streeters that the free economy, via the Invisible Hand, will set the right equilibrium. While CDS and CDO have been widely studied (See Chapter 14) and powerful models have been developed, the modeling of CVA so far is still completely dependent on Monte Carlo, as no closed-form solutions exist and the correlation between counterparty default risk and the risk of the deal is hard to capture.

The easiest way to understand the nature of counterparty risk is a call option diagram in many text books, as follows. The dotted line represents the "moneyness of the deal". If the deal is in-the-money, then there is a positive exposure dollar-for-dollar. Otherwise, i.e. out-of-the-money, the exposure is 0. The solid line is the PV of such exposure. As we can see, the exposure is a call option and we can apply the Black-Scholes model for it.

If a deal has multiple cash flows, then the exposure is a collection (sum) of many such call options. Certainly, the simple sum of call options is incorrect as these

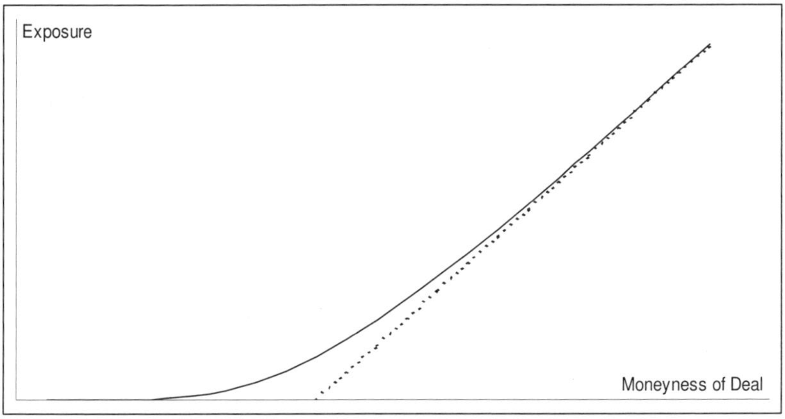

Figure 15.5: P&L Attribution

exposures are inter-dependent. More precisely, later call options are dependent on earlier call options in that if a counterparty defaults at time t then there will be no exposure after t. A rough approximation is the probability-weigh each exposure but such an approximation still ignores inter-dependence. Furthermore, this exposure method ignores the correlation of the counterparty default risk and the underlying risks of the deal.

15.3.4 Risky Funding

The recent 2007-8 crisis has revolutionized what financial modeling has long believed in – the law of one price, a result achieved by no-arbitrage. No-arbitrage was the basic principle Black and Scholes used to derive their option formula. No-arbitrage gave birth to risk-neutral pricing that has dominated the quant finance area for half of a decade. No-arbitrage guarantees the law of one price.

We should recall how Black and Scholes derived the pricing equation (i.e. the partical differential quation). This is known as "price by replication". In other words, Black and Scholes taught us that the price of any security is equal to its cost of hedging (or replicating). Later on, this was shown as the Martingale Representation Theorem (a.k.a. self-financing, which is the base of no-arbitrage). However, to achieve this result, one must have the same borrowing and lending rate. Well, this is not a problem under no-arbitrage, since the only rate relavent under no-arbitrage (or risk-neutral pricing) is the risk-free rate.

Until the 2007-8 crisis, the paradigm seemed to work just fine. Everything is calibrated to LIBOR which is regarded as the risk-free rate by the investment community. The crisis changed everything. During the crisis, deals that were safely collaterized were priced very differently (at much higher prices) from those that were not safely collaterized (either under-collaterized or un-callaterized). Buyers did not have faith in their counterparties's ability to fund the transactions (if the deals were under- or un-collaterized) and consequently demanded for lower prices. LIBOR and OIS rates started to diverge and LIBOR was no longer the risk-free rate anymore. This is the issue called Risky Funding.

There has not been any solution yet for this problem. Proposals have been provided but no conclusion has been drawn. This is a still very live and challenging problem.

Chapter 16

Liquidity Quantification

Liquidity has become the focal point after the recent financial crisis. The largest U.S. bankruptcy – Lehman Brothers, has been largely regarded as a liquidity default. And liquidity regulation hence is the center piece of Basel III and Dodd-Frank regulations. In a new BIS document, not only current liquidity, but all the future potential liquidity shortfalls must be quantified and managed. Basel III also calls for liquidity-risk-weighted regulatory capital. In this chapter, I present a liquidity quantification model (published in the Journal of Fixed Income 2012) I developed that can be used for these regulations.

16.1 Introduction

During the last financial crisis (known as the liquidity crisis), a large number of liquidity-squeezed events had occurred. These incredible events had caused Wall Street to consider abandoning its long proud tradition – marking to market. For as long as Wall Street has existed, it has prided itself in respecting the market prices. Wall Street had always regarded market prices to be a reflection of collective wisdom of the entire investment community. The fact that there is a price reflects an equilibrium. Wall Street is proud to be able to take the full advantage of such equilibriums and develop tools that are consistent with market prices. But during the crisis, such a belief had been put under a strong test. Here are some famous examples.

16.1.1 Some Liquidity Squeeze Examples

Lone Star-Merrill Lynch Deal

On July 28, 2008, a Monday, as the crisis of subprime gradually unfolded, Merrill Lynch, in realizing huge toxic portfolios it owned, would like to sell a particular subprime portfolio that had a face value of $31 billion to a private equity fund owned by Mr. Lone Star. The purchase price was rumored to be 22 cents on the dollar – a 78% discount. On top of that, Merrill Lynch needed to buy back any defaulted asset in the portfolio. In other words, Merrill Lynch was implicitly sold a CDS (for free) to Star, which was worth about 15 cents on the dollar. As a result, Merrill Lynch sold Lone Star a portfolio of financial assets at a discount of 93%!

According to Reuters, the deal had prompted Merrill Lynch to write down all of its toxic assets by at least 75%, an unprecedented price discount seen on Wall Street.

Bear Stearn-JP Morgan Deal

On June 22, 2007, the problems of the two subprime portfolios managed by Bear Stearns Asset Management (BSAM) – Bear Stearns High-Grade Structured Credit Enhanced Leveraged Fund and Bear Stearns High-Grade Structured Credit Fund, started to surface. Its CEO James Cayne initially wanted to keep the two funds separate from the main Bear Stearn bank (as how deals are usually done on Wall Street – so called special purpose vehicle, or SPV). Yet the pressure from the investors and the government forced Bear Stearns to take over. As losses gradually unfolded, Richard A. Marin, a senior executive at Bear Stearns Asset Management responsible for the two hedge funds, was replaced on June 29 by Jeffrey B. Lane, a former Vice Chairman of rival investment bank, Lehman Brothers.

On March 14, 2008, Bear Stearns finally reached to a point where it could no longer operate. Either someone would have to buy it or else it would face liquidation. The Fed called in JP Morgan for the bail out. The price was at $30 per share on March 14, a Friday. After a weekend-long investigation and evaluation of Bear's assets, JP Morgan offered $2 per share. Bear rejected and accused JP Morgan of exploiting the liquidity situation. In fear of a bank run that Bear's bankruptcy could have caused, the Fed yielded to JP Morgan for liquidity concern by agreeing a $30 billion loan to J.P. Morgan. The final deal was $10 per share.

It is a consensus that Bear's default would have been an economic consequence (that is, they lost money in their investments). Yet the bailout amount (from $2 per share to $10 per share) is undoubtedly a consequence of a liquidity squeeze. The liquidity discount here is 80%!

Other Examples

In addition to these two unprecedented liquidity events, there have been numerous examples during the crisis period where asset prices were largely compressed due to urgent needs to unwind those positions (clear evidence of liquidity squeeze).

16.2 Liquidity Discount as a Put Option

Liquidity discount occurs as supply reacts more sensitively to economy than demand does in a non-linear fashion. Liquidity discount reaches the extreme as demand reaches its maximum capacity. Figure 1 depicts this general idea. In Figure 1, we assume that demand and supply of a financial asset jointly determine the equilibrium price of the asset at any given time. The economy is represented by a single state variable (say wealth) symbolized by V. As the economy grows, the supply curve moves to the right and so does the demand curve. To derive the liquidity discount model, we must assume that supply grows faster than demand does in that less elastic demand function is the main reason to cause price discount.

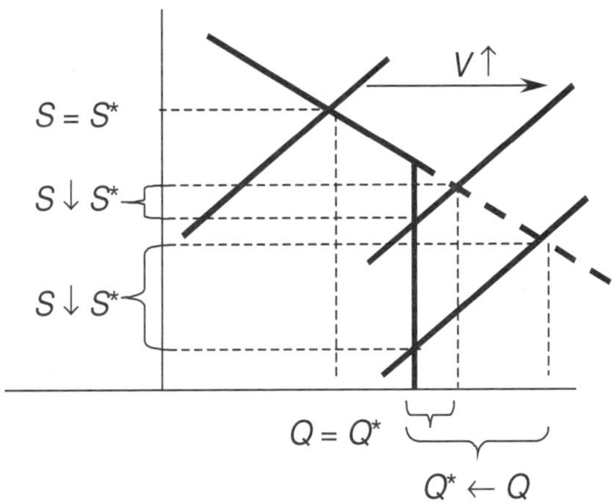

Figure 16.1: Demand-Supply Analysis of Liquidity Discount

As a result, the growth of economy results in an increased equilibrium quantity and a lower price. In the diagram, we assume that the demand is totally insensitive to economy growth and it has a maximum capacity at Q^*. Clearly, our model requires only demand be less sensitive to supply and such an exaggerated demonstration is just for the purpose of easy exposition. At the maximum capacity of the demand function, the quantity can no longer increase and equilibrium can only draw the price down.

In the diagram, the vertical axis represents price (by S) and the horizontal axis represents quantity (by Q). We let the liquidity-constrained price be S^*. In a usual situation, $S = S^*$. In a liquidity-squeezed situation (represented by the situation where demand reaches its maximum capacity Q^*), $S > S^*$. Liquidity discount in this setting is defined as the impact of the slower demand reaction (than supply) and becomes dramatic as the maximum is approached. Based upon the above diagram, we can draw the following conclusion depicted in Figure 2.

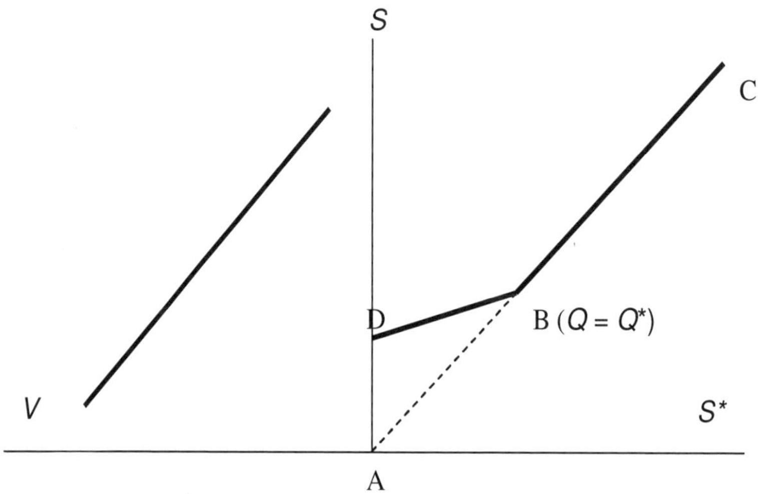

Figure 16.2: Liquidity Discount as a Put

On the left of Figure 16.2, we depict the relationship between the economy (represented by, say wealth V) and perfectly liquidity price S. As shown in Figue 16.1, this is a downward sloping curve. To be shown later, the curvature of S in V must be convex in order to obtain liquidity discount. If the relationship is linear (under which Q^* cannot exist), then there can be no liquidity discount. ,

On the right, we depict the relationship between the perfectly liquidity price S and the liquidity-constrained price S^*. The line $\overline{\text{ABC}}$ is a 45-degree line on which the illiquid price is equal to the liquid price. At point B where the quantity reaches its maximum capacity Q^* the illiquid price starts to decrease rapidly due to problems in liquidity and bends over toward point D. Again, the linear result (by $\overline{\text{BD}}$ line) is just a demonstration. In the next section where we derive the formal model, the graph between points B and D is not linear and has a reflection point.

From the diagram where discount is depicted linearly, it can be seen that liquidity can be explained by a put option. That is,

$$S^* = S - \text{put} \tag{16.1}$$

In the next section, we demonstrate that this is not a simple put due to convexity requirements. But intuitively, the simple put explanation serves the purpose well and can even be used as an approximation, which has already been discussed in the industry. What is offered in this paper is an equilibrium model that is consistent with the Merton model widely used in modeling credit risk. While we do not link our liquidity model yet to credit risk in this paper, it is quite straightforward to do so.

In this section, the demonstration is intuitive but not realistic. In reality, the left and the right panels are related. In other words, how the liquidity price reacts to the economy affects the magnitude of liquidity discount. Furthermore, it is not likely to know the exact amount of Q^*. In the next section, we endogenize the relationships amount the economy V, the perfect liquid price S, and the liquidity-constrained price S^*. We let the liquid price be a convex function of the economy and derive the illiquid price directly. In doing so, we arrive an equilibrium without the specification of Q^*.

16.2.1 The Model

In this section, we develop a formal pricing model for illiquidity which is defined as inability to transact. Inability to transact is the exact description of Figure 16.1 in which lack of demand leads to no transactions. In a perfect world where the Black-Scholes model holds, transactions can take place at any time and hence investors can trade securities and rebalance their portfolios continuously. When transactions are not permitted to be continuous, it presents an extra risk born by the buyer and hence the buyer, in return, should ask for compensation and lower the price of the security.

Once continuous trading is not allowed, market is not (dynamically) complete in the Duffie-Huang sense (1985) and the resulting model is not preference-free. As a result, one must adopt a utility function to gauge the magnitude of the risk premium. We first present a model with the quadratic utility so the standard CAPM can be used. It is straightforward to extend the model to a broader class of utility functions. Using a more complex utility function is certainly better in terms of explaining the reality and providing model flexibility but it loses the closed-form CAPM formula.

The price of an arbitrary security at current time t when no trading is allowed until a future time T can be priced by the most fundamental discounted cash flow method:

$$X(t) = e^{-\xi(T-t)}\mathbb{E}_t[X(T)] \tag{16.2}$$

where $X(t)$ is the cash flow of an arbitrary security at time t, $\mathbb{E}_t[\cdot]$ is the conditional expectation taken at time t, and ξ is the (continuously compounded) risk-adjusted return for the security. Such a discounted cash flow method requires further modeling substances in order to be operational.

To build an explicit model for (16.2), we first assume the Black-Scholes/CAPM model where the underlying economy (represented by a single state variable, say wealth) obeys the following log normal process:

$$\frac{dV}{V} = \mu dt + \sigma dW \tag{16.3}$$

where $W(t)$ is the standard Wiener process and μ and σ are (continuous time) mean and standard deviation of the return of the state variable V.

A perfectly liquid price, S, is a contingent claim on the state variable. The assumption of liquidity discount we make in this paper indicates that S must be a monotonic function in wealth V. In a theorem we shall prove later, the function must be convex in order to arrive at liquidity discount. There are a number of ways to construct such an explicit function. For simplicity, we choose a function that imitates a put option (as opposed to an arbitrary polynomial function) for the following reasons. First, the maturity parameter in the put option can be made equal to the liquidity discount horizon. This provides an extreme convenience in modeling liquidity discount. Secondly, there is a closed form solution for the price of a perfectly liquid asset, which allows us to compare with the price of an equal asset but constrained by liquidity. Lastly, the strike price in the put function ideally characterizes the strength of the liquidity squeeze. The higher is the strike price, the stronger is the liquidity discount and the zero strike price ideally represents perfect liquidity. However, the use of put function does suffer from one drawback. It exists a maximum value for the liquid stock price (at the strike level), which could be unrealistic as economy contracts the price of the security could become unboundedly high. Fortunately, this is the situation where the liquidity discount is small and hence the impact would be small.

Given that S is the price of a security that can be continuously traded, it can be easily computed with the Black-Scholes model when its payoff for a fixed time horizon (time to maturity) mimics a put:

$$\begin{aligned} S(t) &= e^{-r(T-t)}\hat{\mathbb{E}}_t[S(T)] \\ &= e^{-r(T-t)}\hat{\mathbb{E}}_t[\max\{K - V(T), 0\}] \\ &= e^{-r(T-t)}KN(-d_-) - V(t)N(-d_+) \end{aligned} \tag{16.4}$$

where K is the strike price that reflects the strength of the liquidity squeeze, $\hat{\mathbb{E}}_t[\cdot]$ is the risk-neutral expectation, and

$$d_\pm = \frac{\ln V(t) - \ln K + (r \pm {}^1\!/_2\sigma^2)(T-t)}{\sigma\sqrt{T-t}}$$

Note that the adoption of the put payoff is just a convenience to incorporate convexity. Due to the fact that the liquidity-constrained price does not have a closed-form solution (except for the extreme cases), we implement a binomial model to approximate this Black-Scholes result. For the remainder of the paper, when we refer to the Black-Scholes model, it is actually the binomial approximation.

We also note that although the liquidity discount model has no closed form solution, the binomial implementation is actually better in that the model can then be easily augmented to include credit risk proposed by Chen (2002), and Chen, Fabozzi, Pan, and Sverdlove (2008). In this broader model, interactions between liquidity and credit risks can be studied.

Let S^* be the liquidity-constrained price where trading of the security is not permitted until time T. Equation (16.2), as a result, can be replaced with the following equation:

$$X(t) = e^{-\xi(T-t)}\mathbb{E}_t[f(V(T), \Theta), 0\}] \tag{16.5}$$

where Θ represents other parameters needed for the model, $X(t)$ is either S or S^* depending if the equation is used for an instantaneous period or a longer time period, respectively. Note that in an instantaneous period, the physical expectation as in (16.5) is identical to the risk-neutral expectation (details to be explored later in Theorem 1) and hence (16.5) provides the solution to the liquid price. The expected return, ξ, must follow the Capital Asset Pricing Model (CAPM).

Note that S and S^* are identical securities with just one difference: S^* cannot be transacted (or hedged) until time T. The purpose of the paper is to derive the price difference between S and S^*. We shall prove that $S^* < S$ and this is the model of liquidity discount. As we shall demonstrate in an analysis that such discount can be substantial, even under very reasonable assumptions.

We begin our modeling of liquidity discount with the standard CAPM. Note that the Black-Scholes model is consistent with the standard CAPM as follows:

$$\xi = \mathbb{E}\left[\frac{dX}{X}\right] = rdt + \eta\left[\mathbb{E}\left[\frac{dV}{V}\right] - rdt\right] \tag{16.6}$$

where r is the risk-free rate and

$$\eta = \frac{\partial X}{\partial V}\frac{V}{X} \tag{16.7}$$

is the elasticity of the state contingent claim with respect to the underlying economic state variable (similar to an option on its underlying stock). Note that this result is exact only in continuous time. Black and Scholes (1973) prove that the option price is a CAPM result with the underlying stock as the market. Note that $\eta = \beta$ as shown in the following:

$$\begin{aligned}\beta &= \text{cov}\left[\frac{dX}{X}, \frac{dV}{V}\right] \Big/ \text{var}\left[\frac{dV}{V}\right] \\ &= \frac{V}{X}\frac{\text{cov}[X_V dV, dV]}{dV} \\ &= \frac{V}{X}X_V \\ &= \eta\end{aligned} \tag{16.8}$$

where $X_V = \partial X/\partial V$. Re-writing (16.6) in discrete time for a small interval h, we have:

$$\frac{\mathbb{E}[X(t+h)]}{X(t)} = (1+rh) + \beta\left[\frac{\mathbb{E}[V(t+h)]}{V(t)} - (1+rh)\right] \tag{16.9}$$

As a result, we can derive a pricing model for the security as:

$$X(t) = \frac{1}{R(t,t+h)}(\mathbb{E}[X(t+h)] - X_V\{\mathbb{E}[V(t+h)] - RV(t)\}) \tag{16.10}$$

where $R(t,T) = 1 + r(T-t) \approx e^{r(T-t)}$. This result provides an alternative proof of the CAPM argument by Black and Scholes. Note that

$$X_V = \frac{\text{cov}[X(t+h), V(t+h)]}{\text{var}[V(t+h)]} \qquad (16.11)$$
$$= \beta^{\$}$$

is the "dollar beta" (that is, delta is dollar beta). Hence, (16.10) is also an alternative derivation to Jensens model (1972).

It is important to note that CAPM holds only under one of the two assumptions: quadratic utility for the representative agent in the economy or normality for the returns of the risky assets. Consequently, the above CAPM result holds in the Black-Scholes model without any utility assumption must be due to the fact that option returns and stock returns are both normally distributed. It is clear that the stock return is normally distributed, as equation (16.3) postulates. The option return is normally distributed only under continuous time. This is because in continuous time, the option value is linear in stock and must follow the same distribution of that of the stock and hence its return is normally distributed.

In a discrete time where h is large, the option return is no longer normally distributed and the equation (16.10) can no longer hold without the assumption of quadratic utility that guarantees the validity of the CAPM quadratic utility function. The quadratic utility function is a bad assumption in that it has the wrong sign for the relative risk aversion. However, fortunately, the liquidity discount model we derive in this paper suffers very little from the second order effect of the utility function. All the model requires is that investors are risk adverse.

Under quadratic utility, CAPM holds for all securities. Hence, equation (16.10) holds for an h that is not infinitesimally small as follows:

$$X(t) = \frac{1}{R(t,T)}(\mathbb{E}[X(T)] - \beta^{\$}\{\mathbb{E}[V(T)] - R(t,T)V(t)\}) \qquad (16.12)$$

where $R(t,T) = e^{r(T-t)}$ and the dollar beta is computed, for any T, as:

$$\beta^{\$} = \frac{\text{cov}[X(T), V(T)]}{\text{var}[V(T)]} \qquad (16.13)$$

Equation (16.12) is the main result of our model. It states that under quadratic utility, all assets must follow CAPM in determining their values. What we shall demonstrate is that when liquidity discount is present, the value computed by equation (16.12) is less than the perfectly liquid price computed by the Black-Scholes model.

To derive our liquidity discount model, we shall show first that if the relationship between the economy (represented by the state variable V) and the liquid price (S) is linear, then the liquidity discount is nil. Then we show that liquidity discount can exist only if the relationship between the economy and the liquid price is convex.

Theorem 1 When the payoff is linear, then liquidity discount is nil.

Proof Note that if payoff is linear, $X(T) = aV(T) + b$, then the following results hold:

$$\begin{array}{lr} (i) & \mathbb{E}[X(T)] = a\mathbb{E}[V(T)] + b \\ (ii) & \mathrm{var}[X(T)] = a^2\,\mathrm{var}[V(T)] \\ (iii) & \mathrm{cov}[V(T), X(T)] = a\,\mathrm{var}[V(T)] \\ (iv) & \beta^{\$} = a \end{array}$$

where the expectation, variance, and covariance operations are taken under the physical measure. As a result, following (16.12), we have:

$$X(t) = \frac{1}{R(t,T)}\{\mathbb{E}[X(T)] - a(\mathbb{E}[V(T)] - R(t,T)V(t)\} \tag{16.14}$$

where $R(t,T) = e^{r(T-t)}$ and

$$\mathbb{E}[X(T)] - R(t,T)X(t) = a\mathbb{E}[V(T)] - aR(t,T)V(t) \tag{16.15}$$

which gives rise to the following result:

$$X(t) = \frac{1}{R(t,T)}\{aR(t,T)V(t) + b\} \tag{16.16}$$

Note that this is exactly the result of risk neutral pricing (expected value with a "$\hat{\mathbb{E}}$"), i.e.,

$$\begin{aligned} X(t) &= \frac{1}{R(t,T)}\hat{\mathbb{E}}[aV(T) + b] \\ &= aV(t) + b\frac{1}{R(t,T)} \end{aligned} \tag{16.17}$$

This is also the result described by the Martingale Representation Theorem which indicates that any contingent claim under continuous trading can be replicated by the underlying asset and the risk-free asset. The result shown in (16.17) states that equation (16.12) computes also the liquid price, i.e. $X(t) = S(t)$ if the relationship between economy and the asset price is linear. □

In continuous time, there are only two states in every infinitesimal time step (as described in Duffie and Huang (1985)) and hence no liquidity discount can exist. This indicates that if trading is continuous then at each infinitesimal step the payoff is linear and as a result Theorem 1 holds. In other words, continuous trading breaks up a fixed time horizon into small infinitesimal time steps, each of which is a linear payoff and hence liquidity discount does not exist. In the next section, we shall demonstrate this property in a numerical example and Theorem 1 is explicitly demonstrated.

Theorem 2 in the following proves that if the payoff is not linear and is convex, then the equilibrium price is always less than the linear price which, by Theorem 1, is the risk-neutral price where continuous rebalancing is possible.

Although a general proof with any form of convexity is not available, a proof based upon the binomial model that is consistent with our formulation of (16.3) is provided. In particular, as any three points define the convexity, we use a two period binomial model for the proof. It can be referred that with more points (more periods) in the binomial model, the proof stays valid.

Theorem 2 If the payoff is not linear and convex, then $X^{\text{cvx}}(t) < X^{\text{lnr}}(t)$ where $X^{\text{lnr}}(t)$ is defined in [Theorem 1] and identical to $S(t)$ which is the perfectly liquid price and $X^{\text{cvx}}(t)$ is the same as $S^*(t)$ which is the liquidity-constrained price.

Proof We use a two-period binomial model to demonstrate the proof. We let V_{ij} represent the level of the state variable at time period i and state j respectively. In a two-period binomial model, the three state variable values at time 2 are V_{20}, V_{21}, and V_{22} representing low, medium, and high prices respectively. Without loss of generality, we also let $V_{21} = V_0$. Similarly, we also let X_{ij} represent the state contingent claim price at time period i and state j respectively where X_{20}, X_{21}, and X_{22} represent low, medium, and high prices respectively.

We let the convex payoff differ from the linear one by slightly altering the middle one as follows: $X_{21}^{\text{cvx}} = X_{21}^{\text{lnr}} - \varepsilon$ where ε is an arbitrary small positive amount to create the convexity of X in V. The real probability per period is p which is time invariant. Consistent with the notation above, we symbolize compounding at the risk-free rate in two periods as $R(0, 2)$. Our goal is to prove that:

$$\begin{aligned} X_0^{\text{cvx}} &= \frac{1}{R(0,2)}\left\{\mathbb{E}[X_{2\cdot}^{\text{cvx}}] - \beta^{\$}(\mathbb{E}[V_{2\cdot}] - R(0,2)V_0)\right\} \\ &< \frac{1}{R(0,2)}\left\{\mathbb{E}[X_{2\cdot}^{\text{lnr}}] - \beta^{\$}(\mathbb{E}[V_{2\cdot}] - R(0,2)V_0)\right\} \\ &= X_0^{\text{lnr}} \end{aligned} \tag{16.18}$$

where the symbol "2 in the subscript represents the three states in period 2, and cvx or lnr in the superscript represent convex or linear payoffs respectively. In the binomial model, the three physical probabilities are p, $2p(1-p)$, and $(1-p)^2$ for high, medium, and low states respectively. Hence, we have:

$$\begin{array}{rr}
(i) & \mathbb{E}[V_{2\cdot}X_{2\cdot}^{\text{cvx}}] = \mathbb{E}[V_{2\cdot}X_{2\cdot}^{\text{lnr}}] - \varepsilon V_{21}2p(1-p) \\
(ii) & \mathbb{E}[X_{2\cdot}^{\text{cvx}}] = \mathbb{E}[X_{2\cdot}^{\text{lnr}}] - 2\varepsilon p(1-p) \\
(iii) & \mathbb{E}[V_{2\cdot}]\mathbb{E}[X_{2\cdot}^{\text{cvx}}] = \mathbb{E}[V_{2\cdot}]\mathbb{E}[X_{2\cdot}^{\text{lnr}}] - 2\mathbb{E}[V_{2\cdot}]\varepsilon p(1-p) \\
(iv) & \text{cov}[V_{2\cdot}, X_{2\cdot}^{\text{cvx}}] = \text{cov}[V_{2\cdot}, X_{2\cdot}^{\text{lnr}}] + 2\varepsilon p(1-p)(\mathbb{E}[V_{2\cdot}] - V_{21})
\end{array}$$

and then the dollar beta under the convex function can be derived as:

$$\beta^{\text{cvx}} = \beta^{\text{lnr}} + \frac{\varepsilon 2p(1-p)(\mathbb{E}[V_{2\cdot}] - V_{21})}{\text{var}[V_{2\cdot}]} \tag{16.19}$$

As a result,

$$\begin{aligned}
&\beta^{\text{cvx}}\{\mathbb{E}[V_{2\cdot}] - RV_0\} = \\
&\beta^{\text{lnr}}\{\mathbb{E}[V_{2\cdot}] - R(0,2)V_0\} + \frac{2p(1-p)\varepsilon(\mathbb{E}[V_{2\cdot}] - V_{21})(\mathbb{E}[V_{2\cdot}] - R(0,2)V_0)}{\text{var}[V_{2\cdot}]}
\end{aligned} \tag{16.20}$$

and

$$\begin{aligned}
X_0^{\text{cvx}} &= \mathbb{E}[X_{2\cdot}^{\text{cvx}}] - \beta^{\text{cvx}}\{E[V_{2\cdot}] - R(0,2)V_0\} \\
&= \mathbb{E}[X_{2\cdot}^{\text{lnr}}] - 2p(1-p)\varepsilon - \beta^{\text{lnr}}\{\mathbb{E}[V_{2\cdot}] - R(0,2)V_0\} \\
&\qquad - \frac{2p(1-p)\varepsilon(\mathbb{E}[V_{2\cdot}] - V_{21})(\mathbb{E}[V_{2\cdot}] - R(0,2)S_0)}{\text{var}[V_{2\cdot}]} \\
&= X_0^{\text{lnr}} - 2p(1-p)\varepsilon\left\{1 + \frac{(\mathbb{E}[V_{2\cdot}] - V_{21})(\mathbb{E}[V_{2\cdot}] - R(0,2)V_0)}{\text{var}[V_{2\cdot}]}\right\} \\
&< X_0^{\text{lnr}}
\end{aligned} \tag{16.21}$$

This is because, clearly,

$$\mathbb{E}[V_{2\cdot}] > V_{21} = V_0$$

(the expected payoff should be larger than todays value) and

$$\mathbb{E}[V_{2\cdot}] > R(0,2)V_0$$

(the expected return should be more than the risk-free rate) to avoid arbitrage. □

From Theorem 1, we know that $X_0^{\rm lnr} = S(t)$ as the liquid price. Here, $X_0^{\rm cvx} = S^*(t)$ represents the illiquid price. Hence, in summary, $S^*(t) < S(t)$ for all values of finite $R(0, 2)$ and the theorem is proved.

Note that under linearity (between wealth and liquid price), there exists no liquidity discount, which is the same result as continuous trading. As a consequence, the price under linearity $X_0^{\rm lnr}$ is identical to the price under continuous trading $S(t)$. Similarly, when the relationship between wealth and liquidity price is convex, there exists liquidity discount. And the price $X_0^{\rm cvx}$ represents the price under liquidity squeeze, $S^*(t)$.

16.3 Some Analysis

In this section, we provide a numerical example to demonstrate the enormity of liquidity discount. The main model is equation (16.12). While the analysis in this section is based upon an arbitrarily chosen set of parameter values, the result holds in general. While equation (16.12) is closed-form, (13) needs to be computed numerically as:

$$\beta = \frac{\sum_{j=1}^{n} \begin{pmatrix} n \\ j \end{pmatrix} p^j (1-p)^{(n-j)} \{V_i(T) - \bar{V}(T)\}\{X_j(T) - \bar{X}(T)\}}{\sum_{j=1}^{n} \begin{pmatrix} n \\ j \end{pmatrix} p^j (1-p)^{(n-j)} [V_j(T) - \bar{V}(T)]^2} \tag{16.22}$$

where

$$\bar{V}(T) = \sum_{j=1}^{n} \begin{pmatrix} n \\ j \end{pmatrix} p^j (1-p)^{(n-j)} V_i(T)$$
$$\bar{X}(T) = \sum_{j=1}^{n} \begin{pmatrix} n \\ j \end{pmatrix} p^j (1-p)^{(n-j)} X_i(T)$$

are the means of the economic state variable (wealth) and the state contingent claim respectively. In order to demonstrate convergence to the Black-Scholes model in continuous trading, we adopt the binomial framework of Cox, Ross and Rubinstein (1979) with n periods. Given that the binomial model will converge to the Black-Scholes model as n gets large, we shall use a sufficiently large n to represent the limiting Black-Scholes case. In other words, we shall demonstrate that under continuous trading, there is no discount for illiquid trading.

For the sake of easy exposition, we set up the following base case for the binomial model:

Paremeters	
n	100
$T-t$	1 year
σ	0.5
μ	10%
r	5%
V	\$80

To carry out a numerical example, we need to have an explicit functional form for the relationship between the state variable and the state contingent claim that represents either the liquid price (linear payoff) and illiquid price (convex payoff), in the context of the Cox, Ingersoll, and Ross model (1985). As a convenience, we choose a put payoff for the task. The put payoff is convex and negatively monotonic in the underlying economy, and hence can serve the purpose well. We choose an arbitrary strike of 100 to characterize convexity. We shall note that higher is the strike, higher is the convexity.

We first compute the price of the perfect liquid contingent claim, i.e. $X(t) = S(t)$. Note that the liquid price is such that trading takes place continuously and is represented by the Black-Scholes model. In the binomial model where n = 100 the value of the put (representing the perfectly liquid asset price) is almost identical to the Black-Scholes price of \$25.85. We feel that for the sake of computational time, this is minor enough difference for us to demonstrate the value of liquidity discount. This provides us the comfort that n = 100 is a good enough proxy for continuous trading. In the rest of the paper, we shall use the binomial model with n = 100 as the benchmark to examine the properties of the liquidity discount model.

Next we turn to computing the illiquid price, i.e. $X(t) = S^*(t)$. We start our analysis with n = 1 where we demonstrate that in this case, the liquid price (which is computed by the binomial model based upon the risk-neutral probabilities) is identical to the illiquid price (which is computed by the CAPM based upon the physical probabilities). In other words, when rebalancing is permitted at every node in the binomial model, the illiquid price is identical to the liquid price.

In a one period model (i.e. $n = 1$), it is clear that the up and down movements in the binomial model are $u = e^{\sigma\sqrt{\Delta t}} = e^{0.5\times 1} = 1.6487$ and $d = 1/u = 0.6065$. The state variable lattice and the stock payoff are given below (left and right respectively):

State Variable V

	131.90
80	
	48.52

State Contingent Claim Payoff S

	0
28.07	
	51.48

The risk-neutral probabilities are $\hat{p} = \frac{\exp(r\Delta t)-d}{u-d} = \frac{\exp(5\%\times 1)-0.6065}{1.6487-0.6065} = 0.4267$ and $1-\hat{p} = 0.5733$. The liquid price is (call): $0.5733 \times 51.48 \div e^{-5\%} = 28.07$, as in the above binomial tree.

The physical probabilities are $p = \frac{\exp(\mu\Delta t)-d}{u-d} = \frac{\exp(10\%\times 1)-0.6065}{1.6487-0.6065} = 0.4785$ and $1-p = 0.5215$. Following (16.12), we arrive at the same exact price of \$28.07 where the expected level of the state variable is \$88.41; the dollar beta $\beta^{\$}$ is 0.6174; the expected value of the illiquid price is \$26.84; and the risk-free discount is 0.9512.

The fact that the price of the illiquid asset equals the price of the liquid asset suggests that the illiquid price is independent of the physical probability p (and also independent of μ). The reason is that in a single period binomial model, the option payoff is linear in the underlying asset and Theorem 1 applies. This result is extremely crucial in our model in that once we approach continuous trading/rebalancing, the binomial model suggests that the option price within a period is linear in the underlying asset and as a result, the illiquid price must equal the liquid price the boundary condition it must satisfy by definition.

As n becomes large, liquidity discount becomes large. As in the base case where n = 100, the discount is substantial. The binomial value is \$25.86 and the liquidity-constrained price, i.e. (16.12), is \$24.47 representing a 5% discount.

To demonstrate the binomial implementation of our model, we let i be the time index and j be the state index. At the end of the binomial lattice $i = n$. At each time i, the state j is labeled as $0 \leqslant j \leqslant i$. The level of the state variable, the price of the liquid security, and the price of the illiquid security for the economy are then labeled as V_{ij}, S_{ij}, and S^*_{ij} respectively. The risk-neutral probability and the physical probability are defined as usual and given above.

In the following, we present results when the risk preference represented by the Sharpe ratio ranges from 0 (risk free case) to 1.6. Rebalancing frequency is represented by k. And $k = n - 1$ represents perfect rebalancing or continuous trading (that is, for 100 periods, rebalancing 99 times in between is identical to continuous trading).

In the binomial model, the number of periods n must be divisible by one plus the rebalancing frequency, i.e. $k + 1$, to avoid unnecessary numerical errors. In the base case where $n = 100, k$ can be 0, 1, 3, 4, 9, ..., 99. Take $k = 3$ as an example,

Volatility and u and d				
σ	0.8	0.6	0.4	0.2
u	1.083287	1.061837	1.040811	1.020201
d	0.923116	0.941765	0.960789	0.980199

μ (drift)				
0	0.05	0.05	0.05	0.05
0.4	0.37	0.29	0.21	0.13
0.8	0.69	0.53	0.37	0.21
1.2	1.01	0.77	0.53	0.29
1.6	1.33	1.01	0.69	0.37
p (physical probability)				
0	0.4831	0.4892	0.4963	0.5075
0.4	0.5032	0.5092	0.5163	0.5275
0.8	0.5232	0.5293	0.5363	0.5476
1.2	0.5434	0.5494	0.5564	0.5676
1.6	0.5636	0.5695	0.5765	0.5877

Table 16.1: Shapre Ratio and μ and p

rebalancing is allowed at $i = 25$, 50, and 75. At each of these times, equation (16.12) is used to compute the illiquid price at every node at the given time. Specifically, at $i = 75$, equation (16.12) is used to compute values at the nodes that are represented by $j = 0 \sim 75$ (i.e. $S^*_{75,0}$ till $S^*_{75,75}$). Then at $i = 50$, equation (16.12) is again used to compute values where $j = 0 \sim 50$ (i.e. $S^*_{50,0}$ till $S^*_{50,50}$) using the prices from $S^*_{75,0}$ till $S^*_{75,75}$. This process repeats backwards until we reach todays price which is S^*_0.

We compute a number of liquid and illiquid prices under various scenarios. Unless otherwise mentioned, the values of the input variables are taken from the base case. Note that liquidity discount is more severe as investors are more risk averse. To measure the magnitude of risk aversion, we adopt the Sharpe ratio on the underlying state variable, which is excess return scaled by the volatility. We simulate various degrees of Sharpe ratio from 0 (risk free case) to 1.6 with the volatility scenarios from 0.2 to 0.8. At the risk-free rate of 5%, we obtain the required rate of return ($\mu = r + \lambda\sigma$ where λ is Sharpe ratio) from 5% (risk-free case) to 133% ($\lambda = 1.6$).

The results are summarized in Table 16.1. The top panel are binomial parameter values where $u = \exp(\sigma\sqrt{\Delta t})$ and $d = 1/u$ under $n = 100$. The middle panel contains different expected returns) under different volatility scenarios for each Sharpe ratio. The bottom panel presents physical probability values using the binomial formula $p = [\exp(\mu\Delta t) - d]/[u - d]$. Interestingly, each Sharpe ratio corresponds to a physical probability value roughly.

Combine the information of the physical probabilities and other input values, we can then compute liquid and illiquid prices using equation (16.12). Note that when the Sharpe ratio is 0, there is no liquidity discount and the liquid value equals the illiquid value, as proved by Theorem 1. This allows us to examine the magnitude of the liquidity discount as a function of risk preference. Table 2 provides all the liquid (Sharpe ratio is 0) and illiquid prices. The strike price for the result is set at 100 and the state variable is set at 80.

Reported in Table 16.2 are simulated liquid (Sharpe ratio is 0) and illiquid prices (Sharpe ratio is greater than 0). Note that by construction, our model for illiquid prices degenerates to liquid prices as the Sharpe ratio approaches 0. Also as required by the model, when continuous trading is reached ($k = 99$), we obtain liquid prices and risk preference does not matter (bottom panel). Liquidity discount is at maximum when no trading/rebalancing is allowed ($k = 0$).

To visualize the effect, we translate the values in Table 2 from dollar terms to percentage terms. In each case, the liquid price serves as the benchmark (named Black-Scholes value). This is the value consistent with continuous trading. Various comparisons are provided in Figure 16.3 and Figure 4 as follows.

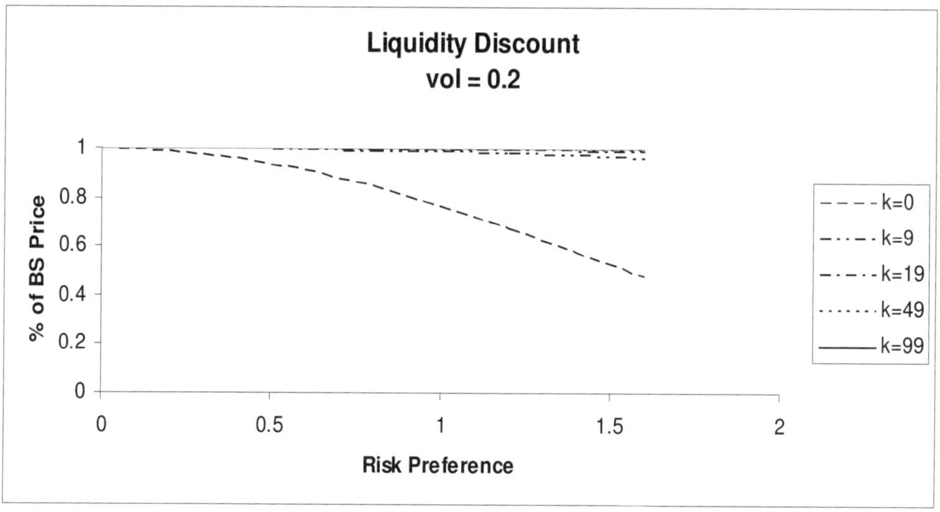

In the upper panel of Figure 16.3, we present the result of liquidity discount under various trading frequencies. The binomial model of $n = 100$ is used as the

σ (volatility)				
$k=0$	0.8	0.6	0.4	0.2
0	35.4303	29.0305	22.7066	16.9860
0.4	29.7751	25.1990	20.5670	16.2768
0.8	21.9833	19.2094	16.5282	14.3871
1.2	14.1540	12.734	11.6111	11.4428
1.6	7.8894	7.2781	7.0476	8.0259
$k=9$	0.8	0.6	0.4	0.2
0	35.4303	29.0305	22.7066	16.9860
0.4	34.8789	28.6789	22.5240	16.9334
0.8	33.9291	28.0265	22.1478	16.8068
1.2	32.5311	27.0385	21.5570	16.5993
1.6	30.6061	25.6581	20.7160	16.2998
$k=19$	0.8	0.6	0.4	0.2
0	35.4303	29.0305	22.7066	16.9860
0.4	35.1880	28.8759	22.6264	16.9630
0.8	34.7790	28.5946	22.4642	16.9087
1.2	34.1950	28.1806	22.2166	16.8222
1.6	33.4223	27.6247	21.8784	16.7021
$k=49$	0.8	0.6	0.4	0.2
0	35.4303	29.0305	22.7066	16.9860
0.4	35.3702	28.9921	22.6867	16.9803
0.8	35.2700	28.9231	22.6469	16.9670
1.2	35.1297	28.8233	22.5871	16.9461
1.6	34.9489	28.6925	22.5073	16.9176
$k=99$	0.8	0.6	0.4	0.2
0	35.4303	29.0305	22.7066	16.9860
0.4	35.4303	29.0305	22.7066	16.9860
0.8	35.4303	29.0305	22.7066	16.9860
1.2	35.4303	29.0305	22.7066	16.9860
1.6	35.4303	29.0305	22.7066	16.9860

Table 16.2: Liquidity Discount Results

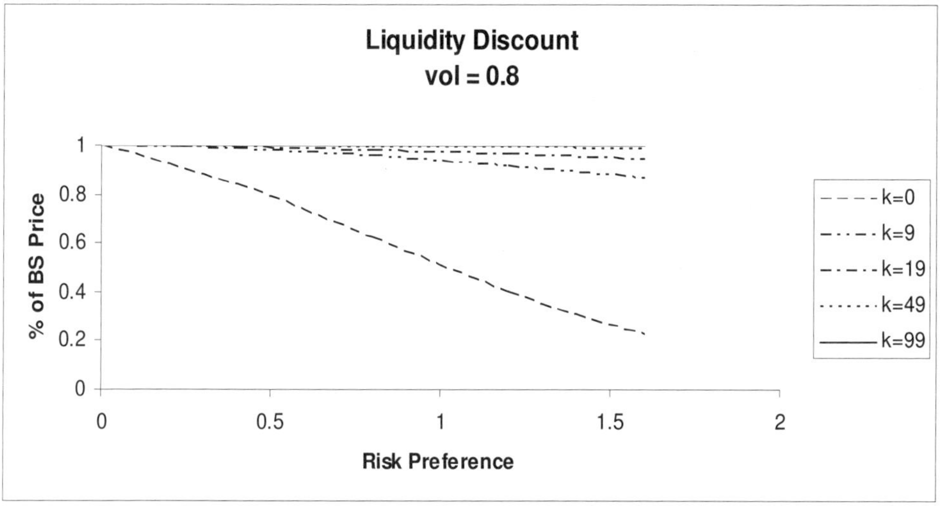

Figure 16.3: A Liquidity Discount Under Various Rebalancing Frequencies

benchmark and regard as the perfectly liquid price which, in the limiting case, converges to the Black-Scholes model. As a result, in the perfectly liquid case where $n = 100$, the number of rebalancing times is $k = 99$. Under no rebalancing, $k = 0$ and this represents the case of extreme illiquidity where investors hold their securities to maturity. Panel A of Figure 16.3 plots the result using $\sigma = 0.2$. The horizontal axis is the expected rate of return of the stock, used to represent risk preference.

It is clear that as the model allows for continuous rebalancing (represented by $k = 99$ in the case of $n = 100$), the price should be the same as the Black-Scholes price where continuous rebalancing is part of the assumption. In Figure 16.3, we do see that the price ratio (of equilibrium over Black-Scholes) is 1 through out the whole range of risk preference. When $k = 0$, the discount because of illiquidity can be severe. We can see from Figure 16.3 that the discount is as bad as 40% as the Sharpe ratio reaches 1.6 and no rebalancing is permitted.

The lower panel of Figure 16.3 is similar to Panel A with a higher volatility value (0.8). As we can see, the liquidity discount is more severe as the volatility is higher. We recall that during the 2007-8 crisis, the volatility was high. For example, in the case of Lehman (see Chen, Chiddi, Imerman, and Soprazetti (2010)) the volatility in many months of 2008 exceeded 100%.

Figure 16.4 presents the same result as Figure 16.3 but examines how various volatility levels affect the liquidity discount. Figure 16.4 sets k to be 0 for the maximum amount of liquidity discount. As we can see from the diagram, the deteri-

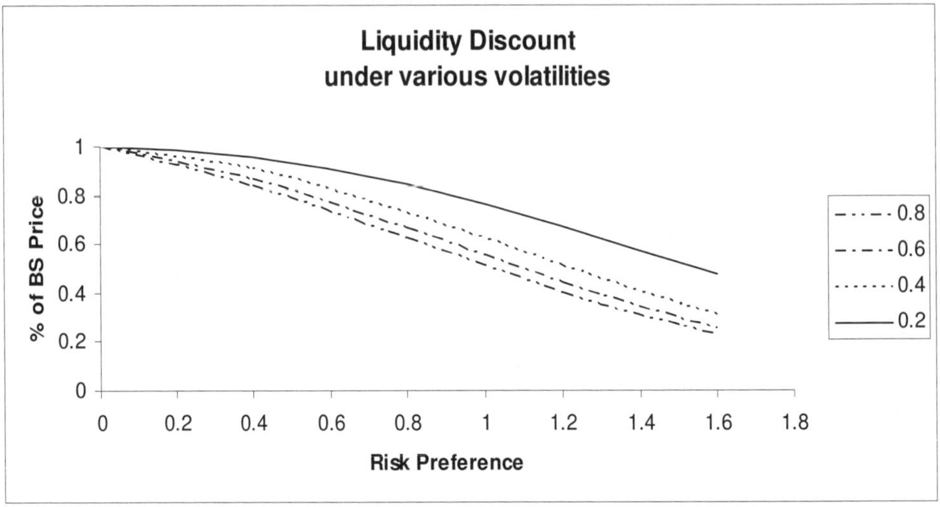

Figure 16.4: A Liquidity Discount Under Various Volatility Levels

oration of the asset price is rather fast. As the volatility is higher, the deterioration is faster.

The next necessary step is to test the model against liquid prices. In other words, when we permit perfect liquidity, i.e. $k = n - 1$, At this situation the equilibrium price must equal to the Black-Scholes (or binomial) price. Figure 16.5 presents the result of convergence under various volatility levels. As we can see, convergence is faster when the volatility is smaller.

In Figure 16.6, we provide the result of our model on the relationship between liquid and illiquid prices. This is the main result of our model which describes the illiquid price (S^*) as a function of the liquid price S. Figure 16.6 is similar to Figure 16.2 but presented with our model. The physical probability is set at 0.6 to exaggerate the result for the visual presentation.

We see in Figure 16.6 that as the liquid price decreases, the illiquid price decreases but at a much faster rate. This is consistent with the description in Figure 16.2 where B-D line bends over to touch the vertical axis.

When the liquid price is high, liquidity discount is small and the two prices are equal to each other. In the numerical example plotted in Figure 16.6, toward the right where the prices are both high, liquidity discount disappears and the curve approaches the 45-degree line asymptotically.

Note that in our model, there is no explicit put option as in Figure 16.2. Our

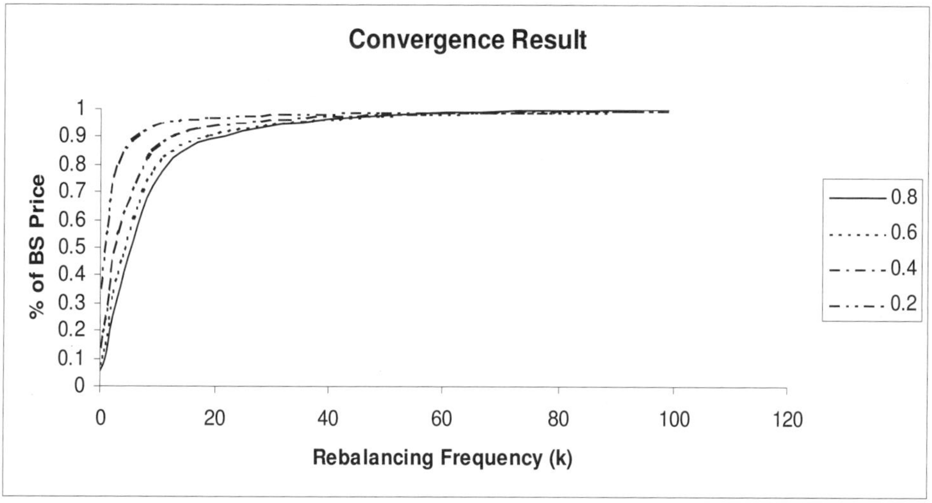

Figure 16.5: A Liquidity Discount Under Various Rebalancing Frequencies

liquidity discount model is derived by limiting trading/rebalancing in a binomial model. Our liquidity discount is computed by assuming a quadratic utility function so that pricing can be achieved via the Capital Asset Pricing Model. Nevertheless, the result of our model describes liquidity discount as a put option.

We also note that the liquidity discount model derived in this paper is closely connected how the relationship between the economy (represented by a single state variable: wealth) and the perfectly liquid price. As a result, the convexity of the relationship determines the severity of the liquidity discount.

16.4 Liquidity Premium

One immediate extension to our model is that we can analyze assets with liquidity premiums. While there are a number of financial assets that suffer from liquidity discount (i.e. lower prices due to limited trading), other assets, not necessarily financial, enjoy liquidity premium. Gold, oil, and even real estate are good examples of assets of such kind.

Symmetrical to the cause of liquidity discount, the cause of liquidity premium is the limited capacity of supply. Parallel to the model of liquidity discount, when an agent is risk adverse and the payoff of a security is concave, limited supply shall cause such an asset to enjoy liquidity premium. The analysis is straightforward as

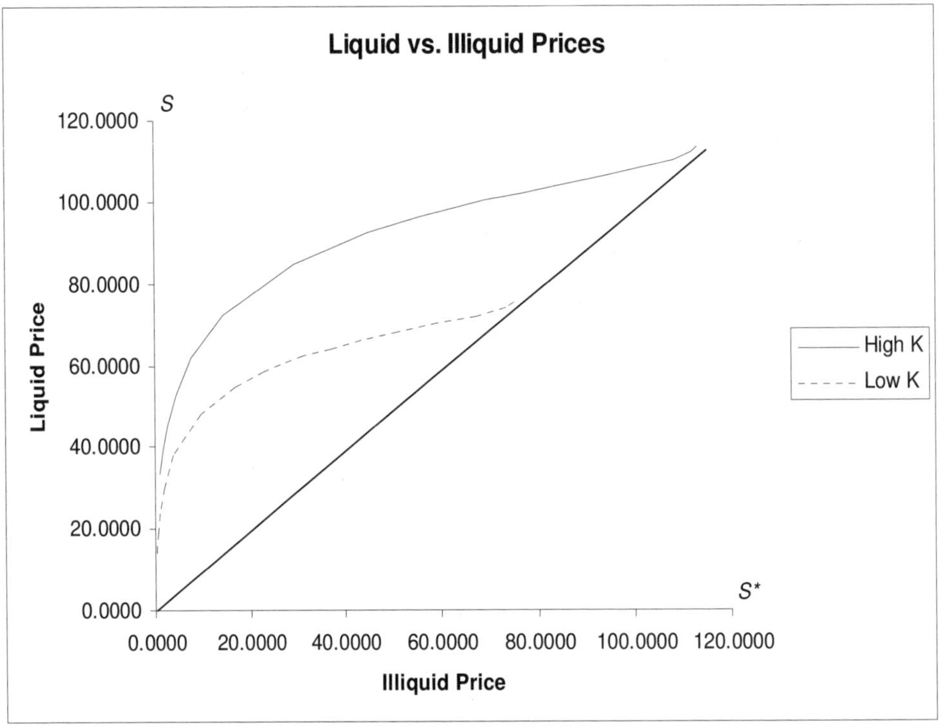

Figure 16.6: Relationship between Liquid and Illiquid Prices

follows.

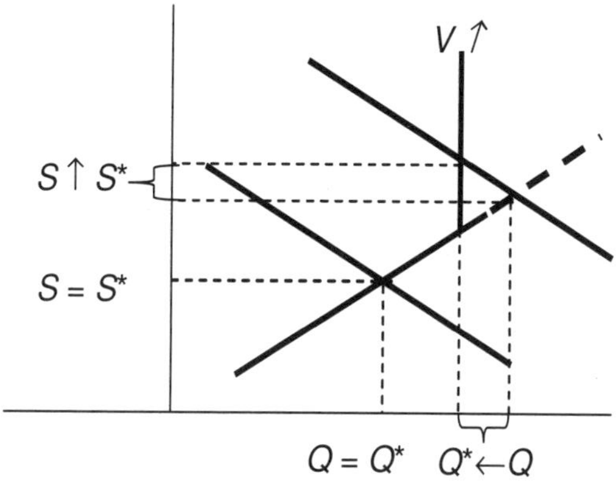

Figure 16.7: Demand-Supply Analysis of Liquidity Premium

Similar to Figure 16.1, Figure 16.7 depicts a situation where demand is more sensitive to economic changes than supply is and supply is bounded by a fixed quantity Q^*. As economy grows, it approaches the maximum capacity and liquidity squeeze (supply-driven) takes place. Contrary to the demand-driven squeeze, now the price under liquidity squeeze is higher with squeeze than without squeeze. In this situation, equilibrium price rises but quantity falls.

A counterpart of Figure 16.2 is shown in Figure 16.8. The left panel of Figure 16.8 describes the relationship between the economy V and perfectly liquid price S. The right panel describes the relationship between liquid price S and illiquid price S^*.

From Figure 16.8, it is clear that liquidity premium can be described as a call option:

$$S^* = S + \text{call} \tag{16.23}$$

As in the liquidity discount case, in our model, there is no need to explicitly model the call option. As long as the liquid price is concave in the state variable of economy, the call-option-like result will be naturally derived.

The following theorem shows that in such a case, the liquidity-squeezed price S^*, is higher than the perfectly liquid price S when the liquid price is a concave function of the economy.

Theorem 3 If the payoff concave, then $S^{\text{lnr}}(t) < S^{\text{cav}}(t)$ where $S^{\text{lnr}}(t)$ is defined in Theorem 1 as $S(t)$ which is the perfectly liquid price and $S^{\text{cav}}(t)$ is the same as

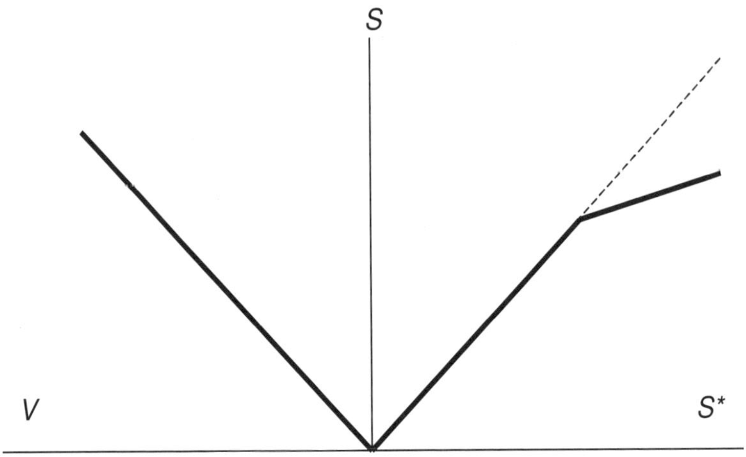

Figure 16.8: Liquidity Premium as a Call

$S^*(t)$ which is the liquidity-constrained price.

Proof Repeat the same procedure of the proof of Theorem 2 and the result follows.

□

The model for liquidity premium is a result that represents the case where supply of the asset is less sensitive than demand to the underlying economy. This is useful in explaining prices of several commodities in the current situation such as gold and oil. These commodities are assets with very limited supply. As the demand of such assets grow stronger, prices rise dis-proportionally to the rest of the economy, resulting in liquidity premiums. Our model argues that if the liquid price is linear in the state variable, then the price of the asset will rise, but there is no liquidity premium. If the liquid price is concave in the state variable, then there is liquidity premium. Similar to liquidity discount, such a liquidity premium can be substantial even the fundamental economy does not change materially.

A counterpart of Figure 16.4 is plotted in Figure 16.9. As the expected return becomes lower, the impact of liquidity premium is more profound. This diagram is generated with the same inputs as the base case with the relationship between the liquid price and the economic state variable as $S = \min\{V, K\}$.

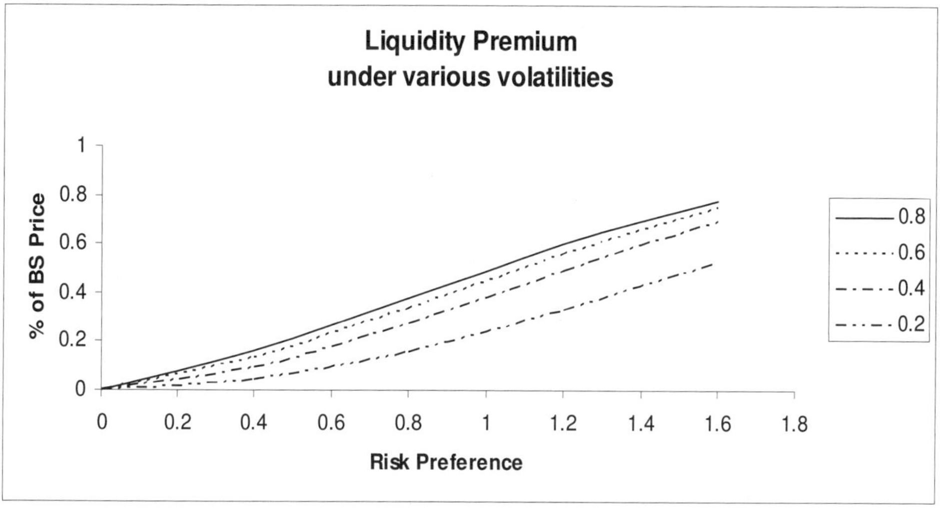

Figure 16.9: A Liquidity Premium Under Various Volatility Levels

Remark

There are a number of applications of the liquidity premium model. Any asset that is under supply squeeze will enjoy a liquidity premium. Obvious recent examples include oil, precious metals like gold, silver and platinum, and agriculture products. In a squeeze situation, these commodity prices deviate from their "fundamental" values and highly inflated prices. As in the analysis for the liquidity discount cases, these situations disappear once the liquidity pressure disappears.

Chapter 17

Bibliography

The bibliography provided here hopefully provides a useful guide for those readers interested in further pursuing quantitative modeling. The list here is by no means exhaustive, but rather what I have read over the years. The classification I provide here is also a reflection of my preference.

17.1 Option Pricing

17.1.1 Basic Classicals - 1997 Nobel Prize of Economics

Black, F. and M. Scholes (1973), The Pricing of Options and Corporate Liabilities, Journal of Political Economy, 81, 637-659.

Merton, R., 1973, "A Rational Theory of Option Pricing", Bell Journal.

17.1.2 Early Work

Samuelson, Paul A., 1965 (Spring), "Rational Theory of Warrant Pricing," Industrial Management Review6, no. 2, p. 13-31.

Sprenkle, Case, 1961, "Warrant Prices as Indications of Expectations," Yale Economic Essays, p. 179-232 (preprinted in Cootner (1967) 412-474).

17.1.3 American Option Pricing

Black, F., 1975, Fact and Fantasy in the use of Options, Financial Analysis Journal 31, 36-41, 61-72.

Kim, I., 1990, The Analytic Valuation of American Options, Review of Financial Studies, 3 (4), 547-72.

Carr, P., 1992, Alternative Characterizations of American Put Options, Mathematical Finance.

Geske, R. and R. Rolls, 1984, On Valuing American Call Options with the Black-Scholes European Formula, Journal of Finance 39, 443-455.

Geske, R., 1979, A Note on An Analytic Valuation Formula for Unprotected American Call Options on Stocks with Known Dividends, Journal of Financial Economics 7, 375-380.

Parkinson, M., 1977, Option Pricing: The American Put," Journal of Business 50, 21-36.

Roll, R., 1977, An Analytical Valuation Formula for Unprotected American Call Option on Stocks with Known Dividends, Journal of Financial Economics 5, 251-58.

Whaley, R., 1981, On the Valuation of American Call Options on Stocks with Known Dividends: An Analytical Valuation Formula for Unprotected American Call Options on Stocks with Known Dividends, Journal of Financial Economics, 9 (2). p 207-11.

17.1.4 Richer Processes

Amin, K.I. and R. Jarrow, 1992, Pricing Options on Risky Assets in a Stochastic Interest Rate Economy, Mathematical Finance 2, 217-237.

Amin, K. (1996), Jump Diffusion Option Valuation in Discrete Time, Journal of Finance, 58, 5.

Cox, J. and S. Ross, 1976, "The Valuation of Options for Alternative Stochastic Processes," JFE.

Cox, J., Notes on Option Pricing I: Constant Elasticity of Diffusions, Unpublished draft, Standford University, September 1975.

Cox, J., The Constant Elasticity of Variance Option Pricing Model, Journal of Portfolio Management, 1996, p.15-17.

Duan, J.C., 1995, The GARCH Option Pricing Model, Mathematical Finance 5(1),

13-32.

Duffie, D., J. Pan, and K. Singleton, Transform Analysis and Option Pricing for Affine Jump-Diffusions, Working Paper, Stanford University.

Heston, S. (1993), A Closed-Form Solution for Options with Stochastic Volatility with Applications to Bond and Currency Options, Review of Financial Studies, 6, 2, 327-343.

Hull, J. and A. White (1987), The Pricing of Options with Stochastic Volatilities, Journal of Finance, 42 (2), 281-300.

Johnson, H. and D. Shanno, 1987, Option Pricing When the Variance is Changing, Journal of Financial and Quantitative Analysis 22, 143-151.

Merton, R., 1976, "Option Pricing when Underlying Stock Returns are Discontinuous," Journal of Financial Economics, May, 125-144.

Scott, L., 1987, "Option Pricing when the Variance Changes Randomly: Theory, Estimation, and an Application," The Journal of Financial and Quantitative Analysis, Vol. 22, No. 4., pp. 419-438.

Scott, L., 1997, Pricing Stock Options in a Jump-Diffusion Model with Stochastic volatility and Interest Rates: Applications of Fourier Inversion Methods, Mathematical Finance, 7 (4), 413-424.

Wiggins, J.B., 1987, Option Values under Stochastic Volatility: Theory and Empirical Estimations, Journal of Financial Economics, 351-372.

17.1.5 Other Underlying Assets

Black, Fischer, 1976, The Pricing of Commodity Contracts, Journal of Financial Economics, 3, 167- 179.

Garmen, B. Mark and Steve W. Kohelgen, 1983, Foreign Currency Option Vaues, Journal of International Money and Finance, 2.

Rabinovitch, Ramon, 1989, Pricing Stock and Bond Options when the Default-Free Rate is Stochastic, Journal of Financial and Quantitative Analysis, 24 (4), 447-457.

Interest rate derivatives see Term Structure of Interest Rates.

17.1.6 Futures

Cox, John C., Jonathan E. Ingersoll, and Stephen A. Ross, 1981, The Relationship Between Forward Prices and Futures Prices, Journal of Financial Economics, 9,

321-346.

Jarrow, Robert A; Oldfield, George S., 1981 (December), Forward Contracts and Futures Contracts, Journal of Financial Economics, 9 (4), 373-82.

17.1.7 Equilibrium Option Pricing

Back, K., 1993, Asymmetric Information and Options, Review of Financial Studies, Vol. 6, No. 3. pp. 435-472.

Bailey, W., and R. Stulz, 1989, The Pricing of Stock Index Options in a General Equilibrium Model, Journal of Quantitative Analysis 24, 1-12.

Bakshi, G.S. and Z. Chen, 1997b, Equilibrium Valuation of Foreign Exchange Claims, Journal of Finance 52, 799-826.

Chen, R., and O. Palmon, 2002, An Empirical Based Option Pricing Model: A Resolution to the Volatility Smile, working paper, Rutgers University.

Naik, V. and M. Lee, 1990, General Equilibrium: Pricing of Options on the Market Portfolio with Discontinuous Returns, Review of Financial Studies 3, 493-521.

17.2 Numerical Methods

Barone-Adesi, G., and R. E. Whaley, 1987, Efficient Analytic Approximation of American Option Values, Journal of Finance, 42, 301-320.

Barone-Adesi, G., and R. J. Elliott, 1989, Free Boundary Problems in the Valuation of Securities, working paper, University of Alberta.

Boyle, P, M. Broadie, and P. Glasserman (1997), Monte Carlo Methods for Security Pricing, Journal of Economic Dynamics and Control, 21, 8-9, 1267-1321.

Boyle, P., 1977, "Options: A Monte Carlo Approach." Journal of Financial Economics 4, 4, 323-338.

Boyle, P., 1988 (March), A Lattice Framework for Option Pricing with Two State Variables, Journal of Financial and Quantitative Analysis, 23 (1), p. 1-12.

Boyle, P., J. Evnine, and S. Gibbs, 1989, Numerical Evaluations of Multivariate Contingent Claims, Review of Financial Studies, 2 (1), p. 241-250.

Chen, Chung, and Yang, 2002 (March), Option Pricing in an Multi-asset, Complete-market Economy, Journal of Financial and Quantitative Analysis, 37 (1), p. 117-136.

Cox, J., S. Ross, and M. Rubinstein, 1979, Option Pricing, A Simplified Approach, Journal of Financial Economics.

Hull, J., and A. White, 1990, Valuing Derivative Securities Using the Explicit Finite Difference Method, Journal of Financial & Quantitative Analysis, 25, 87-100.

Hull, J. and A. White, 1993 (Fall), Efficient procedures for valuing European and American path-dependent derivatives, Journal of Derivatives, 1 (1), p. 21-31.

Hull, J. and A. White, 1994 (Fall), "Numerical procedures for implementing term structure models I: Single-Factor Models," Journal of Derivatives, 2 (1), p. 7-16.

Hull, J. and A. White, 1994 (Winter), "Numerical procedures for implementing term structure models II: Two-Factor Models," Journal of Derivatives, 2 (1), p. 37-48.

Kamrad, B. and P. Ritchken, 1991 (December), Multinomial Approximating Models for Options with k State Variables, Management Science, Vol. 37, No. 12, p. 1640-1652.

Kijima, M. and I. Nagayama, "Efficient numerical procedures for the Hull-White extended Vasicek model," The Journal of Financial Engineering, 3, 4 (September/December 1994) 275-292.

Pelsser, A., "An efficient algorithm for calculating prices in the Hull-White model," ABN-Amro Bank, Derivative Product Research and Development working paper, 1994.

Rendleman, R. and B. Bartter, 1979 (December), Two-State Option Pricing, Journal of Finance.

17.3 Empirical Studies

Babbs, S., and B. Nowman, 1999, Kalman Filtering of Generalized Vasicek Term Structure Models, Journal of Financial and Quantitative Analysis, 34 (1), 115-130.

Bakshi, G., C. Cao, and Z. Chen (1997), Empirical Performance of Alternative Option Pricing Models, Journal of Finance, 52, 2003-2049.

Bates, D., 1991, The Crash of 87': Was It Expected? The Evidence from Options Markets, Journal of Finance 46, 1009-44.

Bates, D., 1996, Jumps and Stochastic Volatility: Exchange Rate Processes Implicit in Deutsche Mark Options, Review of Financial Studies, 9, 69-107.

Black, F. and M. Scholes, 1972, The Valuation of Option Contracts and a Test of

Market Efficiency, Journal of Finance 27, 399-417.

Chen, R. R., and L. 0. Scott, 1993, Maximum Likelihood Estimation of a Multi-Factor Equilibrium Model of the Term Structure of Interest Rates, working paper, University of Georgia.

Chen, R., and L. Scott, 2002, Multi-Factor Cox-Ingersoll-Ross Models of the Term Structure:Estimates and Tests From a Kalman Filter Model, Journal of Real Estate Finance and Economics.

Dumas, B., J. Fleming, and R. Whaley, 1998, Implied Volatility Smiles: Empirical Tests, Journal of Finance 53, 2059-2106.

Emanuel, David C. and James D. Macbeth, Further Results on the Constant Elasticity of Variance Call Option Pricing Model, Journal of Financial and Quantitative Analysis, November 1982, Vol. 17, 533-554.

Longstaff, F., 1995, Option Pricing and the Martingale Restrict, Review of Financial Studies 8, 1091-1124.

MacBeth, J.D. and L.J. Merville, 1979, An Empirical Examination of the Black-Scholes Call Option Pricing Model, Journal of Finance 34, 1173-86.

MacBeth, James D. and Larry J. Merville, Test of the Black-Scholes and Cox Call Option Valuation Models, Journal of Finance, May 1980, v35(2), 285-301.

Rubinstein, M., 1985, Nonparametric Tests of Alternative Option Pricing Models Using All Reported Trades and Quotes on the 30 Most Active CBOE Option Classes from August 23, 1976 through August 31, 1978, Journal of Finance 40, 455-80.

Whaley, R.E., 1981, On the Valuation of American Call Option on Stocks with Known Dividends, Journal of Financial Economics 9, 201-11.

Whaley, R.E., 1982, Valuation of American Call Options on Dividend-Paying Stocks: Empirical Tests, Journal of Financial Economics 10, 29-58.

Whaley, R.E., 1986, Valuation of American Futures Options: Theory and Empirical Tests, Journal of Finance 41, 127-150.

17.4 Default Modeling And Credit Derivatives

Black, F., and J. Cox, 1976, "Valuing Corporate Securities: Some Effects of Bond Indenture Provisions", Journal of Finance.

Chen, R., and J. Huang, 2002, Credit Spread Bounds and Their Implications for Credit Risk Modeling.

Chen, R., and B. Soprazetti, 2002, The Valuation of Default Triggered Credit Derivatives, Journal of Financial and Quantitative Analysis.

Crosbie, P., 2002, Modeling Default Risk, http://www.kmv.com/insight/index.html.

Duffie, D. and K. Singleton, 1997, An Econometric Model of the Term Structure of Interest-Rate Swap Yields, The Journal of Finance, Vol. 52, No. 4, pp. 1287-1321.

Duffie, D. and K. Singleton, 1999, Modeling Term Structures of Defaultable Bonds, The Review of Financial Studies.

Geske, R., 1977, The Valuation of Corporate Liabilities as Compound Options, The Journal of Financial and Quantitative Analysis, Vol. 12, No. 4, pp. 541-552.

Geske, R., and H. Johnson, 1984, The Valuation of Corporate Liabilities as Compound Options: A Correction, The Journal of Financial and Quantitative Analysis, Vol. 19, No. 2, pp. 231-232.

Jarrow, R., D. Lando, S. Turnbull, 1997, A Markov Model for the Term Structure of Credit Risk Spreads, The Review of Financial Studies, Vol. 10, No. 2, pp. 481-523.

Jarrow, R., S. Turnbull, 1995, Pricing Derivatives on Financial Securities Subject to Credit Risk, The Journal of Finance, Vol. 50, No. 1, pp. 53-85.

Leland, H. and Toft, 1996, "Optimal Capital Structure, Endogenous Bankruptcy, and the Term Structure of Credit Spreads", Journal of Finance.

Longstaff, F. and E. Schwartz, 1995, "A Simple Approach to Valuing Risky Fixed and Floating Rate Debt", Journal of Finance.

Merton, R., 1974, "On the Pricing of Corporate Debt: The Risk Structure of Interest Rates", Journal of Finance.

17.5 Continuous Time Mathematics

Harrison, M. and S. Pliska, 1981, "Martingales and Stochastic Integrals in the Theory of Continuous Trading," Stochastic Processes and Their Applications, 11, 215-260.

Harrison, M., and D. Kreps, 1979 (June), "Martingales and Arbitrage in Multiperiod Securities Markets, Journal of Economic Theory. Vol. 2 (3). p 381-408.

Duffie, D., and C-F. Huang, 1985 (November), "Implementing Arrow-Debreu Equilibria by Continuous Trading of Few Long-lived Securities," Econometrica.

Vol. 53 (6). p 1337-56.

Merton, R., 1969, "Lifetime Portfolio Selection under Uncertainty: The Continuous Time Case," REStat.

Merton, R., 1971, "Optimum Consumption and Portfolio Rules in a Continuous Time Model," JET.

17.6 Intertemporal Asset Pricing Models

Merton, R., 1973, "An Intertemporal Asset Pricing Model," Econometrica.

Cox, J. C., J. E. Ingersoll, and S. A. Ross, 1985a, An Interemporal General Equilibrium Model of Asset Prices, Econometrica, 53, 363-384.

Stein, E., and J. Stein, 1991, Stock Price Distributions with Stochastic Volatility: An Analytic Approach, Review of Financial Studies 4, 727-752.

17.7 State Price, Arbitragre

Breeden, D. and R. Litzenberger (1978), "Prices of State-Contingent Claims Implicit in Options Prices," Journal of Business, 51, October, 621-651

Ross, S., 1976, The Arbitrage Pricing of Capital Asset Pricing, Journal of Economy Theory.

17.8 Exotic Options

Boyle, P. P., and S. M. Turnbull, 1989, Pricing and Hedging Capped Options, Journal of Futures Markets, 9, 41-54.

Geske, 1977, "The Valuation of Corporate Liabilities as Compound Options", Journal of Financial and Quantitative Analysis.

Geske, R., 1979, "The Valuation of Compound Options," Journal of Financial Economics. Vol. 7 (1). p 63-81.

Lagnado, R. and S. Osher (1997), "Reconciling Differences," RISK Magazine, April, 79-83

Margrabe, W. "The Value of an Option to Exchange One Asset for Another," The Journal of Finance, Vol. 33, No. 1. (Mar., 1978), pp. 177-186.

Ritchken, P, "On pricing barrier options," Journal of Derivatives, 3, 2 (Winter 1995) 19-28.

Shimko, D., 1993, Bounds of Probability, Risk 6, 33-37.

17.9 Real Option

Moel, A., and P. Tufano, 2002 (Spring), When Are Real Options Exercised? An Empirical Study of Mine Closings, Review of Financial Studies. Vol. 15 (1). p 35-64.

Alvarez, L., and R. Stenbacka, 2001 (Feburary), Adoption of Uncertain Multi-stage Technology Projects: A Real Options Approach, Journal of Mathematical Economics. Vol. 35 (1). p 71-97.

Milne, Alistair; Whalley, A Elizabeth. 'Time to Build, Option Value and Investment Decisions': A Comment.. [Journal Article] Journal of Financial Economics. Vol. 56 (2). p 325-32. May 2000.

Berger, Philip G; Ofek, Eli; Swary, Itzhak. Investor Valuation of the Abandonment Option.. [Journal Article] Journal of Financial Economics. Vol. 42 (2). p 257-87. October 1996.

Majd, Saman; Pindyck, Robert S. Time to Build, Option Value, and Investment Decisions.. [Journal Article] Journal of Financial Economics. Vol. 18 (1). p 7-27. March 1987.

17.10 Risk Neutral Distributions

Ait-Sahalia, Y., Y. Wang, and F. Yared (1998), Do Option Markets Correctly Price the Probabilities of Movements of the Underlying Asset?, Working Paper, University of Chicago

Andersen, L. and R. Brotherton-Ratcliffe (1998), "The Equity Option Volatility Smile: a Finite Difference Approach," Journal of Computational Finance, 1, 2, 5-38.

Brown, G. and K. Toft (1996), Constructing Implied Binomial Trees from Multiple Probability Distributions, Working Paper, University of Texas, Austin.

Chriss, N. (1996), Transatlantic Trees, RISK, July, 45-48.

Derman, E. and I. Kani (1994), "Riding on a Smile," RISK Magazine, February, 32-39

Duan, J.C., 1996, Cracking the Smile," Risk 9, 55-56, 58, 60.

Dupire, B. (1994), "Pricing with a Smile," RISK Magazine, January, 18-20

Dupire, B., 1994, Pricing with a Smile, Risk 7, 18-20.

Eberlein, E., U. Keller, and K. Prause, 1998, New Insights into Smile, Mispricing, and Value at Risk: The Hyperbolic Model, Journal of Business 71, 371-405.

Ederington, L. and W. Guan, 2000 (July), Why Are Those Options Smiling, working paper, University of Oklahoma.

Jackwerth J.C. and M. Rubinstein, 1996, Recovering Probability Distributions from Option Prices, Journal of Finance 51, 1611-1631.

Jackwerth, J. (1996), Generalized Binomial Trees, Working Paper, University of California at Berkeley.

Pea, I., G. Rubio, and G. Serna, 1999, Why Do We Smile? On the Determinants of the Implied Volatility Function, Journal of Banking & Finance 23, 1151-79.

Rubinstein, M., 1994, "Implied Binomial Trees," Journal of Finance, 49, 771-818

17.11 Term Structure Of Interest Rates

Ball, C. and Walter N. Torous 1983 (December), Bond Price Dynamics and Options, Journal of Financial and Quantitative Analysis, 18, No. 4, 517-531.

Black, F. and P. Karasinski, 1991 (July–August), "Bond and option pricing when short rates are lognormal," Financial Analysts Journal, 52-59.

Black, F., E. Derman, and W. Toy, 1990, A One-Factor Model of Interest Rates and Its Application to Treasury Bond Options, Financial Analysts Journal, 46, 33-39.

Brennan, M. J., and E. S. Schwartz, 1979, A Continuous Time Approach to the Pricing of Bonds, Journal of Banking and Finance, 3, 133-155.

Chaplin, G., 1987, The Term Structure of Interest Rates: A Model Based on Two Correlated Stochastic Processes with Closed Form Solutions for Bond and Option Prices, Working Paper 87- 16, University of Waterloo.

Chen, R., 1992, Exact Solutions for Futures and European Futures Options on Pure Discount Bonds, Journal of Financial & Quantitative Analysis, 27, 97-107.

Chen,R. and L. and Scott, 1992, Pricing Interest Rate Options in a Two-Factor Cox–Ingersoll–Ross Model of the Term Structure, Review of Financial Studies, Vol. 5, No. 4, pp. 613-636.

Cheng, S., 1991, On the Feasibility of Arbitrage-Based Option Pricing When Stochastic Bond Price Processes Are Involved, Journal of Economic Theory, 53, 185-198.

Constantinides, George, 1992, A Theory of the Nominal Term Structure of Interest Rates, Review of Financial Studies, Vol.5 (4), 531-552.

Courtadon, G., 1982, The Pricing of Options on Default-Free Bonds, Journal of Financial & Quantitative Analysis, 17, 75-100.

Cox, J. C., J. E. Ingersoll, and S. A. Ross, 1985, A Theory of the Term Structure of Interest Rates, Econometrica, 53, 385-408.

Das, R. and S. Foresi (1996), Exact Solutions for Bond and Option Prices with Systematic Jump Risk, Review of Derivatives Research, 1, 7-24.

Di and K. Singleton, 2000, Specification Analysis of Affine Term Structure Models, Journal of Finance.

Duffie, D., and Kan, 1996, A Yield-Factor Model of Interest Rates, Mathematical Finance.

Dybvig, P. H.. 1989, Bonds and Bond Option Pricing Based on the Current Term Structure, working paper, Washington University in St. Louis.

Dybvig, P. H.. 1997, Bonds and Bond Option Pricing Based on the Current Term Structure, Mathematics of Derivative Securities, Michael A. H. Dempster and Stanley Pliska, eds., Cambridge University Press.

Dybvig, P., J. Ingersoll, and S. Ross, (Jan., 1996), Long Forward and Zero-Coupon Rates Can Never Fall, The Journal of Business, Vol. 69, No. 1. pp. 1-25.

Dybvig, P. and S. Ross, 1996, Long Forward and Zero-Coupon Rates Can Never Fall, Journal of Business, Vol. 69, No. 1, 1-27.

Robert Goldstein, 2000, The Term Structure of Interest Rates as a Random Field, Review of Financial Studies, 13, 365-384

Heath, D., R. Jarrow, and A. Morton, 1990, Bond Pricing and the Term Structure of Interest Rates: A Discrete Time Approximation, Journal of Financial & Quantitative Analysis, 25, 419-440.

Heath, D., R. Jarrow, and A. Morton, 1992, Bond Pricing and the Term Structure of Interest Rates: A New Methodology for Contingent Claims Valuation, Econometrica, 60, 77-105.

Ho, T. S. Y., and S. B. Lee, 1986, Term Structure Movements and Pricing of Interest Rate Claims, Journal of Finance, 41, 1011 -1029.

Hull, J., and A. White, 1990b, Pricing Interest-Rate-Derivative Securities, Review of Financial Studies, 3, 573-592.

Jamshidian, F., 1987, Pricing of Contingent Claims in the One Factor Term Structure Model, Merrill Lynch Capital Markets.

Jamshidian, F., 1989, An Exact Bond Option Formula, Journal of Finance, 44, 205-209.

Leippold, M. and Liuren Wu, 2002, Asset Pricing Under the Quadratic Class, Journal of Financial and Quantitative Analysis, Vol. 37, 271-295.

Longstaff, F. A., and E. S. Schwartz, 1992, Interest Rate Volatility and the Term Structure: A Two-Factor General Equilibrium Model, Journal of Finance, 47, 1259-1282.

Richard, Scott F. An Arbitrage Model of the Term Structure of Interest Rates.. [Journal Article] Journal of Financial Economics. Vol. 6 (1). p 33-57. March 1978

Turnbull, S. M., and F. Milne, 1991, A Simple Approach to Interest-Rate Option Pricing, Review of Financial Studies, 4, 87 -120.

Vasicek, O. A., 1977, An Equilibrium Characterization of the Term Structure, Journal of Financial Economics, 5, 177-188.

17.12 Volatility Forecast

Amin, K. I. and V. Ng, 1997 (Summer), Inferring Future Volatility from the Information in Implied Volatility in Eurodollar Options: A New Approach, Review of Financial Studies, Vol. 10, No. 2. 333-367.

Ball, C. and A. Roma, 1994, Stochastic Volatility Option Pricing, Journal of Financial and Quantitative Analysis 29, 589-607.

Campa, J. and K.H. Chang, 1995, Testing the Expectation Hypothesis on the Term Structure of Volatilities, Journal of Finance 50, 529-547.

Canina, L. and S. Figlewsk, 1993, The Informational Content of Implied Volatility, Review of Financial Studies, Vol. 6, No. 3. 659-681.

Day, T.E. and C.M. Lewis, 1992, Stock Market Volatility and the Information Content of Stock Index Options, Journal of Econometrics 52, 267-287.

Heynen, R., A,G,Z, Kemna, and T. Vorst, 1994, Analysis of the Term Structure of Implied Volatilities, Journal of Financial and Quantitative Analysis 29, 31-56.

Jorion, P., June 1995, Predicting Volatility in the Foreign Exchange Market,Journal of Finance, Vol. 50, No. 2. 507-528.

Kallsen, J. and M. Taqqu, 1998, Option Pricing in ARCH-type Models, Mathematical Finance 8, 13-26.

Lamoureux, C.G. and W.D. Lastrapes, 1990, Forecasting Stock Return Variance: Toward an Understanding of Stochastic Implied Volatilities, Review of Financial Studies 6, 293-326.

17.13 Statistical And Mathematical References

Abramowitz, M., and I. A. Stegun (eds.), Handbook of Mathematical Functions, Tenth Printing, 1972.

Feller, W., An Introduction to Probability Theory and Its Applications, Vol. II, Wiley, New York, 1971.

Glasserman, P., Monte Carlo Methods in Financial Engineering, Springer, 2003.

Johnson, N. L., and S. Kotz, Distributions in Statistics: Continuous Univariate Distributions, Vol. 2, Houghton Mifflin, Boston, 1970.

Karlin, S., and H. Taylor, A First Course in Stochastic Processes, Academic Press

Karlin, S., and H. Taylor, A Second Course in Stochastic Processes, Academic Press

Karatzas, I. and S. Shreve, Brownian Motion and Stochastic Calculus, Springer Verlag, 1991.

Kloeden, P. and E. Platen, Numerical Solution of Stochastic Differential Equations, Springer Verlag, 1992.

Krishnan, V., Nonlinear Filtering and Smoothing, John Wiley and Sons, 1984.

Ludwig, A., Stochastic Differential Equations: Theory and Applications, out of print.

Mitchell, A. and D. Griffiths, The Finite Difference Method in Partial Differential Equations, John Wiley & Sons, 1980.

Pearson, E. S. and H. O. Hartley, Biometrika Tables for Statisticians, Vol I, Cambridge, 1970.

Press, W., S. Teukolsky, W. Vetterling and B. Flannery, Numerical Recipes in C, Cambridge University Press, 1992.

Sankaran, Munuswamy, Approximation to the Non-Central Chi-Square Distribution, Biometrika, June 1963, 199-204.

Schroder, Mark, Computing the Constant Elasticity of Variance Option Pricing

Formula, Journal of Finance, Vol. XLIV. No. 1, March 1989, 211-219.

Spanier, J. and B. An Atlas of Functions, Hemisphere Publishing Corp., 1987.

17.14 Finance Texts

Cochrane, John H., Asset Pricing, Princeton University Press.

Duffie, Darrell, Dynamic Asset Pricing Theory, Princeton University Press.

Duffie, Darrell, Securities Markets: Stochastic Models, Academic Press.

Haugen, Robert, Modern Investment Theory, Prentice Hall.

Huang Chi-Fu, and Robert Litzengerber, Foundations for Financial Economics, North Holland.

Hull, John, 1989, Options, Futures, and Other Derivative Securities, Prentice Hall.

Hull, John, Options, Futures, and Other Derivatives, Prentice Hall.

Ingersoll, Jonathan E., Jr. Theory of Financial Decision Making. Totowa, New Jersey: Rowman & Littlefield.

Kwok, Yue-Kuen, Mathematical Models of Financial Derivatives, Springer.

Merton, Robert, Continuous Time Finance, Blackwell Publisher Inc.

Trigeorgis, Lenos, Real Options, MIT Press.

Index